ANALOG ELECTRONIC CIRCUITS

NOISE

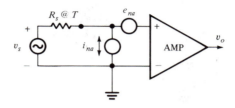

General Amplifier Noise Model

Symbol	Definition	Unit of Measure
e_{na}	Short-circuit input equivalent noise voltage root power spectrum	rmsV/$\sqrt{\text{Hz}}$
i_{na}	Equivalent input current noise root power spectrum	rmsA/$\sqrt{\text{Hz}}$
F	Noise factor: $F = 1 + (e_{na}^2 + i_{na}^2 R_s^2)/4kTR_s$	
NF	Noise figure: $\text{NF} = 10 \log(F)$	
$4kTR$	White noise power density spectrum from a resistor at T K	MSV/Hz

Formulas:

$$S_i(f) \xrightarrow{\quad v_i \quad} \boxed{\mathbf{H}(j\omega)} \xrightarrow{\quad v_o \quad} S_o(f)$$

$$S_o(f) = S_i(f)|\mathbf{H}(j2\pi f)|^2$$

$$v_{on(\text{rms})} = \int_0^\infty S_o(f)df = \int_0^\infty S_i(f)|\mathbf{H}(j2\pi f)|^2 df$$

ANALOG ELECTRONIC CIRCUITS
Analysis and Applications

Robert B. Northrop

UNIVERSITY OF CONNECTICUT

ADDISON-WESLEY PUBLISHING COMPANY

Reading, Massachusetts • Menlo Park, California • New York
Don Mills, Ontario • Wokingham, England • Amsterdam • Bonn
Sydney • Singapore • Tokyo • Madrid • San Juan

This book is in the **Addison-Wesley Series in Electrical and Computer Engineering**
The art on the frontcover is a noninverting, single real-pole low-pass filter (LPF). The art on the backcover illustrates its equivalent circuit.

Credits:
   Figure 8.27 is reprinted courtesy of Joel Cohen, Engineering Manager, Teledyne Crystalonics.
   Figure 12.2 is reprinted courtesy of Hitachi America, Ltd.
   Figure 12.3 is copyrighted 1980 by the RCA Corporation. Reprinted with permission of the Harris Corporation.
   Figure 14.9 is reprinted by permission of Precision Monolithics, Inc.

Library of Congress Cataloging-in-Publication Data

Northrop, Robert B.
   Analog electronic circuits.

   Bibliography: p.
   Includes index.
   1. Linear integrated circuits.   2. Operational
   amplifiers.   3. Electric circuit analysis.   I. Title.
   TK7874.N64   1989   621.381′73   88-22149
   ISBN 0-201-11656-1

Reprinted with corrections March, 1990

BCDEFGHIJ—MA—943210

Contents

Preface

This text is intended to be used in a second (senior) course on analog electronic circuits, and readers are assumed to have had an introductory course on electronic devices and circuits covering not only the basic properties of pn junctions, bipolar junction transistors (BJTs), and junction and MOS field-effect transistors (FETs), but also the physical electronics underlying the behavior of these devices.

In Chapter 1, we review the salient properties of these semiconductor devices at low frequencies as well as the models used to describe their behavior. (There is also a section in Appendix C in which we review the general two-port models for transistors and amplifiers; these models are widely used in circuit analysis.)

Readers are assumed to be skilled in the use of loop and node equations to describe the dynamic behavior of electric and electronic circuits. They are also assumed to be familiar with the use of Thevenin's and Norton's theorems, superposition, Laplace and Fourier transforms, Fourier series, transfer functions, frequency response, and Bode plots. Some experience with the systems analysis techniques of root-locus diagrams (theory, construction, and application) and Mason's signal flow graphs (construction and reduction) is assumed as well. (Appendix A provides a review of signal flow graph techniques.) Section 7.1 presents a summary of the rules for the construction of root-locus diagrams. The root-locus technique is relied on heavily in chapters dealing with the dynamic effects of feedback in electronic systems. Root-locus plots provide a qualitative description of how feedback affects amplifier frequency (and transient) response. They illustrate quantitatively how a system's closed-loop poles move in the s-plane as the scalar loop gain is varied. Although traditionally used with servomechanism and feedback control systems, root-locus plots are well-suited to the analysis and design of analog electronic systems with feedback, including oscillators.

Small-signal linear models are stressed in this text because of their suitability to systems analysis techniques. Chapter 1, as noted, is a review chapter in which the various linear and nonlinear models for electronic devices are presented. Chapter 2 presents several important circuit theorems useful in reducing the complexity of linear electronic circuits when pencil-and-paper analysis is intended.

A review of frequency response techniques is given in Chapter 3; a graphical interpretation of frequency response in the s-plane is presented. Nyquist plots, low-frequency response of reactively coupled amplifiers, and op-amp frequency response are also considered.

Chapter 4 is concerned with the high-frequency response of both discrete and integrated circuit amplifiers; important topics such as the Miller effect, gain–bandwidth product, f_T, and broadbanding techniques are covered.

Chapter 5 deals with the important subject of differential amplifiers, their gain, common-mode rejection ratio, input impedance, and frequency response for difference-mode and common-mode input signals. Common-mode negative feedback is introduced as a means of increasing a differential amplifier's common-mode rejection ratio.

Chapter 6 presents the major effects of feedback on the behavior of electronic circuits; negative and positive voltage feedback and negative and positive current feedback are considered. Op-amp circuits are used to illustrate some of the effects of feedback.

Chapter 7 extends the study of feedback to the frequency domain. Extensive use is made of the root-locus technique to describe how feedback affects the frequency response and stability of amplifiers. The design of oscillators is also approached in Chapter 7 through the use of root-locus methods.

Chapter 8 provides a solid, practical approach to the analysis and design of low-noise amplifiers. The sources of noise in electronic circuits are described, and the means of maximizing an amplifier's output signal-to-noise ratio are presented.

Chapters 9 and 10 are on the design and applications of active filters, including switched-capacitor filters and tracking and adaptive filters. A detailed analysis is given on the Dolby B audio noise reduction system as an example of an adaptive filter system.

Chapter 11 deals with the important topic of phase-lock loops. The component subsystems of a phase-lock loop are described, and applications of PLLs are presented.

Untuned power amplifier designs are considered in Chapter 12. Attention is given to power dissipation and heatsink design as well as power amplifier efficiency. Power op-amps are described.

Chapter 13 covers selected examples of analog electronic systems, including true rms to dc converters, analog multipliers, a self-balancing conductance bridge, and phase-sensitive rectifiers. Chapter 14 treats digital interfaces. Coverage is restricted to R-$2R$ DACs and several important types of ADCs.

Appendix A describes the use and reduction of Mason's signal flow graphs. Signal flow graphs are a convenient means of describing and analyzing lineal feedback systems and can be used effectively to analyze linear electronic systems with both implicit and explicit feedback paths. Appendix B summarizes the features of some of the better-known computer programs for electronic circuit analysis, which run on various IBM PC models and their "clones." A circuit analysis example is given using the Micro-Cap II (Student Version) ECAP. Two-port, linear circuit models are reviewed in Appendix C. These representations are particularly useful for pencil-and-paper analysis of electronic circuits and active filters.

All of the chapters except 13 and 14 have in-depth problem sets that stress analysis, design, and applications. Certain problems are intended to be pencil, paper, and calculator problems, while others are of sufficient complexity that the student must resort to a suitable electronic circuit analysis program to realize a meaningful analysis or a valid design. The numerical answers to selected problems are given in Appendix D. A detailed Solutions Manual for the problems is available for instructors from Addison-Wesley.

This text was written based on the author's experience in teaching a course in electronic circuits and applications for over twelve years in the Electrical and Systems Engineering Department at the University of Connecticut. The author is grateful to his students who have provided much useful feedback for the "fine tuning" of the text and problem sets. I also appreciate helpful comments I have received from teaching colleagues and graduate teaching assistants, specifically, Subhash C. Gulaya and Xiaofeng Qi.

The helpful comments and reviews provided by my editor at Addison-Wesley, Tom Robbins, are greatly appreciated. I am also grateful for the criticisms the following reviewers provided at various stages of the preparation of the manuscript: Artice M. Davis, San Jose State University; Joseph Frank, New Jersey Institute of Technology; John Nyenhuis, Purdue University; J. Alvin Connelly, Georgia Institute of Technology; and William R. Patterson III, Brown University.

Last of all, I must thank my wife, Adelaide, for her moral support, patience, and understanding while I typed my manuscript, revised it, and occasionally searched for mislaid sections.

Chapter *1*

Models for Semiconductor Devices and Integrated Circuit Amplifiers

This chapter reviews large-signal (nonlinear) and small-signal (linear) models used to characterize semiconductor circuit elements, that is, pn diodes, bipolar junction transistors (BJTs), and various types of field-effect transistors (FETs). Low-frequency small-signal models are stressed in this chapter because of their applicability to circuit analysis by pencil and paper as well as by computer, and because of their usefulness in two-port model representations of circuits. (High-frequency models for semiconductor devices are covered in Chapter 4; two-port models are reviewed in Appendix C.)

The relationship of transistor dc operating point (bias conditions) to the numerical values of small-signal model (SSM) parameters is stressed. The mid-frequency small-signal models (MFSSMs) of transistors are used to find the input/output properties of basic one- and two-transistor amplifier configurations. Both BJTs and FETs are considered.

1.1 pn Junction Diodes

An understanding of the electrical properties of pn junction diodes is essential to an appreciation of the behavior of bipolar junction transistors, junction field-effect transistors (JFETs), and other semiconductor devices and systems. In this section, we review the low-frequency terminal characteristics of the pn junction diode and examine various models used to approximate its large- and small-signal behavior.

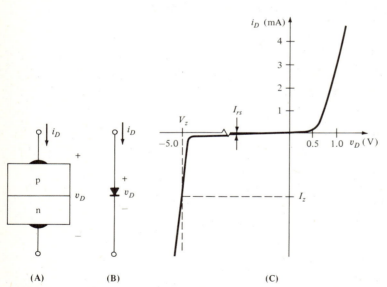

(A)   **(B)**   **(C)**

Figure 1.1   **(A)** "Layer cake" model 1 of a pn junction diode. The diode conducts when $v_D > 0$. **(B)** Symbol for a pn junction diode. **(C)** Current-voltage relation for a pn junction diode showing forward conduction ($v_D > 0$) and avalanche in reverse conduction ($v_D < v_z$).

Figure 1.1(A) illustrates schematically the components of the pn junction diode. A layer of p-doped semiconductor (generally silicon) forms a junction (intimate metallurgical contact) with a layer of n-doped semiconductor. When the p-side of the junction, called the anode, is more positive than the n-side of the junction, termed the cathode, $v_D > 0$ and the diode is said to be forward biased. Under this condition, the diode can conduct relatively large currents, as illustrated in the first quadrant of Fig. 1.1(C). On a linear i_D scale, significant forward conduction of the silicon pn diode does not occur until v_D reaches about 0.65 V.

The volt-ampere curve for the pn diode may be approximated by the well-known relation

$$i_D = I_{rs}(e^{v_D/V_T\eta} - 1) \qquad\qquad 1.1$$

I_{rs} is the reverse saturation current. Its value depends on the diode's geometry, junction type, and semiconductor doping levels. It is theoretically of the order of 1×10^{-15} A for small silicon diodes at room temperature. However, due to practical considerations, I_{rs} is more typically of the order of 1×10^{-9} A. V_T has the dimension of voltage and is given by

$$V_T = \frac{kT}{q} \qquad\qquad 1.2$$

where

k = Boltzmann's constant (1.38×10^{-23} J/K)

T = the junction temperature in kelvins

q = the magnitude of the electron charge (1.6×10^{-19} C)

η is a scaling constant, usually with a value between 1 and 2, that depends on the dominant recombination mechanism in the diode. It is used to force Eq. 1.1 to fit experimental data. At room temperature, $V_T = 0.026$ V.

Under reverse-biased conditions, $v_D < 0$ and $i_D = I_{rs}$ up to V_z, the reverse breakdown voltage. As v_D approaches V_z, minority carriers conducting the reverse i_D acquire enough kinetic energy through collisions to create new hole-electron pairs. These new carriers are in turn accelerated in the high electric field and create still more carriers. This process, called avalanche, can be described empirically by the equation

$$i_D = \frac{I_{rs}}{1 - (v_D/V_z)^n} \qquad\qquad 1.3$$

where the exponent, n, ranges from 3 to 6 and $V_z < v_D < 0$.

Diodes operated in their reverse breakdown regions can be used as voltage sources or references. These diodes are called zener diodes, although the zener reverse breakdown process is different from the avalanche mechanism and generally occurs for $v_D > -6$ V in heavily doped pn junction diodes. Zener diodes are available with various power dissipation values ($P_{max} = V_z I_{D(max)}$), and V_z can range from about 3.2 V to over 100 V.

A typical piecewise-linear model for a zener diode is shown in Fig. 1.2(A). The zener model conducts for $v_D < V_z$. The resistance r_z accounts for the finite slope of the i_D-versus-v_D curve in the breakdown region. Figure 1.2(C) illustrates a typical circuit in which a zener diode supplies a voltage V_z to a load R_L. A dc voltage source, V_s, provides the power to the circuit. To analyze this circuit, we first make a Thevenin equivalent of the port that the zener diode "sees." The R_{eq} of this Thevenin circuit is just R_L in parallel with R_S. The open-circuit voltage is

$$V_{oc} = V_s \frac{R_L}{R_L + R_S} \tag{1.4}$$

and the short-circuit current is

$$I_{sc} = \frac{V_s}{R_S} \tag{1.5}$$

V_{oc} and I_{sc} determine a load line, shown in Fig. 1.2(B), that intersects the zener diode's volt-ampere curve at point Q, the operating point. The reverse diode voltage at Q, $|V_L|$, is the output voltage, V_o. Thus,

$$V_o = |V_L| > 0 \tag{1.6}$$

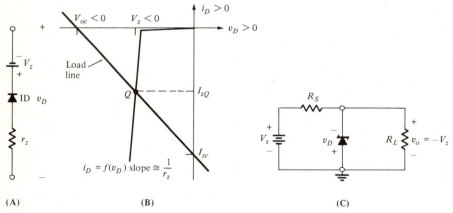

(A)   (B)   (C)

Figure 1.2 **(A)** Piecewise-linear model for pn diode reverse avalanche conduction. ID is an ideal diode with the volt-ampere curve shown in Fig. 1.3(A). **(B)** Graphical solution for the zener (avalanche) diode operating point. The load line is found from the Thevenin (or Norton) equivalent of the dc circuit the zener diode "sees." **(C)** Circuit for a zener diode used as a voltage regulator. R_L is the load resistor; R_S is the series dropping resistor.

The zener diode passes a reverse current, I_{zQ}. By Kirchhoff's current law, V_s must supply a current

$$I_s = I_{zQ} + \frac{V_o}{R_L}$$

1.7

and R_s must be able to dissipate a power of

$$P_s = \left(I_{zQ} + \frac{V_o}{R_L}\right)^2 R_s \text{ W}$$

1.8

When a diode is operated in the region $v_D > V_z$, its behavior can be described by one of several piecewise-linear models, as shown in Fig. 1.3. These models are based on the properties of the ideal diode (ID). The ideal diode is defined by the inequalities

$$i_D > 0 \qquad v_D = 0 \qquad \text{(conducting)}$$

1.9

and

$$i_D = 0 \qquad v_D < 0 \qquad \text{(reverse biased)}$$

1.10

The models of Fig. 1.3(C) and (E) are often used to approximate the base-emitter diode characteristics of a BJT when one sets up quiescent (dc) biasing conditions.

There are obviously many uses of pn junction diodes in electronic circuits, in-cluding: (1) rectification (conversion of a zero-average, time-varying waveform to one with a dc average value); (2) generation of input/output nonlinearities used in signal processing (e.g., peak clipping, log, exponential, dead zone, square law; (3) logic opera-tions (AND and OR gates); (4) temperature compensation of BJT circuits; and (5) current-variable voltage attenuators (for small signals $\ll 0.7$ V).

The latter application will be illustrated here because it is a good way to review the concept of small-signal models of nonlinear circuit elements. A dc current source, I_{DQ}, is used to bias a practical diode as shown in Fig. 1.4(A). The quiescent operating point for the diode is $Q = (V_{DQ}, I_{DQ})$. The capacitor C_s couples a current, $i_s(t)$, to the diode's anode, where it adds to the total diode current, i_D. That is,

$$i_D = I_{DQ} + i_s(t)$$

1.11

As a result of the small sinusoidal component of current, $i_s(t)$, in the diode, a small sinusoidal voltage appears around the dc quiescent voltage. Therefore,

$$v_D = V_{DQ} + v_d(t)$$

1.12

To find the relation between $v_s(t)$, $i_s(t)$, and $v_d(t)$, we note that in the forward-biased region of diode operation around the dc Q-point, the $i_D = f(v_D)$ curve can be

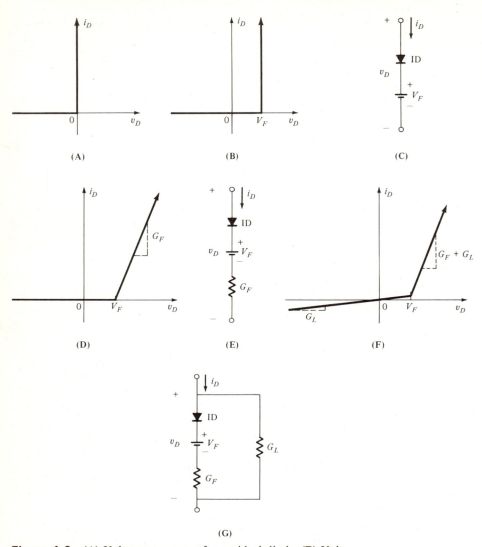

Figure 1.3 **(A)** Volt-ampere curve for an ideal diode. **(B)** Volt-ampere curve for the basic piecewise-linear approximation to a real diode given in **(C)**. **(D)** Volt-ampere curve for the Thevenin piecewise-linear approximation to a real diode given in **(E)**. **(F)** Volt-ampere curve for the Thevenin diode model with reverse leakage conductance, G_L, seen in **(G)**.

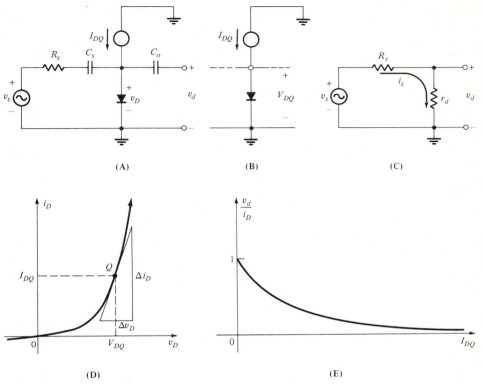

Figure 1.4 **(A)** Circuit for a small-signal, dc current-controlled attenuator using a real diode. C_s and C_o are assumed to have negligible reactance (to be short circuits) for the ac signal, v_s. **(B)** dc equivalent circuit for the current-controlled diode attenuator. **(C)** ac small-signal model for the attenuator. Note that $r_d = f(I_{DQ})$. **(D)** Diode operating point, Q, set by I_{DQ}. **(E)** Attenuation vs. I_{DQ} for the ac signal, v_s.

approximated by a linear line segment having the slope of $i_D = f(v_D)$ at Q such that $v_d = i_s/g_d$. This is

$$g_d = \frac{\partial i_D}{\partial v_D} = I_{rs}e^{(v_D/\eta v_T)}\left(\frac{1}{\eta V_T}\right) = \frac{I_{DQ}}{\eta V_T} \qquad 1.13$$

Equation 1.13 describes the small-signal conductance of the diode at its dc operating point. The capacitor C_o blocks the dc component of v_D, V_{DQ}, so only the ac component of v_D, $v_d(t)$, appears at the output of the attenuator. $v_d(t)$ is related to $v_s(t)$ by the simple voltage divider shown in the ac, small-signal circuit model of Fig. 1.4(C):

$$v_d(t) = v_s(t)\frac{r_d}{R_s + r_d} = \frac{v_s(t)}{R_s g_d + 1} \qquad 1.14$$

Substituting Eq. 1.13 into Eq. 1.14, we obtain

$$v_d(t) = \frac{v_s(t)}{1 + I_{DQ}R_s/\eta V_T}$$

1.15

The attenuation, v_d/v_s, is plotted against the dc diode bias current in Fig. 1.4(E). It should be noted that as long as $|i_s| \ll I_{DQ}$, or $|v_d| \ll V_T$, v_d will contain little harmonic distortion relative to $v_s(t)$.

We have shown that the low-frequency terminal behavior of the pn junction diode can be described by various linear, nonlinear, and piecewise-linear models. Piecewise-linear models are generally used for dc, or large-signal, circuit analysis. Small-signal models are used for ac circuit analysis where there is small departure from the quiescent operating point. The nonlinear diode equation (Eq. 1.1) is used to derive the piecewise-linear and small-signal models; it is seldom used in pencil-and-paper analyses of diode circuits because of its algebraic complexity. Computer circuit analysis programs often use it, however, in detailed analyses.

1.2 The BJT: Large-Signal Behavior and Models

You will never see most of the BJTs that you encounter; they are part of the sealed "innards" of integrated circuits (ICs), and their behavior as circuit elements can be viewed only "in the large" as contributing to the overall input/output characteristics of the ICs. Discrete or individual BJTs are usually encountered in power amplifier circuits, headstages, or input circuits for low-noise amplifiers, and of course in the design of ICs. The reader is presumed to have a background knowledge of the large- and small-signal behavior of BJTs. We review this material in this chapter to provide continuity with material presented in later chapters.

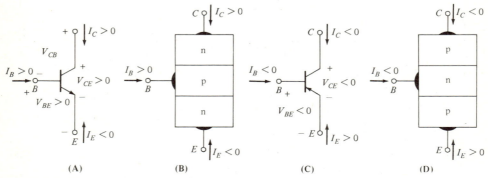

Figure 1.5 **(A)** Symbol and defined voltages and currents for an npn BJT. **(B)** "Layer cake" diagram for an npn BJT. **(C)** Symbol and defined voltages and currents for a pnp BJT. Note the signs. **(D)** Layer cake schematic for a pnp BJT.

There are two basic varieties of BJT: npn and pnp. In describing BJT large-signal behavior, we use the "layer cake" schematic representation of the construction of these devices. Figure 1.5 illustrates the symbols, the defined current directions and voltage polarities, and the layer cake models for npn and pnp BJTs.

A BJT is generally operated in the forward-biased mode in either the grounded-(common) emitter, grounded- (common) base, or common-collector (emitter-follower) configuration. In a modern analog IC such as an operational amplifier (op-amp), it is not uncommon to see all three basic BJT configurations and both npn and pnp devices used. Some manufacturers of BJTs provide volt-ampere curves for the forward-biased grounded-emitter configuration of certain transistors; that is, they provide plots of $I_C = f(V_{CE}, I_B)$. Figure 1.6(C) illustrates a set of these curves for a "typical" silicon small-signal npn BJT. The $I_C = f(V_{CE}, I_B)$ curves for a pnp BJT differ from

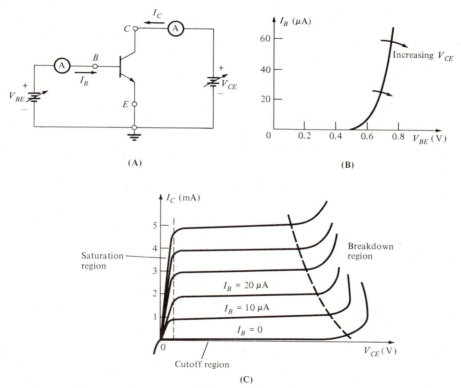

(A)   (B)

(C)

Figure 1.6 **(A)** Circuit used to plot the dc volt-ampere characteristics of an npn BJT. A = ammeter. **(B)** Base-emitter volt-ampere characteristic (at constant V_{CE}) for an npn BJT. **(C)** Collector-emitter volt-ampere characteristics for normal (forward) operation of an npn BJT.

those for an npn device in that the linear operating region is defined for $I_C < 0$, $I_B < 0$, and $V_{CE} < 0$. That is, the dc collector current actually flows out of the pnp transistor's collector; its base current flows out of its base; and so on.

A well-known large-signal nonlinear model for BJT behavior that is useful for computer simulations of transistor circuits is called the Ebers-Moll transistor model. We will describe this model for a grounded-base npn transistor. i_B, i_C, and i_E are defined as positive currents entering the base, collector, and emitter terminals, respectively. From consideration of the base-collector and base-eimitter diodes of the transistor, we have the familiar Ebers-Moll equations:

$$i_C = \alpha_F I_{ES}(e^{\lambda v_{BE}} - 1) - I_{CS}(e^{\lambda v_{BC}} - 1) \qquad\qquad 1.16A$$

$$i_E = -I_{ES}(e^{\lambda v_{BE}} - 1) + \alpha_R I_{CS}(e^{\lambda v_{BC}} - 1) \qquad\qquad 1.16B$$

$$i_B = -(i_C + i_E) \qquad\qquad 1.16C$$

where α_F is the forward alpha of the transistor, typically ranging from 0.9 to 0.998. By definition, $\alpha_F = -I_C/I_E$. The reverse alpha, defined as $\alpha_R = -I_E/I_C$, typically ranges from 0.025 to 0.5, and $\lambda = 1/V_T$. I_{ES} is the base-emitter junction diode's reverse saturation current; I_{CS} is the base-collector junction diode's reverse saturation current. To describe a pnp transistor's behavior by the Ebers-Moll equations, we change the signs of v_{BE}, v_{BC}, I_{ES}, and I_{CS} in Eqs. 1.16A and 1.16B.

It is possible to modify the Ebers-Moll equations and write them in terms of reverse currents calculated under open-circuited conditions. I_{CO} is defined as the reverse collector current calculated with the emitter open circuited ($i_E = 0$) and the base-collector junction reverse biased so that $e^{\lambda v_{BC}} \ll 1$. I_{EO} is the reverse emitter current (i.e., current into the emitter of an npn BJT) calculated with the collector open circuited ($i_C = 0$), $v_{BE} < 0$, and $e^{\lambda v_{BE}} \ll 1$. From the foregoing definitions of I_{CO} and I_{EO}, and using the Ebers-Moll Eqs. 1.16A and 1.16B, we can show that

$$I_{CO} = (1 - \alpha_F \alpha_R)I_{CS} > 0 \qquad\qquad 1.17A$$

$$I_{EO} = (1 - \alpha_F \alpha_R)I_{ES} > 0 \qquad\qquad 1.17B$$

It is also well known (Millman and Grabel, 1987) that

$$\alpha_R I_{CS} = \alpha_F I_{ES} \qquad\qquad 1.18$$

Using Eqs. 1.17A, 1.17B, and 1.18, we find that

$$\alpha_F I_{EO} = \alpha_R I_{CO} \qquad\qquad 1.19$$

If the $(e^{\lambda v_{BE}} - 1)$ term is eliminated from the Ebers-Moll equations and Eq. 1.19 is used, we obtain the useful form

$$i_C = -\alpha_F i_E - I_{CO}(e^{\lambda v_{BC}} - 1) \qquad\qquad 1.20A$$

Similarly, if $(e^{\lambda v_{BC}} - 1)$ is eliminated from the Ebers-Moll equations and Eq. 1.19 is used, we can show that

$$i_E = -\alpha_R i_C + I_{EO}(e^{\lambda v_{BE}} - 1)$$

1.20B

Equation 1.20A for the collector current can also be written in terms of the base current (Eq. 1.16C) and v_{BC}:

$$i_C = \beta_F i_B - I_{CO}(1 + \beta_F)(e^{\lambda v_{BC}} - 1)$$

1.21

where

$$\beta_F = \alpha_F/(1 - \alpha_F)$$
$$1/(1 - \alpha_F) = (1 + \beta_F)$$
$$I_{CEO} = I_{CO}(1 + \beta_F)$$

Using Eqs. 1.20A and 1.20B, we can approximate npn, BJT large-signal dc behavior by a circuit model using pn junction diodes and controlled current sources, as shown in Fig. 1.7(A). When the BJT is normally biased (i.e., when $v_{CB} > 0$, $v_{BE} > 0$, and the device is not cut off or saturated), the model of Fig. 1.7(A) can be reduced to that of Fig. 1.7(B). Note that if the reverse collector current I_{CO} is small compared to $\alpha_F i_E$, then i_C may be given by $\beta_F i_B$, as implied by Eq. 1.21.

When the BJT is saturated, $i_B > i_C/\beta_F$, and both the base-collector and base-emitter junction diodes are forward biased (i.e., $v_{BE} > 0$ and $v_{BC} > 0$). If we assume both v_{BE} and $v_{BC} \gg v_T$, then the Ebers-Moll equations 1.20A and 1.20B can be shown

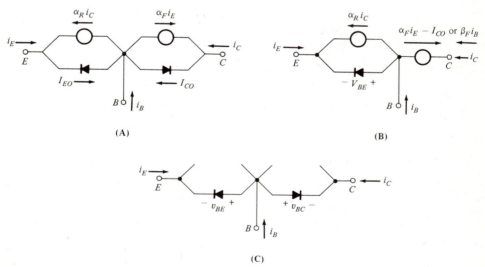

Figure 1.7 **(A)** Ebers-Moll large-signal model for an npn BJT based on Eqs. 1.20A and 1.20B. **(B)** Ebers-Moll circuit model for an npn BJT under normal, forward-biased conditions. **(C)** Saturated BJT modeled by two forward-biased pn junctions.

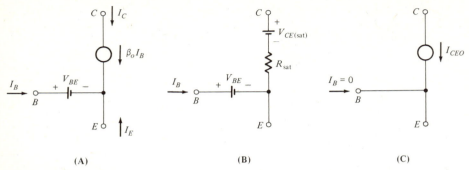

(A)   (B)   (C)

Figure 1.8 **(A)** Simplified large-signal BJT model for normal, forward-biased operation when $I_B > 0$. **(B)** Simplified model for saturation conditions. **(C)** Simplified model illustrating collector-emitter leakage current for $I_B = 0$.

to give (Ghaussi, 1985)

$$v_{CE(sat)} = v_T \ln \frac{(1/\alpha_R) + (0.9\beta_F/\beta_R)}{1 - 0.9} \qquad\qquad 1.22$$

for the condition $i_B = 1.11 i_C/\beta_F$. Typically parameter values for a small-signal silicon BJT yield a $v_{CE(sat)}$ of 0.2 V. Circuit models for a saturated BJT are shown in Fig. 1.7(C) and Fig. 1.8(B).

The large-signal Ebers-Moll model of Fig. 1.7(A) can be simplified for three conditions: normal forward-biased operation, saturation, and cutoff. These simplified piecewise-linear models are shown in Fig. 1.8. The inverted (reverse) mode of operation is seen in the input transistors of TTL logic gates when inputs are high. We will not consider reverse operation of BJTs in this text.

1.3 The BJT: Mid-Frequency Small-Signal Models

Mid-frequency small-signal models (MFSSMs) of BJT behavior are based on the assumption of linear device operation (constant BJT electrical parameters) around a fixed operating (dc biasing or Q-) point. BJT MFSSMs are used to calculate amplifier (and transistor) two-port electrical parameters, such as z-, y-, h-, and hybrid-pi parameters (see Appendix C on two-port models). In simpler terms, amplifier input and output impedance as well as voltage or current gain can be found using linear circuit analysis and one of the equivalent MFSSMs.

Three points should be stressed at this time:

1. MFSSMs do not involve dc.
2. All dc voltage sources are replaced with short circuits.
3. All dc current sources are replaced with open circuits in MFSSM analysis.

In general, the parameters in the MFSSM vary with the transistor's Q-point in (I_C, V_{CE}) space. The same MFSSM is used for both pnp and npn BJTs.

A review of early texts on transistor circuits reveals that a plethora of models were available to characterize the behavior of BJTs as small-signal linear amplifiers. These included "pi," "T," h-parameter, y-parameter, z-parameter, and ABCD-parameter models. In addition, the various models differed for grounded-base, grounded-emitter, and grounded-collector BJT operation. Fortunately, modern electronic engineering practice has reduced the need to work with great numbers of small-signal models. The MFSSMs have been distilled down to the common-emitter and common-base h-parameter models. However, the hybrid-pi model is most often used to characterize BJT small-signal behavior at high frequencies; it is considered in detail in Chapter 4. We will first take a close look at the common-emitter, mid-frequency, h-parameter model.

The common-emitter, h-parameter small-signal model is shown in Fig. 1.9(A). Note that the input (base) portion of this model is a Thevenin equivalent circuit and

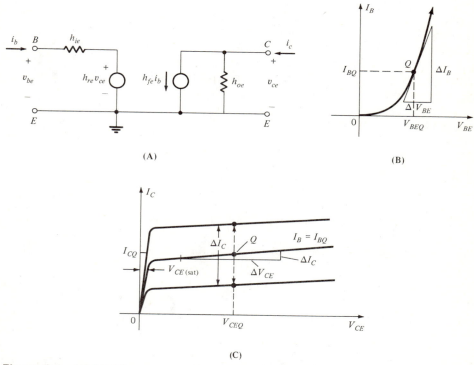

(A)

(B)

(C)

Figure 1.9   **(A)** Mid-frequency small-signal model (MFSSM) for a BJT in the common-emitter configuration using h-parameters. The model is valid for both pnp and npn transistors. **(B)** Base-emitter volt-ampere characteristic for an npn BJT showing how h_{ie} is estimated at the operating point, Q. **(C)** Collector-emitter volt-ampere curves showing how h_{fe} and h_{oe} are estimated at Q.

that the output (collector) portion is a Norton equivalent circuit. The model is valid for small-signal perturbations around the transistor's quiescent operating point. The Q-point is set by dc biasing conditions and is in the linear region of the $I_C = f(V_{CE}, I_B)$ curves. As mentioned previously, dc voltages and currents do not enter the MFSSM nor are they used in any SSM circuit calculations. The currents and voltages used in MFSSM analysis are represented by lowercase letters with lowercase subscripts.

The commonly used common- or grounded-emitter, h-parameter, BJT MFSSM, shown in Fig. 1.9(A), is valid without any sign changes for both pnp and npn transistors. The reason for this common MFSSM is seen in the following expressions defining the common-emitter h-parameters. The negative signs that appear in the numerators and denominators of the defining relations for pnp BJTs cancel, giving positive h-parameters.

The small-signal base input resistance is defined as

$$h_{ie} = \frac{v_{be}}{i_b}\bigg|_{v_{ce}} = \frac{\partial v_{BE}}{\partial i_B}\bigg|_Q \cong \frac{\Delta V_{BE}}{\Delta I_B}\bigg|_{V_{CEQ}} \qquad 1.23$$

h_{ie} can be estimated graphically from the reciprocal of the slope of the transistor's I_B-versus-V_{BE} curve at the Q-point set by the dc biasing circuitry. It also may be found by taking the ratio of the small-signal (or ac) v_{be} to i_b under the condition of $v_{ce} = 0$. v_{ce} can be made zero by connecting a large capacitor with a low reactance at the signal frequency between collector and emitter. Such a capacitor serves as an ac or small-signal collector-emitter short circuit.

Often the $I_B = f(V_{BE}, V_{CE})$ curves are not available, nor are actual measurements of v_{be}/i_b at the Q-point. In these cases, we can approximate h_{ie} by the relation

$$h_{ie} = r_x + r_\pi \qquad 1.24$$

where

r_x = the base spreading resistance, a fixed resistance with a range of 15 to 100 Ω

r_π = the dynamic resistance of the forward-biased, base-emitter pn junction

The value of r_π can be found from the elementary diode equation for the base-emitter junction:

$$i_B = I_{BS}e^{\lambda v_{BE}} \qquad 1.25$$

$$r_\pi = \frac{1}{g_\pi} = \frac{1}{\partial i_B/\partial v_{BE}} = \frac{1}{I_{BS}\lambda e^{\lambda v_{BE}}} \qquad 1.26$$

The dynamic base-emitter junction resistance at the operating point can thus be approximated by

$$r_\pi = \frac{1}{\lambda I_{BQ}} \qquad 1.27$$

where $\lambda = q/kT = 1/V_T$ and I_{BQ} is the dc quiescent base bias current which sets the operating point. λ is typically around $\frac{1}{40}$ V for silicon BJTs at room temperature. The forward current gain, h_{fe} (often called BJT *beta*), is defined by

$$h_{fe} = \frac{i_c}{i_b}\bigg|_{v_{ce}} = \frac{\partial i_C}{\partial i_B}\bigg|_Q \cong \frac{\Delta I_C}{\Delta I_B}\bigg|_Q \qquad\qquad 1.28$$

A graphical estimation of h_{fe} is easily made from the $I_C = f(V_{CE}, I_B)$ curves, as shown in Fig. 1.9(C). h_{fe} can also be measured under ac short-circuit output conditions using a large capacitor across the collector-emitter terminals, as described previously. Manufacturers usually give h_{fe} for several biasing conditions.

The small-signal, hybrid reverse voltage gain, h_{re}, is small for low and mid-frequencies and is generally set equal to zero to simplify the model and analysis. h_{re} is measured under small-signal, open-circuit base conditions. That is, no small-signal (ac) component of base current is allowed to flow in the base; only the dc base bias current, I_{BQ}, flows. h_{re} is given by

$$h_{re} = \frac{v_{be}}{v_{ce}}\bigg|_{i_b = 0} = \frac{\partial v_{BE}}{\partial v_{CE}}\bigg|_Q \cong \frac{\Delta V_{BE}}{\Delta V_{CE}}\bigg|_Q \qquad\qquad 1.29$$

The small-signal, hybrid output conductance, h_{oe}, is also found under small-signal, open-circuit base conditions. It is given by

$$h_{oe} = \frac{i_c}{v_{ce}}\bigg|_{i_b = 0} = \frac{\partial i_c}{\partial v_{CE}}\bigg|_Q \cong \frac{\Delta I_C}{\Delta V_{CE}}\bigg|_Q \qquad\qquad 1.30$$

h_{oe} is small and is often neglected in first-order circuit calculations.

In many instances, BJTs are used as grounded-base amplifiers in which the base is at small-signal ground ($v_b = 0$) and the emitter is driven as the input port. The amplified signal appears at the collector. A congruent set of four h-parameters can be found for the grounded-base (or common-base) BJT configuration. The common-base h-parameter model is shown in Fig. 1.10(A).

$$h_{ib} = \frac{v_{eb}}{i_e}\bigg|_{v_{cb}=0} = \frac{\partial v_{EB}}{\partial i_E}\bigg|_Q \cong \frac{\Delta V_{EB}}{\Delta I_E}\bigg|_Q \qquad\qquad 1.31$$

$$h_{rb} = \frac{v_{eb}}{v_{cb}}\bigg|_{i_e=0} = \frac{\partial v_{EB}}{\partial v_{CB}}\bigg|_Q \cong \frac{\Delta V_{EB}}{\Delta V_{CB}}\bigg|_Q \qquad\qquad 1.32$$

$$h_{fb} = \frac{i_c}{i_e}\bigg|_{v_{cb}=0} = \frac{\partial i_C}{\partial i_E}\bigg|_Q \cong \frac{\Delta I_C}{\Delta I_E}\bigg|_Q \qquad\qquad 1.33$$

$$h_{ob} = \frac{i_c}{v_{cb}}\bigg|_{i_e=0} = \frac{\partial i_C}{\partial v_{CB}}\bigg|_Q \cong \frac{\Delta I_C}{\Delta V_{CB}}\bigg|_Q \qquad\qquad 1.34$$

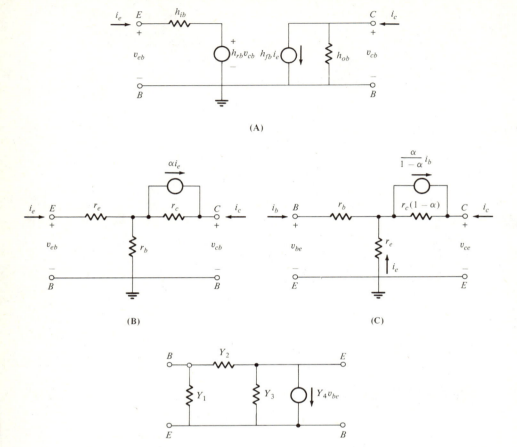

Figure 1.10 **(A)** Common-base, *h*-parameter MFSSM valid for npn and pnp BJTs. **(B)** Common-base, small-signal T model for BJTs. **(C)** Common-emitter, small-signal T model for BJTs. **(D)** Hybrid-pi, common-emitter, small-signal model for BJTs.

In addition to the *h*-parameter small-signal models, T-circuit models are sometimes used. Modern data sheets rarely specify the T-parameters shown in Figs. 1.10(B) and (C) for grounded-base and grounded-emitter BJT configurations, respectively.

Because the same transistor can be modeled using several linear configurations, there are conversion factors relating the four parameters of one two-port model to the four parameters of each of the other two-port models. Table C.1 in Appendix C summarizes some of the more commonly used conversion formulas. The common-collector *h*-parameters have deliberately been omitted from the table because they are seldom used, and circuits using BJTs with their collectors at small-signal ground (i.e., emitter-follower amplifiers) can be easily analyzed using the more frequently encountered common-emitter *h*-parameters.

A linear two-port circuit that is often used to model small-signal BJT behavior is the hybrid-pi model, illustrated in Fig. 1.10(D). This small-signal model is widely used in ac analysis of BJT circuits at high frequencies. It is shown in Appendix C on two-port models that the four parameters of the hybrid-pi model can be expressed in terms of y- and h-parameters. If we assume grounded-emitter operation, the hybrid-pi circuit's admittances can be given in terms of the well-known grounded-emitter h-parameters. Thus,

$$Y_1 = (y_{11} + y_{12}) = \frac{1 + h_{re}}{h_{ie}} \qquad\qquad 1.35$$

$$Y_2 = -y_{12} = \frac{h_{re}}{h_{ie}} \qquad\qquad 1.36$$

$$Y_3 = (y_{22} + y_{12}) = \frac{\Delta h + (-h_{re})}{h_{ie}} = h_{oe} - \frac{h_{re}(1 + h_{fe})}{h_{ie}} = g_o \qquad\qquad 1.37$$

$$Y_4 = (y_{21} - y_{12}) = \frac{h_{fe} + h_{re}}{h_{ie}} = g_m \qquad\qquad 1.38$$

In practice, for pencil-and-paper analyses we often approximate the hybrid-pi model's admittances by $Y_1 = 1/h_{ie}$, $Y_2 = 0$, $Y_3 = 0$, and $Y_4 = g_m = h_{fe}/h_{ie}$. The high-frequency version of the hybrid-pi small-signal model is described in Chapter 4, Sec. 4.1.

Because all SSM parameters are taken at a given operating point, it is important to realize that as the operating point changes, so do the numerical values of the SSM parameters. The reason is that BJT gross characteristics are nonlinear. These variations of SSM parameter values with Q-point can be plotted as normalized curves versus V_{CE} or I_C, as shown in Fig. 1.11 for the common-emitter h-parameters. Note that the base input resistance, h_{ie}, decreases monotonically with increasing collector (or base) current. Also, the forward current gain, h_{fe}, has a broad maximum in the normal I_{CQ} range and falls off for very low and very high I_{CQ}'s.

Because all of the hybrid-pi model's admittances contain h_{ie} in their denominators, they are all strong functions of the quiescent collector current of the BJT. For example, the transconductance of the hybrid-pi small-signal model can be approximated by

$$Y_4 = g_m = \frac{h_{fe}I_{BQ}}{V_T} \qquad\qquad 1.39$$

by substituting Eq. 1.27 into Eq. 1.38.

In conclusion, we observe that the common-emitter and common-base h-parameter MFSSMs are almost univerally used to characterize BJT small-signal, two-port properties. All MFSSMs use four parameters; one or two of these parameters may be set to zero to simplify pencil-and-paper circuit calculations. Because of minus sign cancellations, the same MFSSMs are valid for both pnp and npn transistors. dc voltages and currents do not enter into MFSSM formulations or calculations.

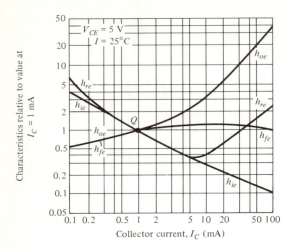

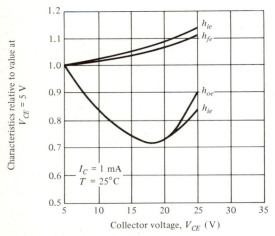

Figure 1.11 **(A)** Normalized plot of the variation of common-emitter *h*-parameters as a function of V_{CE} at constant I_C for a "typical" BJT. **(B)** Normalized plot of the variation of common-emitter *h*-parameters as a function of V_{CE} at constant I_C.

1.4 Analysis of Simple BJT Amplifier Configurations Using Small-Signal Models

In this section, we will review the two-port circuit properties of some common BJT amplifiers using linear MFSSM characterization for the transistors.

Figure 1.12(A) illustrates a simple capacitively coupled grounded-emitter BJT amplifier. Note that the emitter is at small-signal ground at mid-frequencies be-

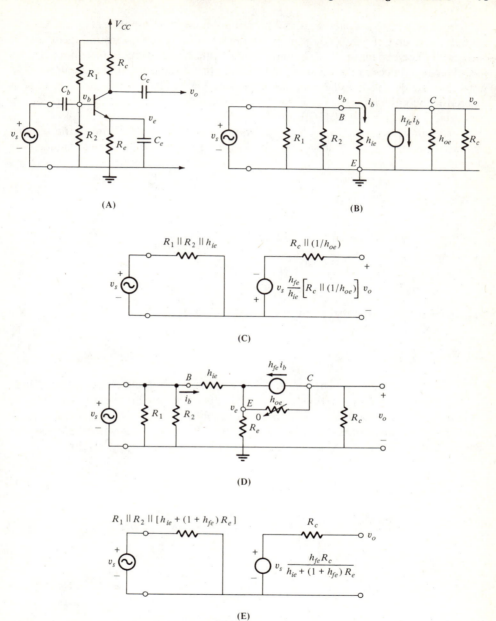

Figure 1.12   (A) Capacitively coupled, grounded-emitter, ac, BJT amplifier. (B) MFSSM for the amplifier of (A). (C) Thevenin model for the amplifier showing R_{in}, R_{out}, and the voltage-controlled voltage source derived from the circuit of (B). (D) MFSSM for the ac amplifier when $C_e = 0$. (E) Thevenin model derived from the amplifier of (D).

cause of the bypass capacitor, C_e. The common-emitter, h-parameter MFSSM is used in the circuit of Fig. 1.12(B) to represent the BJT's behavior around its dc operating point. The capacitors C_b, C_c, and C_e are treated as mid-frequency short circuits (i.e., their reactances are considered to be negligible at the operating frequency). Also note that no dc terms appear in the circuit of Fig. 1.12(B). The V_{CC} node is considered to be at small-signal ground; hence circuit branches going to V_{CC} are shown grounded (R_1, R_c) in Fig. 1.12(B). We have also considered $h_{re} = 0$.

The input resistance seen by v_s is found by inspection; it is simply the parallel combination of R_1, R_2, and h_{ie}, as shown in Fig. 1.12(C). The small-signal base current, i_b, is simply

$$i_b = \frac{v_s}{h_{ie}} \qquad\qquad 1.40$$

The open-circuit voltage at the amplifier's output is

$$v_o = \frac{-h_{fe}i_b}{h_{oe} + (1/R_c)} = -v_s\left(\frac{h_{fe}}{h_{ie}}\right)\left(\frac{R_c}{h_{oe}R_c + 1}\right) \qquad\qquad 1.41$$

and the Thevenin output resistance is simply R_c in parallel with $1/h_{oe}$:

$$R_{eq} = \frac{R_c}{h_{oe}R_c + 1} \qquad\qquad 1.42$$

If the emitter bypass capacitor, C_e, is omitted, the circuit model must include R_e, as shown in Fig. 1.12(D). Because h_{oe} is generally much smaller than G_c, we loose little accuracy in the calculation of the amplifier parameters by setting $h_{oe} = 0$. Now the input resistance that v_s sees is R_1 and R_2 in parallel with the resistance seen looking directly into the base. The latter resistance is v_s/i_b. i_b is simply

$$i_b = \frac{v_s - v_e}{h_{ie}} \qquad\qquad 1.43$$

Now v_e is, by Ohm's law,

$$v_e = i_b(1 + h_{fe})R_e = \frac{v_s}{h_{ie}}(1 + h_{fe})R_e \qquad\qquad 1.44$$

So we have, from Eqs. 1.43 and 1.44,

$$i_b = \frac{v_s - i_b(1 + h_{fe})R_e}{h_{ie}} \qquad\qquad 1.45$$

Equation 1.45 can be solved to get

$$\frac{v_s}{i_b} = h_{ie} + (1 + h_{fe})R_e \qquad\qquad 1.46$$

which is the conductance looking into the base. i_b is greatly reduced by the unbypassed R_e, as is the output open-circuit voltage gain, v_o/v_s. Compare Eq. 1.47 with Eq. 1.41.

$$v_o = -h_{fe}i_b R_c = -v_s \frac{h_{fe}R_c}{h_{ie} + (1 + h_{fe})R_e} \tag{1.47}$$

The Thevenin equivalent output resistance, R_{eq}, is simply R_c, by inspection.

Now let us examine the properties of the grounded-base BJT amplifier using MFSSM analysis. The grounded-base configuration is often used because of its good high-frequency response; it also has noninverting gain. A grounded-base amplifier is shown in Fig. 1.13(A). Figure 1.13((B) illustrates the linear MFSSM for this circuit.

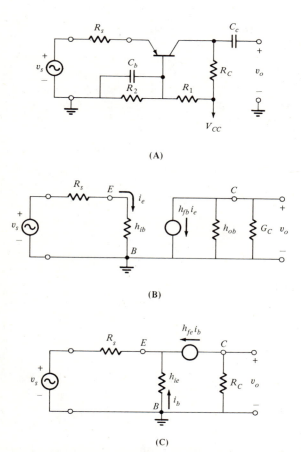

(A)

(B)

(C)

Figure 1.13 **(A)** Grounded-base, pnp, BJT amplifier for ac signals. C_b and C_c are considered to be ac short circuits at the operating frequency. **(B)** Small-signal model for the grounded-base amplifier in (A) using common-base h-parameters. **(C)** Small-signal model for the same amplifier using common-emitter h-parameters.

The common-base h-parameter model for the BJT is used with $h_{rb} = 0$. By inspection, the input resistance is

$$R_{in} = R_s + h_{ib} \tag{1.48}$$

and the Thevenin output resistance is just

$$R_{eq} = \frac{1}{h_{ob} + G_c} \tag{1.49}$$

The Thevenin open-circuit output voltage can be written as

$$v_o = \frac{-h_{fb}i_e}{h_{ob} + G_c} = \frac{-v_s}{R_s + h_{ib}} \frac{h_{fb}}{h_{ob} + G_c} = \frac{v_s|h_{fb}|}{(R_s + h_{ib})(h_{ob} + G_c)} \tag{1.50}$$

Note that the voltage gain for the grounded-base amplifier is noninverting, because h_{fb} is negative and the first term of Eq. 1.45 for v_o also has a minus sign.

Analysis of the grounded-base amplifier is more complex if the common-emitter h-parameter model for the BJT is used (see Fig. 1.13[C]). Analysis is made simpler if we set $h_{oe} = 0$. The node voltage v_e must be found. By Kirchhoff's current law for the unbypassed emitter node,

$$v_e\left(\frac{G_s + 1}{h_{ie}}\right) - h_{fe}i_b = v_sG_s \tag{1.51}$$

i_b is found from Ohm's law:

$$i_b = -\frac{v_e}{h_{ie}} \tag{1.52}$$

Substitution of Eq. 1.52 into Eq. 1.51 yields

$$v_e = \frac{v_sh_{ie}}{R_s(1 + h_{fe}) + h_{ie}} \tag{1.53}$$

Now the input resistance seen by v_s is

$$R_{in} = \frac{v_s}{i_s} = \frac{v_s}{\dfrac{v_s - v_e}{R_s}} = \frac{v_sR_s}{v_s - \dfrac{v_sh_{ie}}{R_s(1 + h_{fe}) + h_{ie}}} \tag{1.54}$$

which easily reduces to

$$R_{in} = R_s + \frac{h_{ie}}{1 + h_{fe}} = R_s + h_{ib} \tag{1.55}$$

The Thevenin open-circuit output voltage is just

$$v_o = -h_{fe}i_bR_c = +h_{fe}R_c\frac{v_e}{h_{ie}}$$

1.56

Substitution of Eq. 1.53 into Eq. 1.56 yields

$$v_o = +\frac{h_{fe}R_c}{h_{ie}}\frac{v_sh_{ie}}{R_s(1 + h_{fe}) + h_{ie}}$$

1.57

Equation 1.57 is equivalent to the gain given by Eq. 1.50 with $h_{ob} = 0$.

The third basic single-BJT amplifier circuit we want to consider is the emitter-follower, or grounded-collector circuit, a version of which is shown in Fig. 1.14(A). Emitter-followers are primarily used as power amplifiers. They have slightly less than unity voltage gain, high input impedance, low output impedance, and very good high-frequency bandwidth. They are often used to drive long transmission lines.

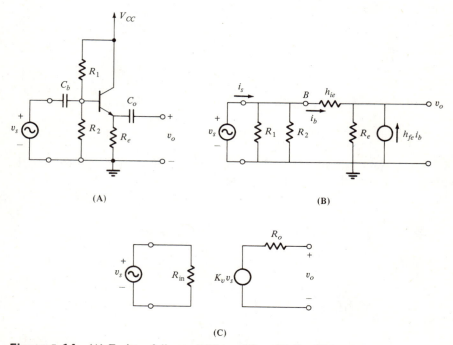

(A)

(B)

(C)

Figure 1.14 **(A)** Emitter-follower BJT amplifier. **(B)** Small-signal model for the emitter-follower; note that C_b and C_o have negligible reactance at the operating frequency. **(C)** Thevenin equivalent model for the emitter-follower found from the circuit of **(B)**.

The MFSSM for the emitter-follower is illustrated in Fig. 1.14(B). Note again that the coupling capacitors C_b and C_o are considered to be short circuits in the operating range of frequencies and that V_{CC} is at small-signal ground potential for R_1 and the BJT's collector.

The open-circuit output voltage is, by Ohm's law and Kirchhoff's current law,

$$v_o = (1 + h_{fe})i_b R_e \tag{1.58}$$

The base current is, by Ohm's law,

$$i_b = \frac{v_s - v_o}{h_{ie}} = \frac{v_s - (1 + h_{fe})i_b R_e}{h_{ie}} \tag{1.59}$$

Equation 1.59 is solved for i_b:

$$i_b = \frac{v_s}{h_{ie} + (1 + h_{ie})R_e} \tag{1.60}$$

The input resistance of the emitter-follower is

$$R_{in} = \frac{v_s}{i_s} = R_1 \| R_2 \| [h_{ie} + (1 + h_{fe})R_e] \tag{1.61}$$

The voltage gain is found by substituting Eq. 1.60 into Eq. 1.58:

$$k_v = \frac{v_o}{v_s} = \frac{R_e(1 + h_{fe})}{h_{ie} + (1 + h_{fe})R_e} \tag{1.62}$$

The Thevenin output resistance (R_o) of the emitter-follower is found by dividing the open-circuit voltage given in Eq. 1.62 by the short-circuit output current given by

$$i_{sc}\Big|_{v_o = 0} = i_b(1 + h_{fe}) = \frac{v_s}{h_{ie}}(1 + h_{fe}) \tag{1.63}$$

Thus, we have

$$R_o = \frac{v_o}{i_{sc}} = \frac{R_e \left(\dfrac{h_{ie}}{1 + h_{fe}} \right)}{R_e + \left(\dfrac{h_{ie}}{1 + h_{fe}} \right)} \tag{1.64}$$

R_o is seen to be R_e in parallel with $h_{ie}/(1 + h_{fe})$. The complete equivalent circuit for the emitter-follower operated under mid-frequency small-signal conditions is illustrated in Fig. 1.14(C), where the parameters R_{in}, K_v, and R_o are given by the expressions just derived.

From the small-signal analysis of the three simple BJT amplifiers discussed here, we may offer some generalizations about driving-point and output resistances. The emitter is a low-resistance point, either as an input or as an output. The base has a medium to high driving-point resistance; it is low for grounded-emitter amplifiers and high for emitter-followers. The collector always appears as a high source resistance regardless of amplifier configuration; it appears as a very high driving-point resistance when the emitter resistance is unbypassed (this property is examined in Chapter 5, Eqs. 5.21 through 5.24). The voltage gains for the grounded-base and grounded-emitter amplifiers are large in magnitude; the gain for an emitter-follower is slightly less than unity.

1.5 The FET: Large-Signal Behavior and Models

Whereas there are only two basic varieties of BJT (npn and pnp), there are seven categories of commercially important field-effect transistors. There are p- and n-channel junction field-effect transistors (JFETs), which are always operated with their gate-channel junctions reverse biased. Metal oxide semiconductor field-effect transistors (MOSFETs) also can be subdivided into p- and n-channel devices, which in turn can be operated in the depletion or enhancement mode. The seventh category of FET utilizes n-doped gallium arsenide as the channel material. The gate is a Schottky metal barrier on the top surface; the drain and source, also on top, are ohmic contacts. Gallium arsenide is used in the substrate of this ultra high frequency MESFET.

In this section, we will examine the large-signal volt-ampere characteristics of certain types of FETs and give algebraic models relating drain current to other circuit parameters.

Typical n-channel JFET volt-ampere curves are shown in Fig. 1.15(A). Note that these curves are subdivided into an ohmic or triode region, a linear saturation or pinchoff region, and a breakdown region.

In the ohmic region, the drain current can be approximated by

$$I_D = I_{DSS}\left[2\left(1 + \frac{V_{GS}}{V_P}\right)\left(\frac{V_{DS}}{V_P}\right) - \left(\frac{V_{DS}}{V_P}\right)^2\right] \quad \text{for} \quad 0 < V_{DS} < |(V_{GS} + V_P)| \qquad 1.65$$

For $V_{GS} \ll V_P$, Eq. 1.65 can be approximated by

$$I_D = I_{DSS}\left[2\left(1 + \frac{V_{GS}}{V_P}\right)\left(\frac{V_{DS}}{V_P}\right)\right] \qquad\qquad 1.66$$

where I_{DSS} is the drain current measured under the conditions of $V_{GS} = 0$ and $V_{DS} > |(V_{GS} + V_P)|$. V_P is the pinchoff voltage; that is, $V_P = V_{GS}$ where $I_D \rightarrow 0$, given $V_{DS} > |(V_{GS} + V_P)|$. (Note that when $V_{DS} > |(V_{GS} + V_P)|$, the FET is the pinchoff region.)

From Eq. 1.66, we see that the $I_D = f(V_{DS}, V_{GS})$ curves go through the origin; their slope is a function of V_{GS} and can be considered to be a voltage-controlled

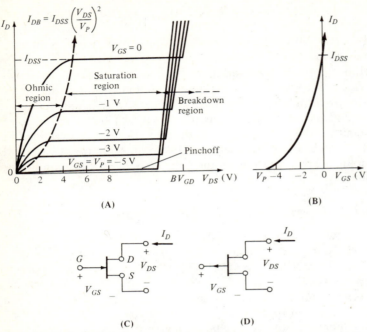

Figure 1.15 **(A)** Drain-source volt-ampere curves for an n-channel JFET. **(B)** I_D-vs.-V_{GS} curve for an n-channel JFET. **(C)** Symbol for an n-channel JFET. **(D)** Symbol for a p-channel JFET.

conductance, g_{ds}:

$$g_{ds} = \frac{\partial i_D}{\partial v_{DS}}\bigg|_{v_{GS}} = I_{DSS}\left[\frac{2\left(1 + \dfrac{V_{GS}}{V_P}\right)}{V_P}\right] \tag{1.67}$$

As long as $V_{DS} \ll |(V_{GS} + V_P)|$ and Eq. 1.66 applies, the drain-source terminals of the JFET can be used as a voltage-controlled conductance in such applications as automatic volume control circuits (AVCs), gain control circuits (attenuators), and voltage-variable active filters.

In the saturation pinchoff region, the drain current is given by

$$I_D = I_{DSS}\left(1 - \frac{V_{GS}}{V_P}\right)^2 \quad \text{for} \quad V_{DS} > |(V_{GS} + V_P)| \tag{1.68}$$

Here the $I_D = f(V_{DS}, V_{GS})$ curves are nearly horizontal for constant V_{GS} in the range $0 > V_{GS} > V_P$. The square-law characteristic of the saturation region of the JFET can pose problems in the design of linear amplifiers because of the generation of even harmonics. However, it makes JFETs good low-noise mixers in superheterodyne radio systems, among others.

The saturation region is separated from the ohmic region by a parabolic boundary; that is, the JFET is in the ohmic region for $I_D > I_{DB}$ at constant V_{DS} and in the saturation region for $I_D < I_{DB}$. I_{DB} is easily found to be

$$I_{DB} = I_{DSS}\left(\frac{V_{DS}}{V_P}\right)^2 \tag{1.69}$$

When p-channel JFETs are considered, I_D and V_{DS} are negative, and normal, reverse-biased V_{GS} (and V_P) are positive, relative to the directions defined for the n-channel JFET shown in Fig. 1.15(C). The transconductance of a JFET, defined by

$$g_m = \left.\frac{\partial i_D}{\partial v_{GS}}\right|_{v_{ds}} \tag{1.70}$$

in the saturation region, can be found from the slope of the I_D-versus-V_{GS} curve (see Fig. 1.15[B]). For the n-channel JFET, this slope is obviously positive over the range $0 > V_{GS} > V_P$. For p-channel JFETs, I_D is negative and V_{GS} is positive over the range $0 < V_{GS} < V_P$. Equation 1.68 for $I_D = f(V_{DS}, V_{GS})$ still applies, however, with I_{DSS} being negative. Thus the I_D-versus-V_{GS} curve also has a positive slope for p-channel JFETs, and JFET transconductance is obviously also positive.

The volt-ampere characteristics of an n-channel enhancement-type MOSFET are shown in Fig. 1.16(A). MOSFET symbols may vary between references; when the substrate (body) is electrically accessible, a terminal B is added as shown in Fig. 1.16(C). The n-channel of this MOSFET exists in a p-substrate because minority carriers (electrons) in the substrate are attracted to the channel region by the positive

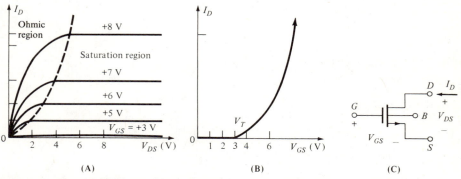

(A)   (B)   (C)

Figure 1.16 **(A)** Drain-source volt-ampere curves for an n-channel enhancement MOSFET. **(B)** I_D-vs.-V_{GS} curve for the n-channel enhancement MOSFET. **(C)** Symbol for the n-channel enhancement MOSFET.

electric field set up by a gate-source potential greater than V_T. V_T is the MOSFET's threshold voltage (analogous to a JFET's V_P). At $V_{GS} > V_T$, the n-channel forms, and if $|V_{DS}| > |(V_{GS} - V_T)|$, the n-channel saturates and a nearly constant $I_D = f(V_{DS}, V_{GS})$ characteristic is observed in the saturation region. I_D is described algebraically in the saturation region by

$$I_D = K(V_{GS} - V_T)^2 = \frac{K}{V_T^2}\left(\frac{V_{GS}}{V_T} - 1\right)^2 \quad \text{for} \quad 0 < |(V_{GS} - V_T)| < V_{DS}| \qquad 1.71$$

The n-channel enhancement MOSFET also has a linear or ohmic channel operating region in which

$$I_D = K[2(V_{GS} - V_T)V_{DS} - V_{DS}^2] \quad \text{for} \quad 0 < |V_{DS}| < |(V_{GS} - V_T)| \qquad 1.72$$

In Eqs. 1.71 and 1.72, K is a constant that is a function of the MOSFET's geometry and certain physical constants. The equation of the line separating the ohmic and saturation operating regions can be shown to be

$$I_{DB} = KV_{DS}^2 \qquad 1.73$$

In obtaining the preceding equations for the n-channel enhancement-type MOSFET, it was assumed that the substrate (B) was tied to the source lead. If the substrate is made positive with respect to the source, V_T is increased, and the I_D-versus-V_{GS} curve is shifted to the right, as shown in Fig. 1.16(B).

In n-channel depletion-type MOSFETs, a lightly doped p-substrate underlies a lightly doped n-channel connecting two heavily doped n-regions for source and drain. In this device, the gate-source transition voltage is a negative pinchoff voltage, and the I_{DSS} symbol may be used to describe the I_D that flows for $V_{GS} = 0$ under saturated channel conditions. Figure 1.17(A) illustrates the volt-ampere characteristics for an n-channel depletion-type MOSFET; Fig. 1.17(B) shows the ID-versus-V_{GS} curve for saturated channel conditions; and the symbol for this type of MOSFET is given in Fig. 1.17(C). In the saturation region,

$$I_D = I_{DSS}\left(1 - \frac{V_{GS}}{V_P}\right)^2 \quad \text{for} \quad V_{DS} > |(V_{GS} + V_P)| > 0 \qquad 1.74$$

And in the ohmic region,

$$I_D = I_{DSS}\left[2\left(1 + \frac{V_{GS}}{V_P}\right)\left(\frac{V_{DS}}{V_P}\right) - \left(\frac{V_{DS}}{V_P}\right)^2\right] \quad \text{for} \quad 0 < V_{DS} < |(V_{GS} + V_P)| \qquad 1.75$$

In analog IC design, we encounter NMOS technology (n-channel MOSFETs) more frequently than PMOS devices (p-channel MOSFETs). There are several reasons for this dominance by NMOS. The mobility of carriers in n-silicon is more than twice that in p-silicon; hence NMOSFETs are faster and, when they are used as switches, their ON resistance is smaller than that of a comparable p-channel device.

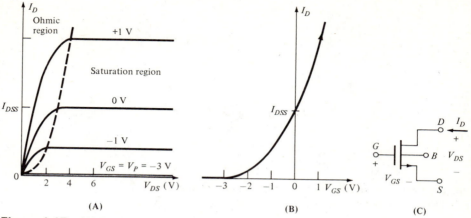

Figure 1.17 **(A)** Drain-source volt-ampere curves for an n-channel depletion MOSFET. **(B)** I_D-vs.-V_{GS} curve for the n-channel depletion MOSFET. **(C)** Symbol for the n-channel depletion MOSFET. Note that it is the same as for the n-channel enhancement MOSFET.

The constant K in Eq. 1.71 is larger for an NMOSFET than for a comparable PMOSFET of the same size. PMOSFETs are frequently used to make complementary symmetry output stages in linear amplifiers and logic circuits. Their use in the latter application has given rise to the acronym CMOS. CMOS architecture is used in the family of logic devices of that name as well as in memories and microprocessors. CMOS logic design and behavior are treated in detail in Taub and Schilling (1977) and in Gray and Meyer (1984).

We have avoided treatment of PMOSFETs in this chapter for reasons of brevity. As in the case of JFETs, the signs of their parameters (I_D, V_{GS}, V_{DS}, V_P, and V_T) are the opposite of the corresponding NMOSFET parameters, which we considered here.

1.6 The FET: Mid-Frequency Small-Signal Models

Although there are at least seven basic varieties of FET, we are indeed fortunate in being able to represent all of them by one MFSSM, illustrated in Fig. 1.18. The low-frequency resistance seen looking into the reverse-biased JFET gate is very high, of the order of teraohms. The low-frequency resistance seen looking into a MOSFET gate is even higher, so we are justified in most cases by representing the gate terminal by an open circuit. In the saturated channel operating region, the drain-source characteristics are well modeled for mid-frequency, small-signal operating conditions by either a Norton or a Thevenin equivalent circuit. In the Norton model, the FET's transconductance, g_m, at the quiescent operating point can be found experimentally by evaluating

$$g_m \cong \left. \frac{\Delta I_D}{\Delta V_{GS}} \right|_{Q, \Delta V_{DS} = 0} \qquad\qquad 1.76$$

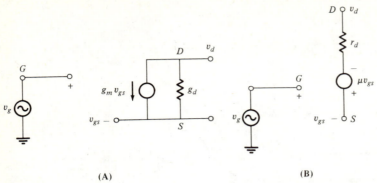

Figure 1.18 **(A)** Current-source (Norton) small-signal model for all FETs. **(B)** Voltage-source (Thevenin) small-signal model for all FETs.

g_m can also be found analytically from differentiation of expressions for $I_D = f(V_{GS})$ in the pinchoff operating region:

$$g_m = \left.\frac{\partial I_D}{\partial v_{GS}}\right|_{Q,v_{ds}} = I_{DSS}\left(-\frac{2}{V_P} + \frac{2V_{GS}}{V_P^2}\right) = I_{DSS}\left[\left(\frac{2}{V_P}\right)\left(\frac{V_{GS}}{V_P} - 1\right)\right]$$

$$= \frac{2I_{DSS}}{|V_P|}\left(1 - \frac{V_{GS}}{V_P}\right) \qquad\qquad 1.77$$

We note from Eq. 1.68 that I_D can be written as

$$\sqrt{I_{DQ}} = \sqrt{I_{DSS}}\left(1 - \frac{V_{GSQ}}{V_P}\right) \qquad\qquad 1.78$$

We can substitute Eq. 1.78 into Eq. 1.77 and obtain

$$g_m = \frac{2I_{DSS}}{|V_P|}\sqrt{\frac{I_{DQ}}{I_{DSS}}} = g_{mo}\sqrt{\frac{I_{DQ}}{I_{DSS}}} \qquad\qquad 1.79$$

Equation 1.79 gives us an algebraic means of estimating the JFET's g_m in terms of the drain current at the operating point, I_{DQ}, and the manufacturer-specified parameters I_{DSS} and V_P.

In addition to the g_m, the MFSSM Norton model contains an output conductance, g_d. g_d accounts for the slight upward slope of the actual I_D-versus-V_{DS} curves in the saturation region. The output conductance may be estimated by

$$g_d \cong \left.\frac{\Delta I_D}{\Delta V_{DS}}\right|_{Q,\Delta V_{GS}=0} \qquad\qquad 1.80$$

If we include a product term as shown in Eq. 1.81, we can modify the saturated channel equation for I_D, given by Eq. 1.68, to include the effect of finite upward slope in the $I_D = |f(V_{DS}, V_{GS})$ curves as V_{DS} increases:

$$I_D = I_{DSS}\left(1 - \frac{V_{GS}}{V_P}\right)^2 \left(1 - \frac{V_{DS}}{V_X}\right) \qquad 1.81$$

where V_X is an intercept voltage, typically 25 V to 50 V in magnitude. V_X is negative for n-channel JFETs and positive for p-channel FETs (see Fig. 1.19[A]).

g_d can be found analytically from Eq. 1.81:

$$g_d = \frac{\partial I_D}{\partial V_{DS}} = I_{DSS}\left(1 - \frac{V_{GS}}{V_P}\right)^2 \left(\frac{-1}{V_X}\right) = -\frac{I_{DQ}}{V_X} > 0 \qquad 1.82A$$

or

$$g_d = \frac{|I_{DQ}|}{|V_X|} \qquad 1.82B$$

An estimate of V_X can be obtained by extrapolating the $I_D = |f(V_{DS}, V_{GS})$ curves in the saturation region back to the negative V_{SD} axis as shown in Fig. 1.19(A).

The Thevenin MFSSM of the JFET's drain-source port provides an alternate description for the FET's electrical behavior around its Q-point. The series resistor is just $r_d = 1/g_d$. The voltage-controlled voltage source is given the gain μ V/V, which can be expressed as

$$\mu = \frac{g_m}{g_d} = \left(\frac{\partial I_D}{\partial V_{GS}}\right)\left(\frac{\partial V_{DS}}{\partial I_D}\right) = \frac{\partial V_{DS}}{\partial V_{GS}}\bigg|_{Q, \Delta I_D = 0} \qquad 1.83$$

Substitution of Eq. 1.79 for g_m and Eq. 1.82(B) for g_d into Eq. 1.83 yields

$$\mu = g_{mo}\sqrt{\frac{I_{DQ}}{I_{DSS}}} \frac{|V_X|}{|I_{DQ}|} = \frac{g_{mo}|V_X|}{\sqrt{I_{DQ}I_{DSS}}} \qquad 1.84$$

μ is seen to vary inversely with I_{DQ} in the saturation region of operation. It is difficult to estimate graphically using the formula given in Eq. 1.83 and is best found using Eq. 1.84.

To illustrate the generality of the MFSSM parameters that we have derived, consider an n-channel enhancement-type MOSFET (see Fig. 1.16). Its drain current is given by Eq. 1.71. Hence its transconductance is found to be

$$g_m = \frac{\partial I_D}{\partial V_{GS}}\bigg|_Q = K(2V_{GS} - 2V_T) = 2\left(\frac{K}{V_T}\right)\left(\frac{V_{GS}}{V_T} - 1\right) = \frac{g_{m(2)}}{\sqrt{I_{D(2)}}}\sqrt{I_{DQ}} \qquad 1.85$$

where $g_{m(2)}$ is the transconductance measured when $V_{GSQ} = 2V_T$ and the transistor channel is saturated. $I_{D(2)}$ is the drain current measured under the same conditions.

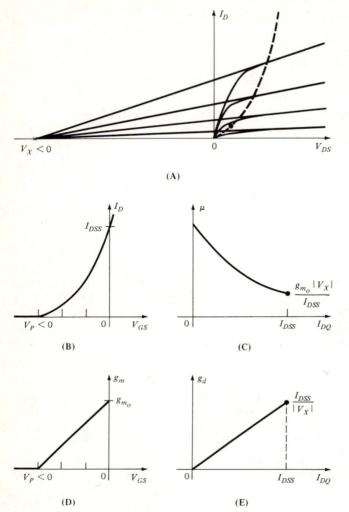

Figure 1.19 **(A)** Drain-source volt-ampere curves for an n-channel JFET showing the intercept voltage, V_X, used to characterize the FET's small-signal drain resistance, r_d. **(B)** I_D-vs.-V_{DS} curve for the JFET of (A). **(C)** Variation of the voltage-controlled voltage source gain, μ, with I_D for the JFET. **(D)** Variation of the voltage-controlled current source transconductance, g_m, with V_{GS} for the JFET.
(E) Variation of the JFET's small-signal drain conductance, g_d, with I_D. Note that g_d is a function of V_X, defined in (A).

Note the similarity of Eq. 1.85 for the g_m of a MOSFET and Eq. 1.79 for the g_m of a JFET; both vary as the square root of the quiescent drain current in the saturation region.

Inspection of the n-channel enhancement MOSFET's $I_D = f(V_{DS}, V_{GS})$ curves in Fig. 1.16 shows that the approximation made in Eq. 1.81 for the JFET will also work for this device:

$$I_D = K(V_{GS} - V_T)^2 \left(1 - \frac{V_{DS}}{V_X}\right) \qquad 1.86$$

(Note that V_X is negative for the n-channel MOSFET.) The output conductance is thus

$$g_d = \left.\frac{\partial I_D}{\partial V_{DS}}\right|_Q = K(V_{GS} - V_T)^2 \left(-\frac{1}{V_X}\right) = \frac{|I_{DQ}|}{|V_X|} \qquad 1.87$$

which is the same result found for the g_d of the JFET.

In summary, a common low- and mid-frequency small-signal model is used for all types of FETs. It can be represented as a Norton or Thevenin equivalent model for the FET's drain-source port. The small-signal parameters μ, g_m, and $g_d = 1/r_d$ vary approximately as shown in Figs. 1.19(C), (D), and (E), respectively.

1.7 Analysis of Simple FET Amplifier Configurations Using Small-Signal Models

Just as we saw in Section 1.4, there are three basic single-transistor amplifier circuits, which we will consider here. A p-channel JFET is shown in a grounded-source amplifier in Fig. 1.20(A). Note that the same small-signal model can be used for n-channel JFETs and all types of MOSFETs. We treat the coupling and bypass capacitors as short circuits at operating frequencies in the small-signal model of this circuit shown in Fig. 1.20(B).

The three parameters characterizing this circuit can be found by inspection. The input resistance, R_{in}, seen by v_1 is just R_g. The Thevenin output resistance, R_o, is

$$R_o = \frac{1}{g_d + G_d} \qquad 1.88$$

The voltage gain, K_v, is

$$K_v = \frac{v_o}{v_1} = -\frac{g_m}{g_d + G_d} \qquad 1.89$$

because

$$v_{gs} = v_g - v_s = v_1 \qquad 1.90$$

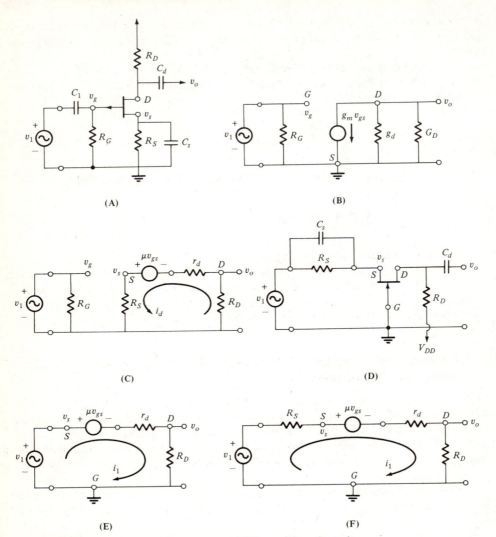

(A)   (B)   (C)   (D)   (E)   (F)

Figure 1.20 **(A)** Grounded-source, ac, JFET amplifier. Capacitors C_1, C_s, and C_d are assumed to be ac short circuits. **(B)** Small-signal model of the grounded-source amplifier. **(C)** Small-signal model of the grounded-source amplifier when $C_s = 0$. Now the source is no longer at ac ground. **(D)** Grounded-gate amplifier using a JFET. **(E)** Small-signal model of the grounded-gate amplifier. **(F)** Small-signal model of the grounded-gate amplifier when $C_s = 0$.

If R_S is left unbypassed ($C_s = 0$), the input resistance is unchanged; however, the voltage gain is more easily found in a pencil-and-paper calculation using the Thevenin small-signal models for FETs in which

$$\mu = g_m r_d \qquad\qquad 1.91$$

By Ohm's law, the mesh current, i_d, is

$$i_d = \frac{\mu(v_1 - v_s)}{R_S + r_d + R_D} \qquad\qquad 1.92$$

If we substitute

$$v_s = i_d R_S \qquad\qquad 1.93$$

into Eq. 1.92 and solve for i_d, we get

$$i_d = \frac{\mu v_1}{r_d + R_D + R_S(\mu + 1)} \qquad\qquad 1.94$$

from which K_v may be easily found:

$$K_v = \frac{v_o}{v_1} = -\frac{\mu R_D}{r_d + R_D + R_S(\mu + 1)} \qquad\qquad 1.95$$

Note that a substantial reduction in K_v can be obtained by not bypassing R_S at signal frequencies.

The output resistance of the grounded-source amplifier can also be found from Fig. 1.20(C) by use of Thevenin's theorem:

$$R_o = \frac{v_{ooc}}{i_{osc}} \qquad\qquad 1.96$$

Using Eq. 1.94 for i_d and setting $R_D = 0$, we find i_{osc} to be

$$i_{osc} = -\frac{\mu v_1}{r_d + R_S(\mu + 1)} \qquad\qquad 1.97$$

Thus R_o can be determined from Eqs 1.95 and 1.97:

$$R_o = \frac{R_D[r_d + R_S(\mu + 1)]}{R_D + r_d + R_S(\mu + 1)} \qquad\qquad 1.98$$

R_o is simply $R_D \| [r_d + R_S(\mu + 1)]$. The second term is the resistance seen looking directly into the drain of the FET.

Next, consider the FET grounded-gate amplifier shown in Fig. 1.20(D). With the assumption that C_s bypasses R_s completely, the MFSSM of the grounded-base amplifier is as shown in Fig. 1.20(E). Because in this model

$$v_{gs} = v_g - v_s = -v_1 \tag{1.99}$$

the voltage across r_d and R_D to ground is just $(1 + \mu)v_1$; hence the voltage gain of this circuit, by the voltage-divider equation, is simply

$$K_v = \frac{v_o}{v_1} = \frac{(\mu + 1)R_D}{R_D + r_d} \tag{1.100}$$

The output resistance is, by inspection,

$$R_o = R_D \| r_d \tag{1.101}$$

and the input resistance that v_1 sees is

$$R_{in} = \frac{v_1}{i_1} = \frac{v_1}{\dfrac{(\mu + 1)v_1}{R_D + r_d}} = \frac{R_D + r_d}{\mu + 1} \tag{1.102}$$

When R_S is left unbypassed, analysis is a bit more complex; the MFSSM for this condition is shown in Fig. 1.20(F). By inspection, we note that v_1 now sees an R_{in} of

$$R_{in} = \frac{v_1}{i_1} = R_1 + \frac{R_D + r_d}{1 + \mu} \tag{1.103}$$

v_s is just

$$v_s = v_1 + i_1 R_S \tag{1.104}$$

Hence v_{gs} is

$$v_{gs} = -(v_1 - i_1 R_S) \tag{1.105}$$

i_1 can be written by Ohm's law:

$$i_1 = \frac{v_1 - \mu v_{gs}}{R_S + v_d + R_D} = \frac{v_1(\mu + 1) - \mu i_1 R_S}{R_S + v_d + R_D} \tag{1.106}$$

This reduces to

$$i_1 = \frac{v_1(\mu + 1)}{r_d + R_D + R_S(\mu + 1)} \qquad\qquad 1.107$$

The voltage gain for this condition is seen to be

$$K_v = \frac{v_o}{v_1} = \frac{(\mu + 1)R_D}{r_d + R_D + R_S(\mu + 1)} \qquad\qquad 1.108$$

and, from the previous example, the Thevenin output resistance can easily be shown to be

$$R_o = R_D \| [r_d + R_S(1 + \mu)] \qquad\qquad 1.109$$

The third basic single-FET amplifier configuration to be treated here is the grounded-drain amplifier, or source-follower, shown in Fig. 1.21(A). Its input resistance is, by inspection,

$$R_{in} = \frac{v_1}{i_1} = R_G \qquad\qquad 1.110$$

and because

$$v_{gs} = v_g - v_s = v_1 - v_o \qquad\qquad 1.111$$

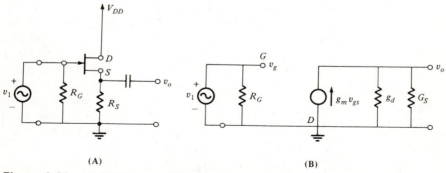

(A)   (B)

Figure 1.21   **(A)** JFET source-follower. **(B)** Small-signal model of the source-follower.

the voltage gain can be shown to be

$$K_v = \frac{v_o}{v_1} = \frac{g_m}{g_m + g_d + G_S} \qquad\qquad 1.112$$

which is clearly positive and less than unity, as it was for the BJT emitter-follower. The source-follower's output resistance can be found from

$$R_o = \frac{v_{ooc}}{i_{osc}} = \frac{K_v v_1}{g_m v_{gs}} = \frac{\dfrac{g_m v_1}{g_m + g_d + G_S}}{g_m v_1} = \frac{1}{g_m + g_d + G_S} \qquad\qquad 1.113$$

R_o is just the parallel combination of r_d, R_S, and $1/g_m$.

In summary, the terminal properties of the three basic FET amplifiers are similar to those of the three basic BJT amplifiers, with the exception that looking into the FET's gate at low- and mid-frequencies, we generally see a very high input resistance in parallel with a large gate *leakage* resistance, R_g. The FET's source generally has a low driving point or source resistance, and the FET's drain is a high-resistance driving point or source.

1.8 Mid-Frequency Small-Signal Analysis of Circuits with Two Transistors

Modern analog ICs typically have from 10 to 100 transistors, are generally direct coupled (i.e., there is a dc conductive pathway from input to output), and may use FETs along with npn and pnp BJTs. Their detailed analysis can be difficult, even by computer circuit analysis programs. However, some insight into large-circuit behavior can be found through analysis of commonly encountered transistor *subassemblies* or modules involving two transistors.

Some typical two-transistor circuit modules are shown in Fig. 1.22. Figure 1.22(A) illustrates the npn version of the well-known Darlington amplifier configuration. The Darlington configuration is used to create an effective single transistor that has a higher short-circuit forward current gain (h_{21}) and a higher input resistance (h_{11}). Darlington configurations are often used as power amplifiers. We will examine the linear two-port properties of the Darlington configuration in the following discussion. The hybrid FET-BJT "Darlington" of Fig. 1.22(B) has similar uses as the BJT version but enjoys the very high input resistance of the FET's gate. A cascode configuration using an FET and a BJT is shown in Fig. 1.22(C). Cascode amplifiers are used in high-frequency applications because of their good high-frequency response (see Chapter 4), and they may also use FET-FET or BJT-BJT architecture. Figure 1.22(D) illustrates a simple BJT differential (difference) amplifier (DA) stage. Differential amplifiers are covered in detail in Chapters 2 and 5.

We first examine the Darlington amplifiers, which are characterized by the MFSSMs shown in Fig. 1.23(A) and (B). In these models, the h_{re}'s have been assumed

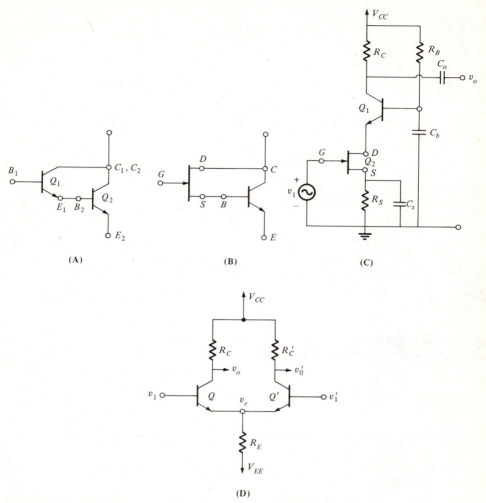

Figure 1.22 **(A)** Darlington connection of two BJTs. **(B)** Hybrid JFET-BJT "Darlington." **(C)** JFET-BJT cascode amplifier. **(D)** BJT differential (difference) amplifier. The circuit is made symmetrical ($Q = Q'$ and $R_C = R_C'$).

to be zero. The four two-port h-parameters, h_{11}, h_{12}, h_{21}, and h_{22}, will be found for these circuits.

The input resistance for the BJT Darlington is easy to find from Fig. 1.23(A):

$$h_{11} = \frac{v_1}{i_1}\bigg|_{v_2=0} = h_{ie(1)} + h_{ie(2)}(1 + h_{fe(1)}) \tag{1.114}$$

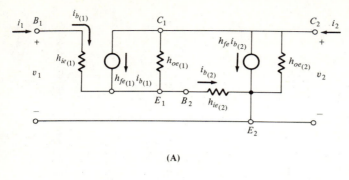

(A)

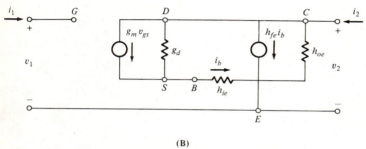

(B)

Figure 1.23 **(A)** Small-signal model of the BJT-BJT Darlington configuration. **(B)** Small-signal model of the JFET-BJT Darlington.

The short-circuit current gain for the BJT Darlington configuration is

$$h_{21} = \left.\frac{i_2}{i_1}\right|_{v_2=0} = \frac{h_{fe(1)}i_{b(1)} + h_{fe(2)}i_{b(2)}}{i_1} = h_{fe(1)} + h_{fe(2)}(1 + h_{fe(1)}) \qquad 1.115$$

The output conductance for the BJT Darlington is found from

$$h_{22} = \left.\frac{i_2}{v_2}\right|_{i_1=0} = h_{oe(2)} + \frac{h_{fe(2)}i_{b(2)} + \dfrac{v_2}{h_{ie(2)} + (1/h_{oe(1)})}}{v_2} \qquad 1.116$$

where

$$i_{b(2)} = \frac{v_2}{h_{ie(2)} + 1/h_{oe(1)}} \qquad 1.117$$

After some algebra, we find

$$h_{22} = h_{oe(2)} + \frac{h_{oe(1)}(h_{fe(2)} + 1)}{1 + h_{ie(2)}h_{oe(1)}} \qquad 1.118$$

Thus the BJT Darlington's input resistance is significantly higher than that of either BJT taken alone; its current gain is greater than the product of the two individual BJT h_{fe}'s; and its output conductance is higher than that of either BJT taken alone.

The input resistance of the hybrid FET-BJT Darlington is infinite, so $i_1 = 0$. Clearly, the h-parameter model cannot be used, because h_{21} is a current-controlled current source. Instead, we use a y-parameter model described by the equations

$$i_1 = y_{11}v_1 + y_{12}v_2 \qquad\qquad 1.119$$

$$i_2 = y_{21}v_1 + y_{22}v_2 \qquad\qquad 1.120$$

By inspection of the MFSSM for the hybrid Darlington, $y_{11} = 0$ and $y_{12} = 0$. y_{21} is the hybrid Darlington's transconductance and is given by

$$y_{21} = \left.\frac{i_2}{v_1}\right|_{v_2=0} \qquad\qquad 1.121$$

in which the output short-circuit current, i_2, must be calculated. First we note that

$$v_{gs} = v_1 - v_s \qquad\qquad 1.122$$

v_s is found by writing a node equation on the source node:

$$v_s\left(gd + \frac{1}{h_{ie(2)}}\right) - g_m(v_1 - v_s) = 0 \qquad\qquad 1.123$$

which yields

$$v_s = \frac{v_1 g_m}{g_m + g_d + 1/h_{ie(2)}} \qquad\qquad 1.124$$

$i_{b(2)}$ is then

$$i_{b(2)} = \frac{v_s}{h_{ie(2)}} = \frac{v_1 g_m/h_{ie(2)}}{g_m + g_d + (1/h_{ie(2)})} \qquad\qquad 1.125$$

The desired short-circuit current, i_2, is thus

$$i_2 = h_{fe(2)}i_{b(2)} + g_m v_{gs} - v_s g_d \qquad\qquad 1.126$$

Substitution of Eq. 1.125 for $i_{b(2)}$ and Eqs. 1.122 and 1.124 into Eq. 1.126 yields, after some algebra,

$$i_2 = \frac{v_1 g_m(h_{fe(2)} + 1)}{h_{ie(2)}(g_m + g_d) + 1} \qquad\qquad 1.127$$

From Eq. 1.127 for i_2, it is easy to write the transconductance, y_{21}, as

$$y_{21} = \frac{j_2}{v_1} = \frac{g_m(h_{fe(2)} + 1)}{h_{ie(2)}(g_m + g_d) + 1} \qquad \text{1.128}$$

The output conductance of the hybrid Darlington is

$$y_{22} = \frac{v_2}{i_2}\bigg|_{v_1 = 0} \qquad \text{1.129}$$

v_2 can be found by writing node equations for v_2 and v_s using Kirchhoff's current law. It is left as a chapter problem to find the expression for y_{22} for the hybrid Darlington (note that v_{gs} and $i_{b(2)}$ can be expressed in terms of v_s, and the conductance $h_{oe(2)}$ appears in parallel with the rest of the output conductance and thus can be added to it).

The JFET-BJT cascode circuit of Fig. 1.22(C) can be analyzed using the MFSSM of Fig. 1.24. First, we will find the open-circuit voltage gain. Note that

$$v_{gs} = v_g - v_s = v_1 \qquad \text{1.130}$$

$$v_o = -h_{fe}i_bR_c \qquad \text{1.131}$$

$$i_b = \frac{-v_e}{h_{ie}} \qquad \text{1.132}$$

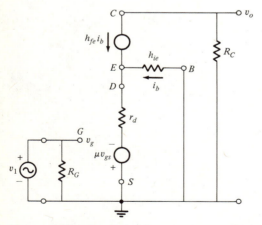

Figure 1.24 Small-signal model for the JFET-BJT cascode amplifier. C_b, C_s, and C_o are treated as ac short circuits, and h_{oe} is assumed to be zero. The Thevenin JFET small-signal model is used.

Therefore,

$$v_o = \frac{+v_e(h_{fe}R_e)}{h_{ie}} \qquad\qquad 1.133$$

v_e is found by writing a node equation:

$$v_e\left(g_d + \frac{1}{h_{ie}}\right) + v_1\mu g_d - h_{fe}\frac{-v_e}{h_{ie}} = 0 \qquad\qquad 1.134$$

Solving for v_e, we find

$$v_e = \frac{-\mu g_d v_1}{g_d + \left(\dfrac{1 + h_{fe}}{h_{ie}}\right)} \qquad\qquad 1.135$$

Substitution of Eq. 1.135 for v_e into Eq. 1.133 yields, after some algebra,

$$K_v = \frac{v_o}{v_1} = -\frac{\mu h_{fe}R_c}{h_{ie} + r_d(1 + h_{fe})} \qquad\qquad 1.136$$

The cascode's input resistance is essentially R_g, and its output resistance is R_c, by inspection.

The differential amplifier of Fig. 1.22(D) is treated in detail in Chapter 5 and will not be considered here.

In this section, we have examined four commonly encountered two-transistor circuits and have used their MFSSMs to characterize the two-port behavior of these circuits (excluding the differential amplifier) at low and mid-frequencies.

SUMMARY

In this chapter, we have reviewed the important two-port models for semiconductor devices; both large-signal (nonlinear) and small-signal (linear) models are used to describe the two-port behavior of single transistors, amplifiers using single transistors, and amplifiers using two transistors.

Linear small-signal models for transistors have been stressed because they are applicable in most situations where ac frequency analysis is desired. An important property of small-signal models is the strong dependence of certain SSM parameters on the transistor operating point. Of special note is the dependence of FET g_m on $\sqrt{I_{DQ}}$ and BJT g_m on I_{BQ}. Nonlinear circuit representations are used primarily in computing a transistor's dc operating point and the static input/output characteristics of amplifiers. Electronic circuit analysis programs on computers are the best way to handle circuits with nonlinear models or with large numbers of nodes.

Characteristics of linear two-port networks are reviewed in Appendix C.

PROBLEMS

The first seven problems involve the dc biasing of simple transistor stages.

1.1   Find R_B and R_C required to make $V_{CEQ} = -7$ V and $I_{CQ} = -1.2$ mA. Let $h_{FE} = 150$, $I_{CBO} = 0$, and $V_{BEQ} = -0.65$ V.

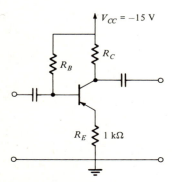

Figure P1.1

1.2   Find R_B and R_C such that the Q-point is at $V_{CEQ} = 5$ V and $I_{CQ} = 1$ mA. Let $h_{FE} = 99$, $I_{CBO} = 0$, and $V_{BEQ} = 0.65$ V.

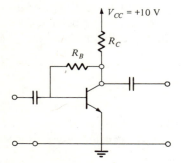

Figure P1.2

1.3   Find I_{CQ}, I_{BQ}, V_{EQ}, V_{CEQ}, and R_1 such that $V_{CQ} = 8$ V. Let $h_{FE} = 99$, $I_{CBO} = 0$, and $V_{BEQ} = 0.65$ V. See Fig. P1.3.

1.4   A BJT emitter-follower is shown in Fig. P1.4. Find I_{EQ}, I_{BQ}, R_B, and the quiescent power dissipation of the BJT and in R_E. Assume $V_{EQ} = 0$, $h_{FE} = 49$, $V_{BEQ} = 0.7$ V, and $I_{CBO} = 0$.

1.5   A simple JFET grounded-source stage with fixed bias is shown in Fig. P1.5. Find R_D such that $V_{DSQ} = 7.5$ V. Assume $V_{GS} = 1.5$ V, $I_G = 0$, $I_{DSS} = 8$ mA, and $V_P = -3.5$ V.

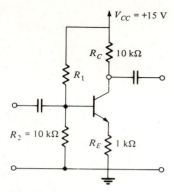

Figure P1.3

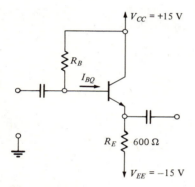

Figure P1.4

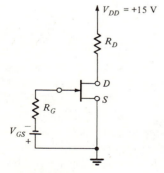

Figure P1.5

1.6   The JFET source-follower stage shown in Fig. P1.6 is operated with fixed bias. Find the V_G required to make $V_{SQ} = 5$ V ($V_{DSQ} = -5$ V), and also I_{DQ} and V_{GSQ}. Assume $I_{DSS} = 4$ mA and $V_P = +2$ V.

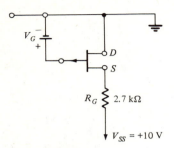

Figure P1.6

1.7   The JFET amplifier shown in Fig. P1.7 is operated with self-bias so that $V_{DQ} = 7$ V (to ground). Find the R_S required and also I_{DQ}, V_{DSQ}, and V_{GSQ}. Assume $I_{DSS} = 10$ mA and $V_P = -4$ V.

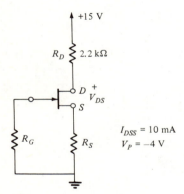

$I_{DSS} = 10$ mA
$V_P = -4$ V

Figure P1.7

Problems 1.8 through 1.27 involve mid-frequency small-signal models of BJTs and FETs. Some can also be solved using an ECAP such as Micro-Cap II if suitable transistor dc parameters are chosen and appropriate resistor and dc source values are used. Micro-Cap dc analysis can be used to find the mid-frequency (or dc) gain at the circuit's operating point.

1.8   Use the node equations of the general y-parameter two-port model (see Appendix C) to show that the mid-frequency, hybrid-pi, BJT two-port model shown in Fig. 1-10(D) has admittances $Y_1 = (y_{11} + y_{12})$, $Y_2 = -y_{12}$, $Y_3 = (y_{22} + y_{12})$, and $Y_4 = (y_{21} - y_{12})$.

1.9   Express the four y-parameters of the hybrid-pi, BJT two-port model (see Problem 1.8) in terms of the grounded-emitter h-parameters, h_{ie}, h_{re}, h_{oe}, and h_{fe}. (*Hint:* Use Table C.1 in Appendix C.) Note that Y_3 may be negative.

1.10 Use the mid-frequency, grounded-emitter h-parameter model for the BJT to find expressions for the four general two-port h-parameters for the BJT amplifier shown. Assume $h_{oe} = h_{re} = 0$.

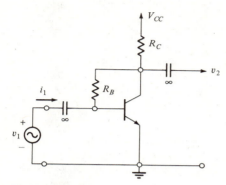

Figure P1.10

1.11 The BJT circuit shown in Fig. P1.11 is a class A, complementary symmetry power amplifier output stage. Both BJTs have identical h_{fe}'s and h_{ie}'s; their h_{oe}'s and h_{re}'s = 0. Find the R_{in} seen by v_1, the Thevenin output resistance, and the open-circuit voltage gain, $K_v = v_o/v_1$. The capacitors are treated as short circuits at mid- and high frequencies.

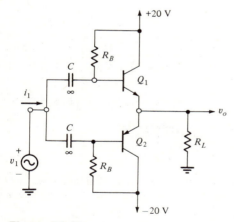

Figure P1.11

1.12 (a) The two-transistor circuit shown in Fig. P1.12(a) is not a Darlington pair. Derive expressions for the general two-port parameters of the circuit (h_{11}, h_{12}, h_{21}, and h_{22}) in terms of the BJT h-parameters, $h_{ie(1)}$, $h_{fe(1)}$, $h_{ie(2)}$, $h_{fe(2)}$, and $h_{oe(2)}$. Let $h_{re(1)} = h_{oe(1)} = h_{re(2)} = 0$.

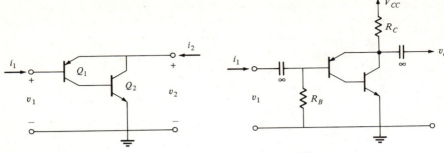

Figure P1.12(a)   **Figure P1.12(b)**

(b) Assume the feedback pair is connected as shown in Fig. P1.12(b). Find numerical values for the mid-frequency gain, $K_v = v_o/v_1$, and for R_{out}. Assume $h_{ie(1)} = 2$ kΩ, $h_{ie(2)} = 1$ kΩ, $h_{fe(1)} = 100$, $h_{fe(2)} = 10$, $h_{oe(2)} = 5 \times 10^{-5}$ S, and $R_C = 100$ Ω.

1.13 (a) The feedback pair of Problem 1.12 is oriented as shown in Fig. P1.13(a). Find expressions for the two-port circuit's h_{11}, h_{12}, h_{21}, and h_{22} in terms of $h_{ie(1)}$, $h_{fe(1)}$, $h_{ie(2)}$, $h_{fe(2)}$, and $h_{oe(2)}$. Let $h_{re(1)} = h_{oe(1)} = h_{re(2)} = 0$.

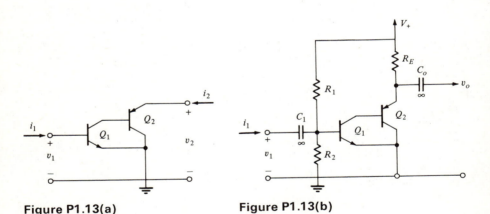

Figure P1.13(a)   **Figure P1.13(b)**

(b) The circuit is connected as shown in Fig. P1.13(b) to make an amplifier. Use the h-parameter model found in part (a) to derive an expression for the amplifier's gain, $K_v = v_o/v_1$.

1.14 A Darlington amplifier is to be used in the grounded-base configuration. Assume $h_{re(1)} = h_{re(2)} = h_{oe(1)} = h_{oe(2)} = 0$.

(a) Find expressions for the general two-port h-parameters for the circuit shown in Fig. P1.14(a) in terms of $h_{ie(1)}$, $h_{fe(1)}$, $h_{ie(2)}$, and $h_{fe(2)}$.

(b) The Darlington is connected as shown in Fig. P1.14(b) to make an amplifier. Find expressions for $K_v = v_o/v_1$, R_{in}, and R_{out}. Assume mid-frequency operation where

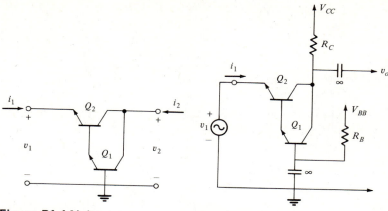

Figure P1.14(a)

Figure P1.14(b)

capacitors are shorts. Use the *h*-parameter model from part (a) to obtain your results.

1.15 A JFET-BJT "Darlington" amplifier is shown in Fig. P1.15. Use the MFSSMs for Q_1 and Q_2 to derive an expression for $K_v = v_o/v_1$ and R_{out} of the amplifier. Assume the Norton small-signal model for the JFET; let $g_d = 0$. Let $h_{re} = h_{oe} = 0$ for the BJT.

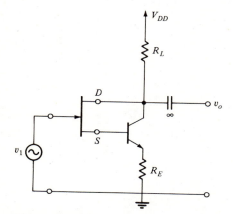

Figure P1.15

1.16 The amplifier shown in Fig. P1.16 may be thought of as a MOSFET source-follower driving a BJT grounded-base stage. Use an ECAP to obtain the dc transfer curve for this amplifier. The FET is characterized by BETA factor = 0.015, threshold voltage = 4, drain resistance = source resistance = 0, GAMMA = LAMBDA = 0, and PHI = 0.6. The BJT is a 2N2369 with a forward β of 100. Scan V_1 from -5 V to $+5$ V. Find the gain at $V_1 = 0$ V. Treat C_s as an open circuit at dc. (Note that small-signal analysis

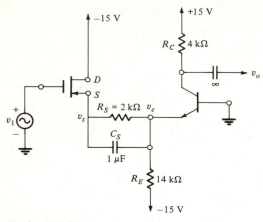

Figure P1.16

of this circuit requires determination of the operating point so that the FET's g_m and g_d and the BJT's h_{ie} can be found.)

1.17 A discrete, capacitively coupled, grounded-base/emitter-follower amplifier stage is shown in Fig. P1.17. Use the common-emitter, h-parameter small-signal models of the BJT to find expressions for the amplifier's $K_v = v_o/v_1$, R_{in}, and R_{out}. Assume Q_1 and Q_2 are identical with $h_{re} = h_{oe} = 0$. All capacitors are short circuits at operating frequencies, and R_1 is in parallel with $R_2 = R_B$.

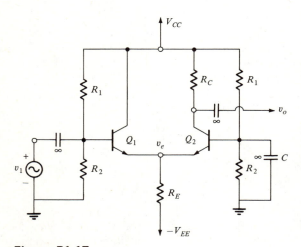

Figure P1.17

1.18 A BJT cascode amplifier is shown in Fig. P1.18. Assume Q_1 and Q_2 are identical, $h_{oe} = h_{re} = 0$, and capacitors are short circuits at operating frequencies.

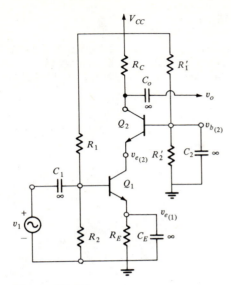

Figure P1.18

(a) Draw the MFSSM for the amplifier.

(b) Find expressions for R_{in}, R_{out}, and $K_v = v_o/v_1$ at mid-frequencies.

(c) Compare the K_v of the cascode amplifier to that of Q_1 with just R_C as load.

1.19 A JFET-JFET cascode amplifier is shown in Fig. P1.19 in which Q_1 has $g_{m(1)}$ and $g_{d(1)}$ and Q_2 has $g_{m(2)}$ and $g_{d(2)}$. The capacitors are short circuits at operating frequencies.

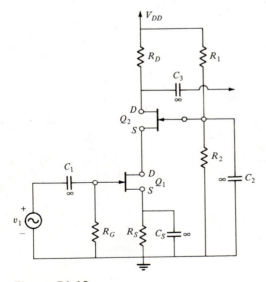

Figure P1.19

(a) Draw the MFSSM for the amplifier.

(b) Find an expression for $K_v = v_o/v_1$ at mid-frequencies.

1.20 The circuit shown in Fig. P1.20 is a bootstrapped, headstage FET-BJT amplifier. Assume for the BJT that $h_{oe} = h_{re} = 0$, $h_{fe} = 99$, and $h_{ie} = 2$ kΩ, and for the JFET that $g_d = 0$ and $g_m = 4000$ μS. Use $R_1 = R_4 = 2$ kΩ, $R_2 = 500$ Ω, and $R_3 = 10$ kΩ.

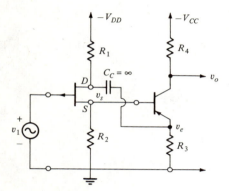

Figure P1.20

(a) Draw the MFSSM for the amplifier. Assume C_c is a short circuit at signal frequencies.

(b) Find a numerical value for $K_v = v_o/v_1$. (*Hint:* Write node equations on v_s and v_e.)

1.21 A low-noise JFET is used to make a headstage for an ideal op-amp (IOA), as shown

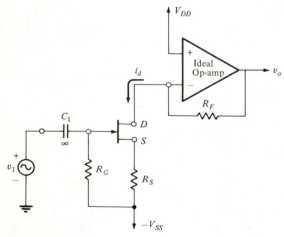

Figure P1.21

in Fig. P1.21. Use MFSSM analysis to find an expression for $K_v = v_o/v_1$. Assume the JFET has g_m and $g_d > 0$.

1.22 The circuit shown in Fig. P1.22 is a JFET-BJT "Darlington" emitter-follower. Assume the FET's $g_d = 0$ and the BJT's $h_{oe} = h_{re} = 0$. Find the amplifier's voltage gain and Thevenin output resistance.

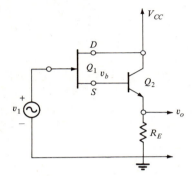

Figure P1.22

1.23 Find the voltage gain and output resistance for the amplifier shown in Fig. P1.23. Assume identical JFETs with g_m and $g_d > 0$.

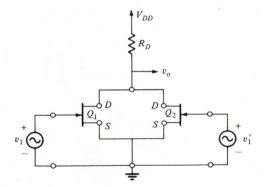

Figure P1.23

1.24 Use MFSSM analysis to find expressions for the voltage gain and Thevenin output resistance for the JFET-BJT feedback-pair amplifier shown in Fig. P1.24. Assume the FET has g_m and $g_d > 0$ and the BJT has $h_{re} = h_{oe} = 0$.

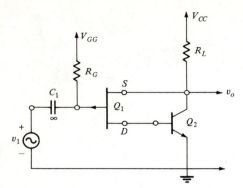

Figure P1.24

1.25 Use MFSSM analysis to find expressions for the voltage gain and Thevenin output resistance of the JFET-BJT feedback-pair amplifier shown in Fig. P1.25. Assume the FET has g_m and $g_d > 0$ and the BJT has $h_{re} = h_{oe} = 0$.

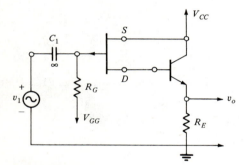

Figure P1.25

1.26 Derive an expression for the small-signal resistance seen looking into the FET's drain.

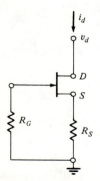

Figure P1.26

1.27 Derive an expression for the small-signal resistance looking into Q_2's drain. Assume Q_1 has $g_{m(1)}$ and $g_{d(1)}$ and Q_2 has $g_{m(2)}$ and $g_{d(2)}$.

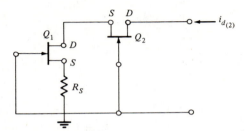

Figure P1.27

Problems 1.28 through 1.37 are concerned with linear active two-port theory. They are provided for review of this material, which is covered in Appendix C.

1.28 Find numerical values for the four h-parameters of the linear active circuit shown in Fig. P1.28. Draw the h-model and label its components.

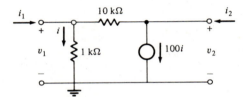

Figure P1.28

1.29 Find the y-parameters for the linear active two-port circuit shown in Fig. P1.29. Draw the y-model and label its components.

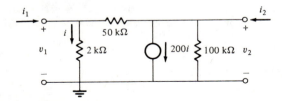

Figure P1.29

1.30 Find the z-parameters for the circuit of Problem 1.29. Draw the z-model and label its components.

1.31 Find the y-parameters for the two-port linear active circuit shown in Fig. P1.31.

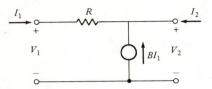

Figure P1.31

1.32 Find the input resistance, the current gain, and the voltage gain for the terminated two-port linear active circuit shown in Fig. P1.32.

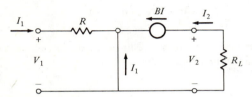

Figure P1.32

1.33 Give the h-parameter two-port model for an ideal transformer characterized by the equations

$$v_2 = nv_1 \qquad i_1 = -ni_2$$

where n is the secondary-to-primary turns ratio. An ideal transformer is a lossless two-port, that is, power in = power out. Hence $v_1 i_1 = v_2 i_2$.

1.34 An ideal transformer characterized by the equations in Problem 1.33 has a capacitor C connected across its port 2 terminals. Find an expression for $Y_1(s) = V_1/I_1$.

1.35 Repeat Problem 1.34 for the case of a resistor R_L connected to port 2.

1.36 This is the classic power transfer problem. An ideal transformer characterized by the equations in Problem 1.33 has port 1 connected to a Thevenin equivalent circuit. Port 2

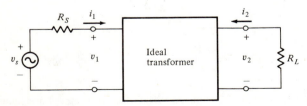

Figure P1.36

is connected to a load resistor, R_L. Find an expression for the value of n that will maximize the power dissipation in R_L.

1.37 A *gyrator* two-port is defined by the equations

$$v_1 = -Gi_2 \qquad v_2 = Gi_1$$

where G is the gyration resistance. Port 2 of the gyrator is connected to a capacitor, C_L. Find an expression for the impedance seen looking into port 1.

Chapter 2

Network Theorems Used in the Analysis of Linear Active Circuits

The network theorems presented in this chapter enable relatively complex electronic circuit models (having more than two nodes) to be reduced to simpler formats that are amenable to pencil-and-paper analysis (not every circuit we encounter needs to be analyzed on a computer). The theorems include the substitution theorem, the reduction theorem, the bisection theorem, and the Miller theorem. The reader should already be quite familiar with Thevenin's and Norton's circuit theorems; they will be used as needed.

2.1 The Substitution Theorem

The substitution theorem, under certain conditions, allows the replacement of an active, current-controlled voltage source (CCVS) with a resistor or the replacement of a voltage-controlled current source (VCCS) with a conductance. This theorem is illustrated in Fig. 2.1

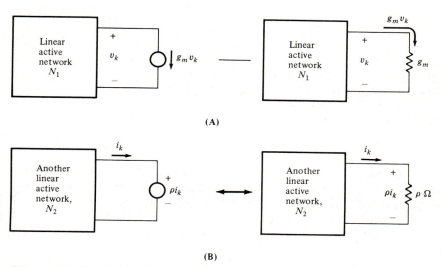

(A)

(B)

Figure 2.1   **(A)** Current form of the substitution theorem. The proof lies in the node equations for the circuit, which are unchanged. **(B)** Voltage form of the substitution theorem. The proof lies in the loop equations, which are the same.

EXAMPLE 2.1

Application of the Current Form of the Substitution Theorem

Consider the FET source-follower at mid-frequencies shown in Fig. 2.2(A). The MFSSM for this amplifier is given in Fig. 2.2(B). We note that $v_{gs} = (v_1 - v_s)$, so the voltage-controlled current source (VCCS) in Fig. 2.2(B) can be split (the principle of

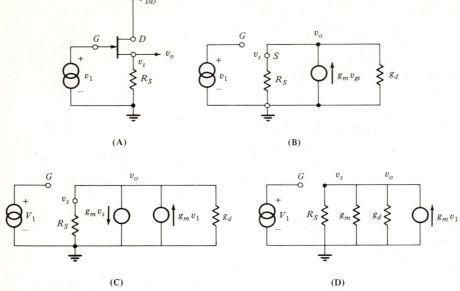

Figure 2.2 **(A)** FET source-follower. **(B)** MFSSM of the FET source-follower. **(C)** MFSSM with the source split (superposition). **(D)** Application of the substitution theorem to replace the $g_m v_s$ current source with a conductance, g_m.

superposition), as shown in Fig. 2.2(C). Since the controlled source, $g_m v_s$, lies between the v_s node and ground, the v_s node equation is unchanged if we replace it with a passive conductance of value g_m. The MFSSM is redrawn one final time to illustrate that one independent VCCS, $v_1 g_m$, develops $v_s = v_o$ across the three conductances in Fig. 2.2(D). Thus, by Ohm's law, the FET source-follower gain is written by inspection:

$$A_v = \frac{v_o = v_s}{v_1} = \frac{g_m}{g_m + g_o + G_S} \qquad 2.1$$

As you can see, this gain is positive (noninverting) and slightly less than unity. Other examples of the substitution theorem will be encountered in the problems section.

⬤

2.2 The Reduction Theorem

The reduction theorem is a fairly complex circuit theorem which has two voltage and two current forms. It involves the scaling of component values in one of two parts of the linear active circuit (LAC); this scaling allows the removal of a certain dependent CCCS or the removal of a dependent VCVS. The LAC must have the nodal geometry shown in Fig. 2.3(A) for the current forms of the reduction theorem to be applied. Figure 2.3(B) illustrates the circuit loop geometry required for the application of the voltage forms of the reduction theorem. Note that the two parts of the LAC, N_1 and N_2, must be separated such that the only links between N_1 and N_2 are the ones containing the dependent source, as shown in either Fig. 2.3(A) or (B). No current may enter or leave N_1 or N_2 through any other coupling links.

Proof of the reduction theorem is complex and is not given here; it involves impedance scaling of partitioned node or loop equations for the two parts of the LAC. A discussion of the proof is given in Angelo (1969).

First, let us consider the *current form* of the reduction theorem. It may be stated as follows:

> All *voltages* in N_1 and N_2 (in Fig. 2.3[A]) *remain unchanged* if the dependent current source, βi_k, is removed and
>
> > If each R, L, and $1/C$ (impedance) in N_1 is *divided* by $(\beta + 1)$ and the current of each current source in N_1 is *multiplied* by $(\beta + 1)$. (N_2 is untouched.)
>
> *or*
>
> > If each R, L, and $1/C$ (impedance) in N_2 is *multiplied* by $(\beta + 1)$ and the current of each current source in N_2 is *divided* by $(\beta + 1)$. (N_1 is untouched.)

The *voltage form* of the reduction theorem may also be stated:

> All *currents* in N_1 and N_2 (in Fig. 2.3[B]) *remain unchanged* if the dependent voltage source, μv_j, is replaced with a short circuit and
>
> > If each R, L, and $1/C$ and the voltage of each voltage source in N_1 is *multiplied* by $(\mu + 1)$. (N_2 is untouched.)
>
> *or*
>
> > If each R, L, and $1/C$ and the voltage of each voltage source in N_2 is *divided* by $(\mu + 1)$. (N_1 is untouched.)

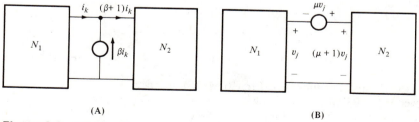

(A) (B)

Figure 2.3 **(A)** Circuit geometry necessary to apply the current form of the reduction theorem. **(B)** Circuit geometry required to apply the voltage form of the reduction theorem.

The instructions for using this theorem are deceptively simple; we will illustrate its effectiveness for simplifying circuits with two examples.

EXAMPLE 2.2

Application of the Current Form of the Reduction Theorem

Consider the Darlington emitter-follower circuit in Fig. 2.4(A); its unreduced MFSSM

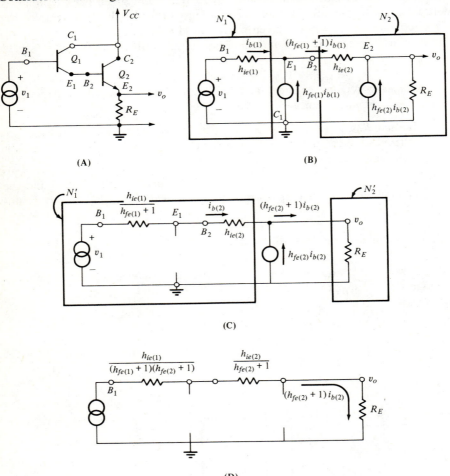

(A)

(B)

(C)

(D)

Figure 2.4 **(A)** Darlington emitter-follower. **(B)** MFSSM of the emitter-follower. **(C)** MFSSM after applying the reduction theorem to N_1. New N_1' and N_2' are defined for a second application of the theorem. **(D)** MFSSM after two successive applications of the reduction theorem. Because voltages in N_1 and N_2 are unchanged, it is easy to find v_o.

is given in Fig. 2.4(B). The reduction theorem is applied to N_1 in the MFSSM, yielding the once-scaled circuit of Fig. 2.4(C). New networks N_1' and N_2' are defined, and again the reduction theorem is applied, this time to N_1'. The final result is shown in Fig. 2.4(D). Because the reduction theorem transformations took place to the left of R_E, the circuit of Fig. 2.4(D) can be used to find the gain of the Darlington emitter-follower by the voltage-divider method:

$$A_v = \frac{v_o}{v_1} = \frac{R_E}{R_E + \dfrac{h_{ie(2)}}{\beta_2 + 1} + \dfrac{h_{ie(1)}}{(\beta_2 + 1)(\beta_1 + 1)}}$$

$$= \frac{R_E(\beta_1 + 1)(\beta_2 + 1)}{R_E(\beta_1 + 1)(\beta_2 + 1) + h_{ie(2)}(\beta_1 + 1) + h_{ie(1)}} \qquad 2.2$$

The emitter-follower's output resistance can also be found by inspection of Fig. 2.4(D). It is simply R_E in parallel with the series transformed resistances:

$$R_{\text{out}} = R_E \left\| \left(\frac{h_{ie(2)}}{\beta_2 + 1} + \frac{h_{ie(1)}}{\beta_1\beta_2 + \beta_2 + \beta_1 + 1} \right) \right. \qquad 2.3$$

Note that the input resistance "seen" by v_1 cannot be found using the N_1-scaled circuit of Fig. 2.4(D). The reason is that the scaling of N_1 and N_1' did not preserve $i_{b(1)}$. Two successive applications of the reduction theorem on the N_2's shown in Fig. 2.5(A) yield the reduced MFSSM shown in Fig. 2.5(C). The input impedance can thus be written by inspection:

$$R_{\text{in}} = \frac{v_1}{i_{b(1)}} = h_{ie(1)} + h_{ie(2)}(\beta_1 + 1) + R_E(\beta_1\beta_2 + \beta_2 + \beta_1 + 1) \qquad 2.4$$

●

EXAMPLE 2.3 ●

Application of the Voltage Form of the Reduction Theorem

Consider the FET-FET cascode amplifier shown in Fig. 2.6(A). Figure 2.6(B) illustrates the unreduced MFSSM for the cascode amplifier, with parts N_1 and N_2 defined. The reduction theorem is applied to N_1 to find the gain of the circuit. Figure 2.6(C) illustrates the simple series circuit remaining after the reduction theorem is used. Note that i_d is unchanged in the voltage form of the theorem, so v_o will be

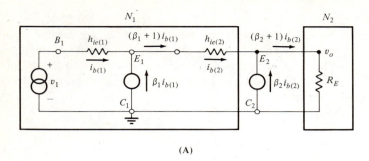

(A)

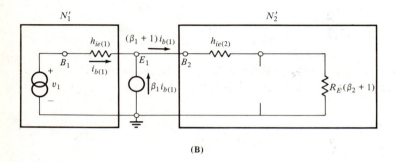

(B)

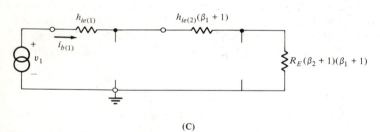

(C)

Figure 2.5 **(A)** MFSSM of the Darlington emitter-follower with N_1 and N_2 defined to preserve $i_{b(1)}$. **(B)** MFSSM after the reduction theorem transformation of N_2 in (A). **(C)** MFSSM after the second reduction theorem transformation on N_2 in (B). $i_{b(1)}$ is preserved; hence the circuit of (C) can be used to find R_{in} by inspection.

$-i_d R_D$. Thus we can write an expression for the FET-FET cascode amplifier's mid-frequency gain by inspection of Fig. 2.6(C):

$$A_v = \frac{v_o}{v_1} = -\frac{\mu_1(\mu_2 + 1)R_D}{R_D + r_{d(2)} + r_{d(1)}(\mu_2 + 1)}$$

2.5

●

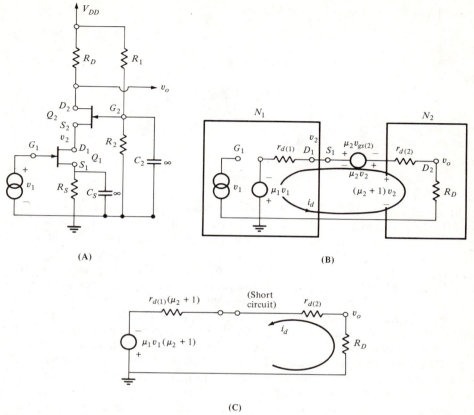

Figure 2.6 **(A)** FET-FET cascode amplifier. The bypass capacitors are considered to be short circuits at mid-frequencies. **(B)** MFSSM for the cascode amplifier. Note that $v_{gs(2)} = -v_2$, and the redefinition of the second VCVS. N_1 and N_2 are defined. **(C)** The reduction theorem is applied to N_1 in (B) to yield this reduced small-signal model. i_d is unchanged.

Other examples of the application of the reduction theorem to simplify circuit analysis are given in the problems at the end of the chapter.

2.3 The Bisection Theorem

The bisection theorem is used to simplify the analysis of symmetrical circuits, such as differential amplifier stages (treated in detail in Chapter 5). A differential amplifier has two inputs, v_1 and v_1'. v_1 and v_1' can be broken down into two forms of symmetrical input to the differential amplifier: the common-mode and difference-mode inputs. These inputs are defined as follows:

$$v_{1(c)} \triangleq \frac{v_1 + v_1'}{2} \qquad\qquad 2.6$$

$$v_{1(d)} \triangleq \frac{v_1 - v_1'}{2} \qquad\qquad 2.7$$

From these definitions, we note that

$$v_1 = v_{1(c)} + v_{1(d)} \quad \text{and} \quad v_1' = v_{1(c)} - v_{1(d)} \qquad\qquad 2.8$$

$v_{1(c)}$ is the average value of v_1 and v_1'. If $v_1 = v_1'$, then $v_{1(c)} = v_1$ and $v_{1(d)} = 0$. If $v_1 = -v_1'$, then $v_{1(c)} = 0$ and $v_{1(d)} = v_1$. If $v_1 > 0$ and $v_1' = 0$, then $v_{1(c)} = v_1/2$ and $v_{1(d)} = v_1/2$. We often discuss the behavior of differential amplifiers under conditions of purely common-mode or difference-mode inputs, as will be seen in Chapter 5.

The bisection theorem has two different forms, one for a symmetrical linear active circuit (SLAC) under purely difference-mode inputs, the other for the case where the SLAC has purely common-mode inputs.

The circuit for difference-mode excitation is illustrated in Fig. 2.7(A). Using superposition, we first apply $v_1 = v_{1(d)}$, letting $v_1' = 0$. In general, potentials v_k and v_j will occur between the three cross-links shown. (There is no limit to the number of cross-links between the symmetrical halves of the SLAC.) Next, we continue the use of superposition by applying $v_1' = -v_{1(d)}$ with $v_1 = 0$. Because of symmetry, the voltages between the cross-links will now be $-v_k$ and $-v_j$. The superposition principle allows us to sum the component voltages to obtain the net result of simultaneous inputs by a difference-mode signal. It is obvious that when the summing is done, $[v_k + (-v_k)] = 0$, and so forth. This means that for difference-mode excitation of the SLAC, all potential differences between the cross-links on the axis of symmetry are zero and, in particular, the potential difference between any cross-link and the ground cross-link is zero. Thus, when a symmetrical circuit has difference-mode inputs, all cross-links on the axis of symmetry can be shorted together and to ground without disturbing either half of the SLAC. Now either half of the SLAC can be analyzed for difference-mode gain, difference-mode input impedance, and so on. Obviously, the analysis will involve about half the nodes of the complete circuit.

The circuit for common-mode excitation is shown in Fig. 2.8(A). Again using superposition, we first apply $v_1 = v_{1(c)}$, letting $v_1' = 0$. This causes currents i_j, i_k, and i_p to flow in the cross-links. Next, we apply $v_1' = v_{1(c)}$ with $v_1 = 0$. Now, due to symmetry, currents i_j', i_k', and i_p' flow in the opposite directions in their respective links. The magnitudes of the current pairs are obviously equal by symmetry. Superposition allows us to sum the current pairs on the cross-links to find the net effect of common-mode excitation. The superposition of the current pairs sums to zero on each cross-link; if the links carry zero current, they can be cut on the axis of sym-

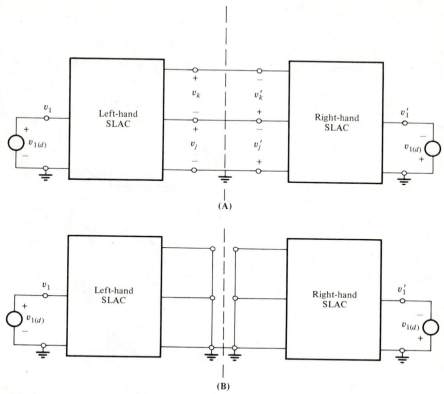

Figure 2.7 **(A)** Schematic of a symmetrical linear active circuit (SLAC). The dashed line is the axis of symmetry. The circuit is assumed to be under difference-mode excitation, where the input to the left-hand side is a positive voltage, $v_1 = v_{1(d)}$, and the input to the right-hand side is $v_1' = -v_{1(d)}$. **(B)** Reduced circuit after application of the bisection theorem, given difference-mode excitation. All links crossing the axis of symmetry are grounded; either half-circuit can be used for analysis.

metry without disturbing the analysis of either half of the SLAC, given common-mode inputs. This situation is shown in Fig. 2.8(B).

Often symmetrical circuits have components that lie on the axis of symmetry. These on-axis components appear to present a problem in implementing the bisection theorem. They can be moved aside, however, to merge with the symmetrical halves of the SLAC. Figure 2.9 illustrates the disposition of frequently encountered on-axis components.

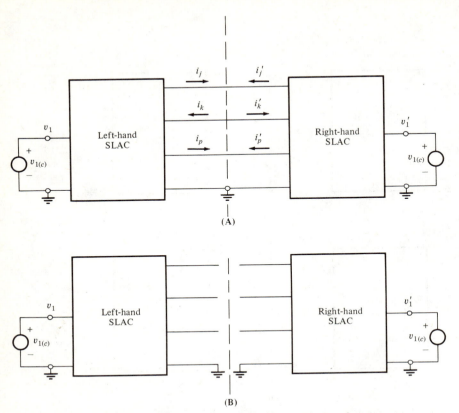

Figure 2.8 **(A)** SLAC under common-mode excitation. Note that $v_1 = v'_1 = v_{1(c)}$. **(B)** Bisected SLAC under common-mode excitation. Following the application of superposition, we can show that it is possible to cut all cross-links on the axis of symmetry without disturbing either half of the SLAC. Analysis for common-mode inputs can then be done on either half-circuit.

EXAMPLE 2.4

Application of the Bisection Theorem

Consider the symmetrical JFET differential amplifier circuit at mid-frequencies shown in Fig. 2.10. Note that in Fig. 2.10(C), where the SLAC has difference-mode excitation, the method illustrated in Fig. 2.7(B) is used, while in Fig. 2.10(D), where the SLAC has common-mode excitation, the procedure shown in Fig. 2.8(B) is applied. The output for difference-mode inputs is easily calculated from Fig. 2.10(C):

$$v_o = -i_d R_D = -\frac{\mu v_{1(d)} R_D}{R_D + r_d} \qquad 2.9$$

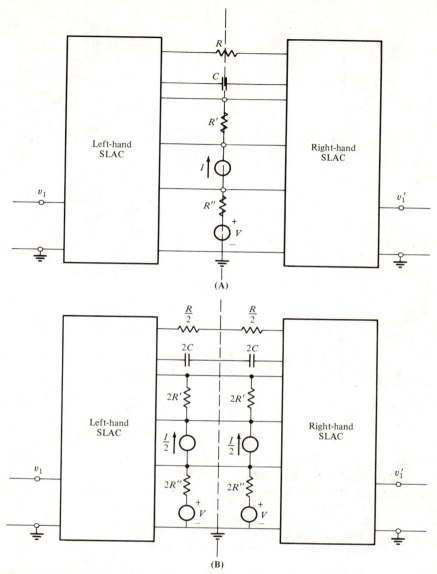

Figure 2.9 **(A)** Components on the axis of symmetry of a SLAC. **(B)** Disposition of on-axis components to permit application of the bisection theorem.

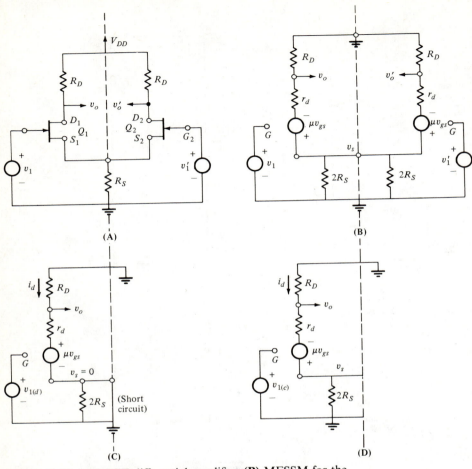

Figure 2.10 **(A)** FET differential amplifier. **(B)** MFSSM for the FET differential amplifier. Note that R_S on the axis of symmetry is treated according to the protocol in Fig. 2.9(B). **(C)** Left half of the differential amplifier after the bisection theorem has been applied for difference-mode inputs. **(D)** Left half of the differential amplifier after the bisection theorem has been applied for common-mode inputs.

The output for common-mode inputs is made more complicated by the negative feedback from v_s:

$$v_{gs} = v_{1(c)} - v_s \qquad\qquad 2.10$$

$$v_s = i_{d(2)} R_S \qquad\qquad 2.11$$

$$i_d = \frac{\mu v_{gs}}{R_D + r_d + 2R_S}$$ 2.12

$$v_o = -i_d R_D$$ 2.13

Equations 2.10 and 2.11 are substituted into Eq. 2.12, which is solved for i_d and substituted into Eq. 2.13 for v_o, giving

$$v_o = -\frac{v_{1(c)}\mu R_D}{R_D + r_d + 2R_S(\mu + 1)}$$ 2.14

It is interesting to observe that $v_o/v_{1(d)}$ given by Eq. 2.9 is much greater than $v_o/v_{1(c)}$ given by Eq. 2.14. More will be made of this inequality between common-mode and difference-mode gains in Chapter 5 on differential amplifiers.   ●

2.4 The Miller Theorem

The Miller theorem is useful in the simplified analysis of a certain class of feedback circuits which includes high-frequency, inverting-gain voltage amplifiers that have capacitive coupling between input and output nodes. A two-port circuit relevant to this discussion of the Miller theorem is shown in Fig. 2.11(A). An admittance Y_f is connected between the input (V_1) and output (V_2) nodes. The output voltage, v_2, is assumed to be proportional to the input voltage, V_1; that is,

$$V_2 = A_v V_1$$ 2.15

This relation implies that an ideal VCVS with gain A_v is connected between the V_2 node and ground. The current flowing into Y_f from the V_1 node, I_f, can be written from Ohm's law and Eq. 2.15:

$$I_f = Y_f(V_1 - A_v V_1) = Y_f(1 - A_v)V_1$$ 2.16

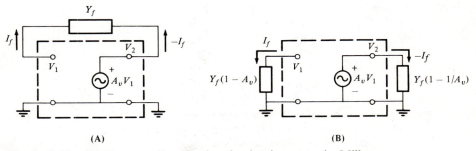

(A)   (B)

Figure 2.11   **(A)** Two-port linear active circuit relevant to the Miller theorem. **(B)** Circuit of (A) modified by the Miller theorem.

The current flowing into Y_f from the V_2 node is simply $-I_f$; it can be written in terms of V_2 using Eq. 2.15 and Ohm's law:

$$-I_f = Y_f\left(V_2 - \frac{V_2}{A_v}\right) = Y_f\left(1 - \frac{1}{A_v}\right)V_2 \qquad 2.17$$

Thus the summed branch currents at nodes 1 and 2 are not changed if the circuit of Fig. 2.11(A) is redrawn as shown in Fig. 2.11(B) with admittance $Y_m = Y_f(1 - A_v)$ from node 1 to ground and admittance $Y'_m = Y_f(1 - 1/A_v)$ from node 2 to ground.

EXAMPLE 2.5

Application of the Miller Theorem

Consider the op-amp circuit of Fig. 2.12(A). Because the op-amp is ideal, the inverting gain from V_1 to V_2 is given by the well-known relation

$$\frac{V_2}{V_1} = A_v = -\frac{R_2}{R_1} \qquad 2.18$$

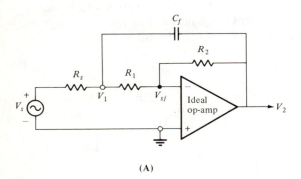

(A)

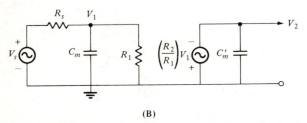

(B)

Figure 2.12 **(A)** Op-amp circuit to illustrate application of the Miller theorem. **(B)** Reduced circuit showing the Miller capacitance $C_m = C_f(1 + R_2/R_1)$. Note that the output capacitance, C'_m, does not affect the circuit's gain, because of the VCVS. Resistance R_1 appears to ground because $V_{sj} = 0$ (virtual ground) for the ideal op-amp.

The feedback admittance between the V_2 and V_1 nodes is $Y_f = sC_f$ in Laplace notation. From the preceding discussion, we see that C_f can be removed from between the V_2 and V_1 nodes and replaced with admittance Y_m from node 1 to ground. Y_m can be written simply as

$$Y_m = sC_f(1 - A_v) = sC_f\left(1 + \frac{R_2}{R_1}\right) = sC_m \qquad 2.19$$

The capacitance $C_m = C_f(1 + R_2/R_1)$ is called the Miller capacitance; the Miller capacitance and Miller effect are treated in more detail in Chapter 4, Sec. 4.3. If $R_2/R_1 > 1$, then the Thevenin source (V_s, R_s) sees an enlarged input capacitance, C_m, to ground, which acts with R_s to form a low-pass filter. The output admittance, Y'_m, does not affect the circuit's gain in this case. It is left as an exercise to show that the overall gain of the circuit of Fig. 2.12(A) is given by

$$\frac{V_2}{V_s}(s) = -\frac{R_2/(R_s + R_1)}{sC_fR_s\dfrac{(R_1 + R_2)}{(R_1 + R_s)} + 1} \qquad 2.20$$

●

Note that application of the Miller theorem is valid only if the output voltage, V_2, is determined by an ideal VCVS.

If the output of the two-port is describable by a Thevenin equivalent circuit (a VCVS in series with a source resistance), as shown in Fig. 2.13(A), then analysis by use of the Miller theorem is not valid. Figure 2.13(B) illustrates an incorrect application of the Miller theorem to the analysis of the circuit of Fig. 2.13(A).

Inspection of the circuit of Fig. 2.13(B) yields an incorrect overall gain given by

$$\frac{V_2}{V_s}(s) = \frac{-\mu}{[sC_f(\mu + 1)R_1 + 1][sC_f(1/\mu + 1)R_o + 1]} \qquad 2.21$$

Valid analysis of the circuit of Fig. 2.13(A) can be done by writing the node equations for V_1 and V_2 and solving for V_2 by Cramer's rule. This process yields the correct transfer function,

$$\frac{V_2}{V_s}(s) = \frac{-\mu[1 - sC_fR_o(1/\mu)]}{sC_f[R_1(\mu + 1) + R_o] + 1} \qquad 2.22$$

Note that the incorrect gain expression, Eq. 2.21, reduces to the correct gain expression, Eq. 2.22, if $R_o \to 0$ in Eq. 2.21. Of course, when $R_o = 0$, V_2 is determined by an ideal VCVS, under which condition application of the Miller theorem is valid.

In conclusion, we remark that the Miller theorem is perhaps the least useful of the circuit theorems discussed in this chapter, for the reason that one rarely finds a two-port circuit in which V_2 is determined by a pure VCVS.

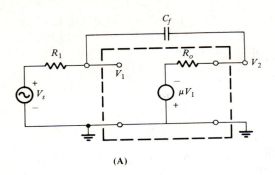

(A)

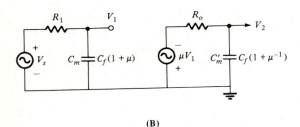

(B)

Figure 2.13 **(A)** Two-port circuit with a Thevenin source driving the V_2 node. **(B)** Incorrect reduction of the circuit of (A) through misapplication of the Miller theorem.

SUMMARY

In this chapter, we have seen how the use of certain circuit theorems can simplify the analysis of electronic systems, often reducing the analysis to the solution of one or two simultaneous equations. Although use of electronic circuit analysis programs has generally made application of simplifying circuit theorems unnecessary (the computer easily handles the complexity), these theorems are useful for dealing with suitable circuits when using a pencil, paper, and a calculator.

PROBLEMS

2.1 Use the substitution theorem to simplify the grounded-gate JFET amplifier's MFSSM and write an expression for its input resistance, output resistance, and mid-frequency gain. Assume g_m, g_d for JFET small-signal model; $g_d \cong 0$.

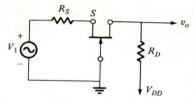

Figure P2.1

2.2 The JFET-BJT hybrid "Darlington" in Fig. P2.2 is to be analyzed using the substitution theorem (on the FET's VCCS) and the reduction theorem (on the BJT's CCCS). Assume MFSSMs:

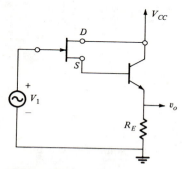

Figure P2.2

JFET: g_m, v_{gs}, g_d

BJT: h-parameter common-emitter model with $h_{re} = h_{oe} = 0$

(a) Find an expression for V_o/V_1.
(b) Find the Thevenin resistance that R_E sees.

2.3 Use two repeated applications of the reduction theorem (each time on N_2) to find R_{in} and V_o/V_1 for the two-BJT circuit shown in Fig. P2.3. Assume common-emitter, h-parameter MFSSMs in which $h_{re(1)} = h_{re(2)} = h_{oe(1)} = h_{oe(2)} \cong 0$. (*Note*: $i_{b(2)} = -\beta_1 i_{b(1)}$.)

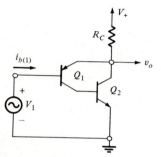

Figure P2.3

2.4 The MOSFETs, shown in Fig. P2.4 are used to drive the armature of a small dc motor. The circuit is balanced, so that when $V_1 = 0$, $I_m = 0$. Assume the following:

$$Q_1 = Q_2$$

g_m, g_d (Norton) model

Motor armature resistance, $R_A \approx 0$

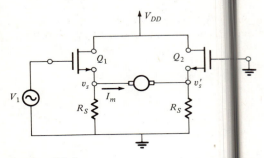

Figure P2.4

(a) Draw the MFSSM for the circuit.

(b) Use the substitution theorem to eliminate all controlled sources in the circuit except the VCCS controlled by V_1.

(c) Assuming $R_A = 0$, derive an expression for $G_m = I_m/V_1$, the circuit's transconductance. What source conductance (Norton) does the motor see?

2.5 A differential amplifier with common-mode negative feedback is shown in Fig. P2.5. Assume:

JFETs: $Q_1 = Q_2$ g_m, g_d (Norton) MFSSM; $g_d = 0$

BJT: Common-emitter h-parameter model with $h_{re} = 0$

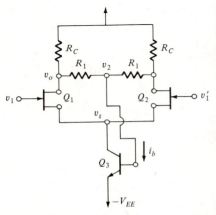

Figure P2.5

(a) Draw the complete MFSSM for the circuit. Show symmetry.
(b) Draw the left-half, difference-mode small-signal model using the bisection theorem.
(c) Draw the left-half, common-mode small-signal model using the bisection theorem.

2.6 The matched FET pair shown in Fig. P2.6 allows an inexpensive op-amp to be used to make a low-noise differential amplifier. Assume $Q_1 = Q_2$ with g_m and $g_d > 0$.

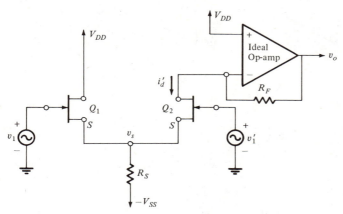

Figure P2.6

(a) Draw the MFSSM for the amplifier. Note that the op-amp's summing junction is at small-signal ground.
(b) Find an expression for $v_s = f(v_1, v_1')$.
(c) Find expressions for the differential gain, $A_D = v_o/v_{1(d)}$, and the common-mode gain, $A_c = v_o/v_{1(c)}$, where $v_{1(d)} = (v_1 - v_1')/2$ and $v_{1(c)} = (v_1 + v_1')/2$. Use the bisection theorem.

2.7 Use the Miller theorem to find the transfer function, in time-constant form, of the ideal op-amp circuit in Fig. P2.7.

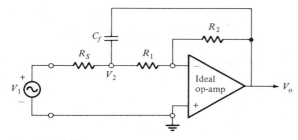

Figure P2.7

2.8 Two ideal op-amps are used to make a *Miller integrator*, as shown in Fig. P2.8. Use the Miller theorem to derive an expression for V_o/V_1 in time-constant form. Assume $R_3/R_2 \gg 1$.

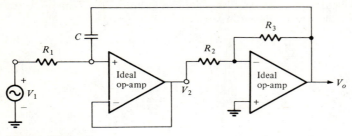

Figure P2.8

Chapter 3

Review of Frequency Response Analysis of Amplifiers

In terms of frequency response, there are two broad categories of linear broadband amplifiers: direct-coupled (DC), and reactively coupled (RC). Direct-coupled amplifiers use resistive (conductive) elements to couple the input to the amplifier, the amplifier's stages to one another, and the output stage to the load. They can amplify zero-frequency (dc) signals as a result of their direct coupling, and their frequency response function is generally flat down to zero frequency. All IC op-amps are DC amplifiers, as are IC differential instrumentation amplifiers. The design of DC amplifiers presents special problems such as the need to compensate for thermally induced dc offset voltage drift and the requirement to match dc quiescent voltage levels between their gain stages.

Reactively coupled amplifiers generally have their input, output, and gain stages isolated by dc-blocking circuit elements (capacitors or transformers). Modern practice in the design of RC amplifiers uses capacitors to couple non-zero-frequency signals. In the early days of electronics, interstage coupling in RC amplifiers made extensive use of transformers. This practice was soon abandoned, however, because it was found that transformers were bigger, heavier, more expensive, and had poorer frequency response than simple resistor-capacitor interstage coupling. Transformers are still used in some power amplifier applications to couple the load to the power output stage more efficiently.

RC amplifiers are bandpass amplifiers; that is, their frequency response rises from zero (scalar) at zero frequency to a constant value in the mid-band frequency region, and then falls off to zero (scalar) at high frequencies (see Fig. 3.4). If we are interested in amplifying signals with no average (dc) value, then we can use RC amplifier design (or DC amplifiers, for that matter). Examples of signals with no dc component are audio (speech, music, etc.), double-sideband/suppressed-carrier signals used in instrumentation and control, and RF (radio-frequency) signals. Signals having dc or very low frequency components arise in instrumentation (measurement of pH, light intensity) and in control (dc motor speed).

3.1 The Transfer Function and Frequency Response

The dynamic behavior of most linear systems (including electronic amplifiers) can be described by an nth-order linear differential equation with constant coefficients or, alternatively, by a set of n state equations, If $x(t)$ is the amplifier's input, and $y(t)$ is its output, the describing differential equation may be written in the form

$$K(b_m \dot{x}^m + b_{m-1} \dot{x}^{m-1} + \cdots + b_1 \dot{x} + b_0 x) = \dot{y}^n + a_{n-1} \dot{y}^{n-1} + \cdots + a_1 \dot{y} + a_0 y$$

$$3.1$$

where $n \geq m$ and $\dot{x}^k = d^k[x(t)]/dt^k$. If we take the Laplace transform of Eq. 3.1, setting initial conditions equal to zero, we obtain

$$KX(s)(s^m + b_{m-1}s^{m-1} + \cdots + b_1 s + b_0) = Y(s)(s^n + a_{n-1}s^{n-1} + \cdots + a_1 s + a_0)$$

$$3.2$$

which is easily put in the form of a transfer function:

$$\frac{Y(s)}{X(s)} = A_v(s) = K \frac{s^m + b_{m-1}s^{m-1} + \cdots + b_1 s + b_0}{s^n + a_{n-1}s^{n-1} + \cdots + a_1 s + a_0}$$
  3.3

The numerator polynomial (mth-order) has m roots (i.e., m complex s values which, when substituted into the numerator, make it equal to zero). The m roots of the numerator are known as the zeros of $A_v(s)$. Similarly, the denominator polynomial has n roots, which are the poles of $A_v(s)$. They are called poles because when the denominator goes to zero, $A_v(s)$ goes to infinity.

If the numerator and denominator polynomials are factored to reveal the poles and zeros, $A_v(s)$ can be written in *Laplace format*:

$$A_v(s) = K \frac{(s + z_1)(s + z_2) \cdots (s + z_m)}{(s + p_1)(s + p_2) \cdots (s + p_n)}$$
  3.4

In Eq. 3.4, it is understood that the zeros are at $s = -z_1, s = -z_2, \ldots, s = -z_m$, and the poles are at $s = -p_1, s = -p_2, \ldots, s = -p_n$. The poles and zeros of $A_v(s)$ can be real, lying on the real axis in the s-plane, or occur in complex-conjugate pairs. Real and complex-conjugate zeros can lie in either the left-half or right-half s-plane. The poles of $A_v(s)$ must lie in the left-half s-plane if the amplifier is to be stable. Because the complex-conjugate roots occur in pairs, they can be written in the form:

$$(s + z_1)(s + z_1^*) = s^2 + 2\zeta\omega_n s + \omega_n^2 = s^2 + 2as + a^2 + b^2$$
  3.5

The parameter ζ in Eq. 3.5 is called the damping factor, and ω_n, the undamped natural frequency. Parameters ζ, ω_n, b, and a have geometrical interpretation in the s-plane, as illustrated in Fig. 3.1. It can be shown that $\zeta = \cos\theta$ and $\omega_n =$ the radial distance from the origin to the poles (or zeros). It is evident that for complex-conjugate roots, $0 < \zeta < 1$. If $\zeta > 1$, the roots are no longer complex-conjugate; they lie on the real axis.

The transfer function $A_v(s)$ given by Eq. 3.4 in factored Laplace format can be used to find the amplifier's time response to impulses, steps, ramps, and so on at the input using classical Laplace transform techniques:

$$y(t) = \mathcal{L}^{-1}[Y(s) = A_v(s)X(s)]$$
  3.6

$A_v(s)$ can also be used to determine the system's steady-state sinusoidal frequency response when $x(t) = X \sin(\omega t)$. The steady-state $y(t)$ will also be a sine wave of the same frequency but will have a different amplitude and phase than $x(t)$. That is,

$$y(t) = Y \sin(\omega t + \psi)$$
  3.7

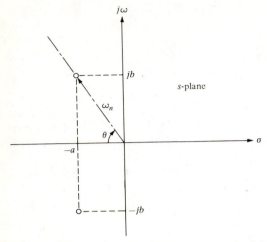

Figure 3.1   Complex-conjugate roots in the left-half s-plane.

If we substitute $s = j\omega$ in $A_v(s)$, we obtain the system's steady-state frequency response:

$$A_v(j\omega) = \frac{\mathbf{Y}}{\mathbf{X}}(j\omega) = \frac{Y\,\underline{/\psi}}{X\,\underline{/0°}}(j\omega) \qquad\qquad 3.8$$

It is often desirable to plot $A_v(j\omega)$, versus frequency. The customary way to do this is by the Bode plot technique, in which we plot

$$20\log|A_v(j\omega)| \text{ vs. } \omega \qquad \text{(log scale)}$$

and

$$\mathrm{ARG}\ A_v(j\omega) \text{ vs. } \omega \qquad \text{(log scale)}$$

These functions are generally plotted on semilogarithmic graph paper, the log axis being used for ω in both cases. It is most convenient to do a Bode magnitude plot when $A_v(j\omega)$ is in time-constant format, illustrated in the example of Eq. 3.9:

$$A_v(j\omega) = K\frac{(j\omega + z_1)(j\omega + z_2)}{(j\omega + p_1)[(j\omega)^2 + j\omega2\zeta\omega_n + \omega_n^2]} \qquad \text{(Laplace form)} \qquad 3.9\text{A}$$

$$= K_v\frac{\left(\dfrac{j\omega}{z_1} + 1\right)\left(\dfrac{j\omega}{z_2} + 1\right)}{\left(\dfrac{j\omega}{p_1} + 1\right)\left[\left(\dfrac{-\omega^2}{\omega_n^2} + 1\right) + j\omega\dfrac{2\zeta}{\omega_n}\right]} \qquad \text{(time-constant form)} \qquad 3.9\text{B}$$

where the dc gain is

$$\mathbf{A}_v(j0) = K_v = \frac{K z_1 z_2}{p_1 \omega_n^2}$$ 3.10

3.2 Bode Plots

As we have noted, Bode plots are a convenient and widely used means of describing a linear system's steady-state frequency response.

EXAMPLE 3.1

Bode Plot of a Real-Pole Transfer Function

To illustrate some of the mechanics of Bode plotting, we use the following frequency response function, which has two real poles (see Fig. 3.2).

$$\mathbf{A}_v(j\omega) = \frac{-200}{(j\omega/10^3 + 1)(j\omega/10^4 + 1)}$$ 3.11

The phase angle of this transfer function is written as

$$\psi = -180° - \left[\tan^{-1}\left(\frac{\omega}{10^4}\right) + \tan^{-1}\left(\frac{\omega}{10^3}\right) \right]$$ 3.12

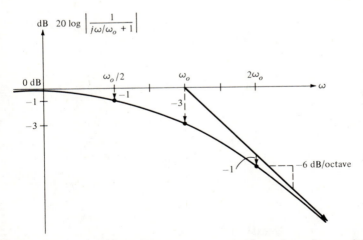

Figure 3.2 Standard Bode magnitude plot for a denominator single real-pole term.

The $-180°$ comes from the minus sign in $\mathbf{A}_v(j\omega)$ (this denotes a $-180°$ phase shift at low frequencies). The arctan functions come from the two denominator vector terms in $\mathbf{A}_v(j\omega)$; the vector terms have unity real parts and imaginary parts of $\omega/10^3$ or $\omega/10^4$. The angle of each term is thus the arctan of the rise/run. The denominator angle terms are summed negatively, the numerator angle terms with positive signs.

The units of $20|\log \mathbf{A}_v(j\omega)|$ are decibels, abbreviated dB. For the example of Eq. 3.11,

$$
\begin{aligned}
20 \log|\mathbf{A}_v(j\omega)| &= 20 \log(200) - \left(20 \log\left|\frac{j\omega}{10^3} + 1\right| + 20 \log\left|\frac{j\omega}{10^4} + 1\right| \right) \\
&= 20 \log(200) - 20 \log\sqrt{\frac{\omega^2}{10^3} + 1} - 20 \log\sqrt{\frac{\omega^2}{10^4} + 1}
\end{aligned}
\qquad 3.13
$$

Thus the dB gain of $|\mathbf{A}_v(j\omega)|$ is plotted as the algebraic sum of logarithmic terms. To make plotting easier, we construct asymptotes for the frequency-dependent terms. We consider each frequency-dependent term at frequencies much lower than its corner frequency, at its corner frequency, and well above its corner frequency. The results of this procedure are illustrated in Fig. 3.3 for poles of the form $(j\omega/\omega_o + 1)$. Note that at frequencies well below the break frequency, ω_o, the real pole's asymptote subtracts 0 dB from the overall transfer function's dB plot. Note also that it is easy to show that the actual dB curve for one real pole is down -1 dB from the 0 dB asymptote at $\omega = \omega_o/2$; it is down -3 dB from the 0 dB asymptote when $\omega = \omega_o$; and it is down -1 dB from the -20 dB/decade asymptote at $\omega = 2\omega_o$.   ●

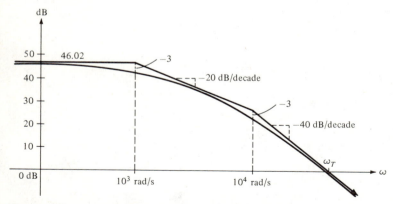

Figure 3.3   Bode magnitude plot for the example of Eq. 3.11 in Example 3.1.

EXAMPLE 3.2

Bode Plot of a Bandpass Transfer Function

Now consider a bandpass frequency response example:

$$\mathbf{A}_v(j\omega) = \frac{-K(j\omega)}{(j\omega/\omega_1 + 1)(j\omega/\omega_2 + 1)} \qquad 3.14$$

where $\omega_1 \ll \omega_2$. The phase of $\mathbf{A}_v(j\omega)$ given in Eq. 3.14 is

$$\psi = -180° + 90° - \left[\tan^{-1}\left(\frac{\omega}{\omega_1}\right) + \tan^{-1}\left(\frac{\omega}{\omega_2}\right)\right] \qquad 3.15$$

As before, the $-180°$ term is from the minus sign in the numerator of $\mathbf{A}_v(j\omega)$, the $+90°$ comes from the j term in the numerator (recall that $j\omega$ is one notation for $\omega 90 = \omega e^{j90°}$ and the negative arctan terms come from the poles.

To plot $20 \log|\mathbf{A}_v(j\omega)|$ for the bandpass transfer function, we first find the mid-band gain, $A_{v(\text{mid})}$. Because $\omega_1 \ll \omega_m \ll \omega_2$ in the mid-band, we can write

$$A_{v(\text{mid})} = |\mathbf{A}_v(j\omega_m)| = \frac{K\omega_m}{(\omega_m/\omega_1)(1)} \qquad 3.16$$

At ω_2, the dB plot begins decreasing at -20 dB/decade. The net dB magnitude plot for this $\mathbf{A}_v(j\omega)$ is shown in Fig. 3.4.

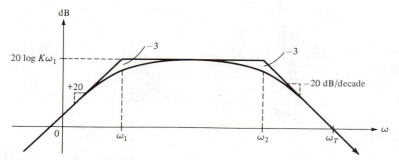

Figure 3.4 Bode magnitude plot for the frequency response function of Example 3.2. The mid-frequency band lies between $2\omega1$ and $\omega_2/2$. The frequency response $= 0$ dB at ω_r.

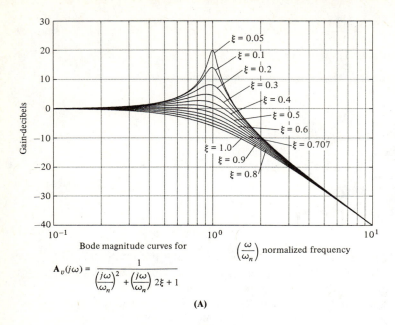

Bode magnitude curves for $\left(\dfrac{\omega}{\omega_n}\right)$ normalized frequency

$$\mathbf{A}_v(j\omega) = \dfrac{1}{\left(\dfrac{j\omega}{\omega_n}\right)^2 + \left(\dfrac{j\omega}{\omega_n}\right) 2\xi + 1}$$

(A)

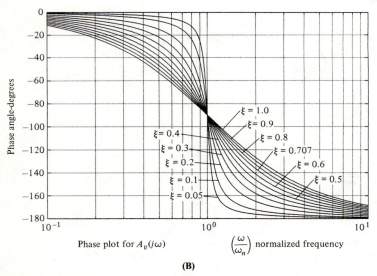

Phase plot for $A_v(j\omega)$ $\left(\dfrac{\omega}{\omega_n}\right)$ normalized frequency

(B)

Figure 3.5   **(A)** Bode magnitude plot for a quadratic complex-conjugate pole pair. **(B)** Phase plot for the complex-conjugate pole pair.

EXAMPLE 3.3 •

Bode Plot of a Transfer Function with a Complex-Conjugate Pole Pair

As a final review example of Bode plot construction, consider the quadratic low-pass transfer function given by Eq. 3.17. This $A_v(j\omega)$ has a complex-conjugate pole pair.

$$A_v(j\omega) = \frac{K_{mid}}{(j\omega)^2/\omega_n^2 + j\omega_2\zeta/\omega_n + 1}$$ 3.17

The phase of the frequency response function is

$$\psi = -\tan^{-1}\left(\frac{2\zeta\omega/\omega_n}{1 - \omega^2/\omega_n^2}\right)$$ 3.18

It is plotted in Fig. 3.5(B). The dB magnitude plot of Eq. 3.17 exhibits some curious characteristics as the damping factor ranges from 0 to 1. These characteristics are shown in Fig. 3.5(A). Note that as $\zeta \to 0$, the complex-conjugate poles approach the $j\omega$ axis of the s-plane, and the system becomes more resonant—its Bode plot shows an increased peak at $\omega = \omega_n$. At $\omega = \omega_n$, it is easy to show that

$$|A_v(j\omega_n)| = \frac{K_{mid}}{2\zeta}$$ 3.19

•

3.3 Graphical Interpretation of Frequency Response in the s-Plane

To illustrate how frequency response may be obtained graphically, consider the amplifier's transfer function in factored Laplace form as shown in Eq. 3.4. Because s can have values only on the $j\omega$ axis in the s-plane, we can interpret $A_v(j\omega)$ as a ratio of *vector* differences in the s-plane. For example, consider the following transfer function, in vector form:

$$A_v(\mathbf{s}) = K\frac{\mathbf{s} - \mathbf{z}_1}{(s - \mathbf{p}_1)(s - \mathbf{p}_2)} \qquad \text{where } \mathbf{s} = j\omega$$ 3.20

Here, $\mathbf{z}_1 = -z_1$ (negative real zero), and $\mathbf{p}_1 = -p_1$ and $\mathbf{p}_2 = -p_2$ (negative real poles). In the s-plane, $A_v(j\omega)$ is evaluated by drawing the vector differences, as shown in Fig. 3.6. From Fig. 3.6, we see that

$$\text{ARG } A_v(j\omega) = \phi_1 + (\theta_1 + \theta_2)$$ 3.21

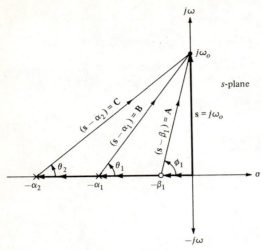

Figure 3.6   Use of vector differences in the s-plane to evaluate the frequency response function of Eq. 3.20. $\omega = \omega_o$ in the example.

and

$$|\mathbf{A}_v(j\omega)| = K \frac{|j\omega - \mathbf{z}_1|}{|j\omega - \mathbf{p}_1||j\omega - \mathbf{p}_2|} = \frac{KA}{BC} \qquad 3.22$$

A, B, and C are the magnitudes of the vector differences.

If a system has a narrow bandpass frequency response function described by

$$\mathbf{A}_v(\mathbf{s}) = K \frac{\mathbf{s}}{(\mathbf{s} - \mathbf{p}_1)(\mathbf{s} - \mathbf{p}_1^*)} \qquad \text{where } \mathbf{s} = j\omega \qquad 3.23$$

Where, $\mathbf{p}_1^*$ is the conjugate of $\mathbf{p}_1$, and $\mathbf{p}_1 = -a + jb$; then we can see how $\mathbf{A}_v(j\omega)$ behaves near $\omega_n = \sqrt{a^2 + b^2}$ using the graphical construction of Fig. 3.7. The phase of this $\mathbf{A}_v(j\omega)$ is simply

$$\text{ARG } \mathbf{A}_v(j\omega) = \phi_1 - (\theta_1 + \theta_2) \qquad 3.24$$

However, $\theta_1 = 90°$ in this case for all ω. The magnitude of $\mathbf{A}_v(j\omega)$ is

$$|\mathbf{A}_v(j\omega)| = K \frac{\omega_o}{|j\omega_o - \mathbf{p}_1||j\omega_o - \mathbf{p}_1^*|} = \frac{K\omega_o}{AB} \qquad 3.25$$

Note that as $\mathbf{s} = j\omega$ approaches the pole at $(-a + jb)$, the vector $|A| = |j\omega - \mathbf{p}_1|$ becomes short, causing a peak in $|\mathbf{A}_v(j\omega)|$.

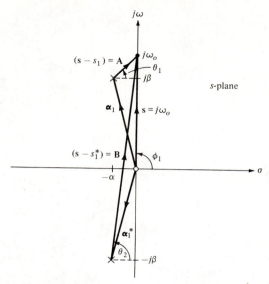

Figure 3.7 Graphical construction of $H(j\omega)$ for a bandpass transfer function.

3.4 Nyquist Plots

Nyquist plots are mentioned briefly for background; we do not use them in this text. A Nyquist plot is a polar plot of frequency response magnitude (scalar) plotted radially versus the phase angle. Nyquist plots can be used to predict amplifier stability under closed-loop feedback conditions.

EXAMPLE 3.4

Nyquist Plot

Assume an amplifier has the following frequency response function:

$$A_v(j\omega) = \frac{-100}{(j\omega/10^6 + 1)(jw/10^7 + 1)}$$

3.26

Its phase is

$$\psi = -180° - \left[\tan^{-1}\left(\frac{\omega}{10^6}\right) + \tan^{-1}\left(\frac{\omega}{10^7}\right) \right]$$

3.27

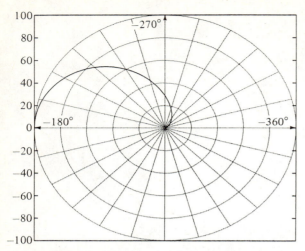

Figure 3.8 Nyquist plot of $\mathbf{H}(j\omega)$ for the two-pole transfer function of Eq. 3.26 in Example 3.4. Note that the phase starts at $-180°$ at $\omega = 0$.

and its magnitude is

$$|\mathbf{A}_v(j\omega)| = \frac{100}{\sqrt{\left(\dfrac{\omega}{10^6}\right)^2 + 1} \; \sqrt{\left(\dfrac{\omega}{10^7}\right)^2 + 1}}$$ 3.28

When we plot Eqs. 3.27 and 3.28 on polar coordinates, we obtain the Nyquist diagram shown in Fig. 3.8. Notice that the Nyquist plot starts out at zero frequency (dc) with a gain magnitude of 100 and a phase of $-180°$ and goes to zero gain magnitude at a phase of $-360°$ at $\omega = \infty$. ●

3.5 Low-Frequency Behavior of RC Amplifiers

Cascaded stages of reactively coupled (RC) amplifiers introduce high-pass behavior in the frequency response function. Consider the capacitively coupled, grounded-emitter BJT amplifier in Fig. 3.9(A). Figure 3.9(B) shows the low- and mid-frequency model for this BJT amplifier. The first step in frequency analysis of this circuit is to consider C_1 and C_2 to have zero reactance and to find the mid-frequency gain, $A_{v(mid)}$. From Fig. 3.9(B), we see that

$$v_o = -\frac{h_{fe}i_b}{G_C + G_L}$$ 3.29

$$i_b = \frac{v_s}{h_{ie}}$$ 3.30

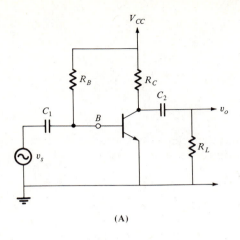

(A)

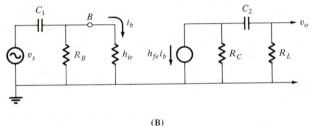

(B)

Figure 3.9 **(A)** Capacitively coupled, grounded-emitter BJT stage. Two coupling capacitors, C_1 and C_2, are used. Their effects must be considered at low frequencies but can be neglected at mid- and high frequencies. R_L is the input resistance of the next stage. **(B)** Low- and mid-frequency small-signal model of the transistor amplifier in (A). We assume $h_{oe} \ll G_C$.

Substituting Eq. 3.30 into Eq. 3.29, we obtain

$$A_{v(\text{mid})} = \frac{v_o}{v_s} = -\frac{h_{fe}}{h_{ie}(G_C + G_L)} \qquad 3.31$$

The low-frequency effects of the coupling capacitors, C_1 and C_2, are now considered. An expression for v_o is written after doing a Norton-to-Thevenin transformation on the collector circuit of Fig. 3.9(B):

$$v_o = -h_{fe}i_bR_C\left(\frac{R_L}{R_L + R_C + \dfrac{1}{sC_2}}\right) = -h_{fe}i_bR_c\left[\frac{sC_2R_L}{sC_2(R_L + R_C) + 1}\right] \qquad 3.32$$

$$i_b = \frac{v_b}{h_{ie}}$$
3.33

$$v_b = v_s \left(\frac{\dfrac{R_B h_{ie}}{R_B + h_{ie}}}{\dfrac{R_B h_{ie}}{R_B + h_{ie}} + \dfrac{1}{sC_1}} \right) = v_s \left[\frac{sC_1 \left(\dfrac{R_B h_{ie}}{R_B + h_{ie}} \right)}{sC_1 \left(\dfrac{R_B h_{ie}}{R_B + h_{ie}} \right) + 1} \right]$$
3.34

Substituting Eq. 3.34 into Eq. 3.33, and then that result into Eq. 3.32, we arrive at an expression for the low-frequency gain of the BJT amplifier:

$$A_v(s) = -\left(\frac{h_{fe}R_C}{h_{ie}} \right) \frac{(sC_2 R_L)[sC_1(R_B \| h_{ie})]}{[sC_2(R_L + R_C) + 1][sC_1(R_B \| h_{ie}) + 1]}$$
3.35

Observe that $A_v(s)$ given by Eq. 3.35 has two zeros at the origin of the s-plane; they arise from the high-pass filters formed by c_1 and c_2. The low-frequency poles lie at $\mathbf{p}_1 = -1/C_2(R_C + R_L)$ rad/s and $\mathbf{p}_2 = -1/C_1(R_B \| h_{ie})$ rad/s. A Bode magnitude plot of the frequency response $|A_v(j\omega)|$ is shown in Fig. 3.10. There is no special reason why ω_2 should be greater than ω_1. ω_2 can be equal to or less than ω_1, depending on the choice of parameter values. As a practical rule, every series coupling capacitor used in an RC amplifier adds a zero at the origin and a low-frequency pole to the system's overall transfer function.

Another source of low-frequency poles and zeros in an amplifier's overall transfer function arises from the use of emitter resistors and bypass capacitors in BJT *grounded-emitter* amplifiers (or, equivalently, the use of source resistors and bypass capacitors in FET *grounded-source* amplifiers). Such an FET circuit is illustrated in Fig. 3.11(A). Notice that series coupling capacitors have been omitted to simplify

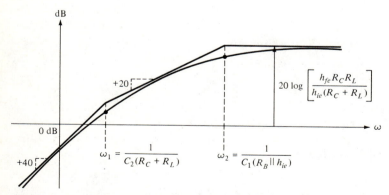

Figure 3.10   Bode magnitude plot of the low- and mid-frequency response of the BJT amplifier of Fig. 3.9(A).

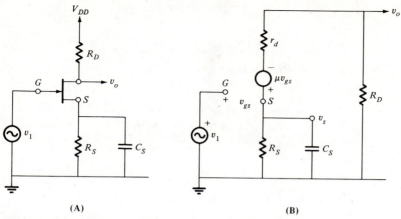

Figure 3.11 **(A)** FET grounded-source amplifier with bypassed source resistor. **(B)** Low- and mid-frequency small-signal model for the FET amplifier.

calculations in this circuit. As in the preceding case, we first find the mid-frequency gain, where C_s is assumed to have zero reactance and $v_s = 0$. By inspection of the series circuit, we find

$$A_{v(\text{mid})} = \frac{v_o}{v_1} = -i_d R_D = -\frac{\mu R_D}{R_D + r_d} \qquad 3.36$$

At low frequencies, C_s is assumed to have significant reactance, and $v_s = 0$. Now we can write

$$v_{gs} = v_1 - v_s \qquad 3.37$$

(Equation 3.37 is an implicit statement of negative feedback in this amplifier.) Also,

$$v_s = \frac{i_d}{G_S + sC_S} = \frac{i_d R_S}{sR_S C_S + 1} \qquad 3.38$$

$$v_o = -i_d R_D \qquad 3.39$$

and

$$i_d = \frac{\mu v_{gs}}{R_D + r_d + \dfrac{1}{G_S + sC_S}} \qquad 3.40$$

Equations 3.37 and 3.38 are substituted into Eq. 3.40, and the resulting expression for i_d is put into Eq. 3.39 to give, after some algebra, an expression for the low-frequency gain in time-constant form for the FET amplifier:

$$A_v(s) = \frac{V_o}{V_1} = -\frac{-\mu R_D(1 + sR_SC_S)}{[R_D + r_d + R_S(\mu + 1)]\left[s\,\dfrac{C_SR_S(R_D + r_d)}{R_D + r_d + R_S(\mu + 1)} + 1\right]} \qquad 3.41$$

This gain transfer function has a real zero at $Z_1 = -1/R_SC_S$ rad/s and a real pole at $p_1 = -[R_D + r_d + R_S(\mu + 1)]/C_SR_S(R_D + r_d)$ rad/s in the s-plane. The frequency of the pole is set by the resistance C_S "sees" looking up the FET's source, in parallel with R_S. That is, C_S sees an R_{eq} of

$$R_{eq} = \frac{\left(\dfrac{R_D + r_d}{\mu + 1}\right)R_S}{\dfrac{R_D + r_d}{\mu + 1} + R_S} \qquad 3.42$$

The Bode magnitude plot for the FET amplifier's low- and mid-frequency response is shown in Fig. 3.12. Unlike the RC high-pass coupling treated in the previous example for the BJT amplifier, the FET amplifier is seen to have finite gain at zero frequency.

It is easily seen that an RC amplifier with several stages, whether BJT or FET, will have a complex low-frequency behavior, having one real pole and zero for every bypassed emitter (or source) resistor, and one zero at the origin and one low-frequency pole for every series capacitive coupling circuit used. Fortunately, modern IC amplifiers use direct coupling, hence avoiding the complexity of low-frequency poles and zeros. Low-frequency poles and zeros also can create stability problems when overall negative voltage feedback is applied around an RC amplifier.

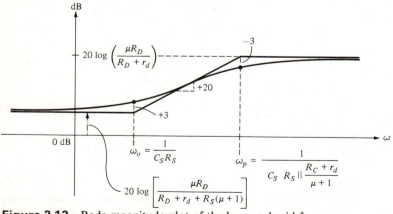

Figure 3.12 Bode magnitude plot of the low- and mid-frequency response of the FET amplifier of Fig. 3.11(A).

3.6 Operational Amplifier Frequency Response

As a final topic in this chapter, we examine the frequency response of typical *compensated* IC op-amps. Op-amps are designed to be used with large amounts of negative feedback transmitted through a variety of feedback transfer functions. To ensure stability over a wide range of conditions, the op-amp's frequency response is compensated (internally or with external capacitors) to be of the form

$$A_v(s) = \frac{V_o}{V_i - V_i'} \cong \frac{Kv}{\tau_a s + 1} \qquad\qquad 3.43$$

This open-loop $A_v(s)$ has a one real-pole, low-pass frequency response characteristic. Of special note is the dc gain, K_v, which is usually in the range 10^4 to 10^8, depending on design and application. Such a high K_v is the result of two or more cascaded dc gain stages; hence we would expect to see two or more high-frequency poles in the op-amp's open-loop transfer function. Compensation for stability inserts a dominant, low-frequency real pole at $\omega_a = 1/\tau_a$ such that $|A_v(j\omega)| < 1$, before the effects of the other natural high-frequency poles are felt in the op-amp's frequency response. Thus Eq. 3.43 is a good approximation to the gain of a frequency-compensated op-amp.

An important small-signal parameter of op-amps is their gain-bandwidth product (GBWP), found from Eq. 3.43 by multiplying the op-amp's dc gain by its -3 dB bandwidth in hertz. Thus,

$$\text{GBWP} = \frac{K_v}{2\pi\tau_a} \qquad\qquad 3.44$$

Note that for a single real-pole transfer function, the GBWP is also equal to f_T, the frequency at which $|A_v(j2\pi f_T)| = 1$, or 0 dB.

More will be said in Chapter 4 about the special problems of high-frequency response in amplifiers.

SUMMARY

In this chapter, we have stressed the use of the Bode plot to describe amplifier frequency response. Because most modern IC amplifiers are direct-coupled (DC), we have been concerned mostly with the details of mid- and high-frequency sinusoidal response. Properties of the complex-conjugate (quadratic) pole pair are covered, and the vector (graphical) interpretation of the rational polynomial transfer function in factored form is given. The vector form is used in root-locus analysis, found in Chapters 7 and 11.

Low-frequency analysis has also been treated. We have seen how capacitive bypassing of transistor amplifier emitter (or source) resistors affects the low-frequency portion of the transfer functions, and also how low-frequency response is attenuated by resistor-capacitor interstage coupling circuits.

The frequency-compensated op-amp transfer function was shown to be approximately a real pole with very high dc gain.

PROBLEMS

3.1 Sketch dimensioned Bode magnitude and phase plots for the following transfer functions. Where applicable, show asymptote slopes, dB corrections of the actual magnitude frequency response curve at break frequencies, and ω_T values.

(a) $H(s) = \dfrac{-10^8}{s + 100}$

(b) $H(s) = \dfrac{s - 10^3}{s + 10^3}$

(c) $H(s) = \dfrac{100}{\dfrac{s^2}{10^9} + \dfrac{s}{9.0909 \times 10^3} + 1}$

(d) $H(s) = \dfrac{-100}{\dfrac{s^2}{10^9} + \dfrac{s}{1.581 \times 10^5} + 1}$

(e) $H(s) = \dfrac{100(s/1.581 \times 10^5)}{\dfrac{s^2}{10^9} + \dfrac{s}{1.581 \times 10^5} + 1}$

(f) $H(s) = \dfrac{-s + 10^4}{(s + 10)(s + 10^4)}$

(g) $H(s) = \dfrac{1 - e^{-s}}{s}$

(h) $H(s) = \dfrac{s^2 + 10^8}{s^2 + s(4 \times 10^3) + 10^8}$

(i) $H(s) = \dfrac{s + 50}{s}$

(j) $H(s) = \dfrac{-10^3 e^{-0.1s}}{s + 3}$

3.2 Make a dimensioned Nyquist (polar) plot for the following transfer functions:

(a) $A_L(s) = \dfrac{-10^6}{\dfrac{s}{100} + 1}$

(b) $A_L(s) = \dfrac{-s(4 \times 10^3)}{s^2 + s(4 \times 10^3) + 10^8}$

3.3 Low-Frequency Response

(a) Use SSM analysis to make a dimensioned Bode magnitude plot for the low- and

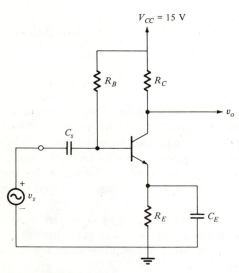

Figure P3.3

mid-frequency response, $\mathbf{H}(j\omega) = (V_o/V_s)(j\omega)$, of the BJT circuit in Fig. P3.3. Assume $R_C = 7\ \text{k}\Omega$, $R_E = 1\ \text{k}\Omega$, $h_{re} = h_{oe} = 0$, $h_{ie} = 2.6\ \text{k}\Omega$, $h_{fe} = 100$, $R_B = 1.335\ \text{M}\Omega$, $C_S = 0.1\ \mu\text{F}$, and $C_E = 33\ \mu\text{F}$.

(b) Use an ECAP such as Micro-Cap II to plot the frequency response required in part (a). Assume the BJT is a 2N2369. Compare your result with that of part (a), in particular, low- and mid-frequency gain and the frequencies of the pole and zero.

3.4 Low-Frequency Response

Use SSM analysis to make a dimensioned Bode magnitude plot of the low- and mid-frequency response of the FET amplifier shown in Fig. P3.4. Assume $R_D = 5.6\ \text{k}\Omega$, $R_S = 1.2\ \text{k}\Omega$, $R_G = 10\ \text{M}\Omega$, $C_1 = 1\ \text{nF}$, $C_S = 4.2\ \mu\text{F}$, $g_{mo} = 6 \times 10^{-3}\ \text{S}$, $I_{DSS} = 3\ \text{mA}$, $I_{DQ} = 2\ \text{mA}$, and $r_d = 50\ \text{k}\Omega$.

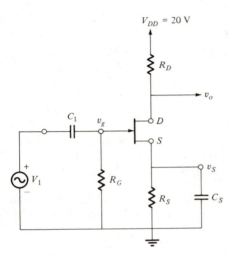

Figure P3.4

3.5 A certain op-amp is specified as having a dc differential gain of 6×10^5 and a unity-gain frequency of 25 MHz. Find the amplifier's gain-bandwidth product and its τ_a (see Eq. 3.43).

3.6 N low-pass systems, each with gain

$$H(s) = \frac{K_v\omega_o}{s + \omega_o}$$

are cascaded. Derive an expression for the -3 dB frequency, ω_b, for the N cascaded low-pass systems.

3.7 N high-pass systems, each with gain

$$F(s) = \frac{K_v s}{s + \omega_o}$$

are cascaded. Derive an expression for the -3 dB frequency, ω_b, for the N cascaded high-pass systems.

3.8 Make a dimensioned Bode magnitude plot for the low- and mid-frequency response of the FET source-follower shown in Fig. P3.8. Assume $g_m = 3500\ \mu S$, $r_d = 40\ k\Omega$, $R_S = 1\ k\Omega$, $R_2 = 5\ k\Omega$, $C_2 = 0.1\ \mu F$, $C_1 = 0.1\ \mu F$, and $R_G = 100\ k\Omega$.

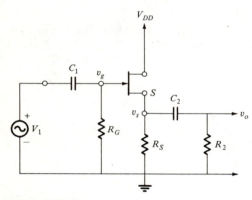

Figure P3.8

3.9 Low-Frequency Response

(a) Use SSM analysis to make a dimensioned Bode magnitude plot for the low- and mid-frequency response of the BJT emitter-follower shown in Fig. P3.9. Assume $h_{re} = h_{oe} = 0$, $h_{fe} = 100$, $h_{ie} = 525\ \Omega$, $R_E = 1\ k\Omega$, $R_2 = 10\ k\Omega$, $R_B = 87.8\ k\Omega$, $C_1 = 0.1\ \mu F$, and $C_2 = 1\ nF$.

(b) Use an ECAP such as Micro-Cap II to make the Bode magnitude plot required in part (a). Use a 2N2369 BJT; compare your results with those obtained in part (a).

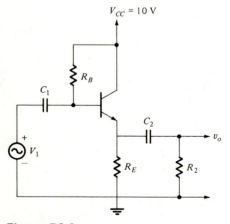

Figure P3.9

Chapter

4

High-Frequency Behavior of Simple Amplifier Stages

In this chapter, we will examine the fundamental limitations to amplifier high-frequency response. These limitations will be discussed in the context of single active device stages using BJTs and FETs. High-frequency small-signal models are presented which approximately describe the behavior of BJTs and FETs at high frequencies; these high-frequency small-signal models are used to predict circuit behavior.

Broadbanding, the science (and sometimes art) of extending amplifier high-frequency response, is introduced. Some broadbanding strategies use circuit architecture to reduce effective input capacitance, whereas others improve bandwidth by trading off mid-band gain for bandwidth through the use of negative feedback (discussed in detail in Chapters 6 and 7).

We stress that high-frequency amplifier performance can generally be considered separately from low- and mid-frequency response in broadband amplifiers, hence separation of the treatment of high-, and low- and mid-frequency, response between this chapter and Chapter 3.

4.1 The Hybrid-Pi High-Frequency Model for the BJT

The hybrid-pi high-frequency small-signal (HFSSM) for the BJT is now almost universally used to model and predict high-frequency BJT circuit behavior. It is shown in Fig. 4.1 with the notation we use in this text. Parameter notations vary between authors of electronics texts, and some of the aliases are given in the figure caption. The transconductance of the hybrid-pi small-signal model, g_m, is defined under short-

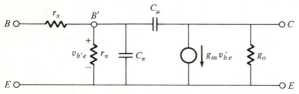

Figure 4.1   BJT, hybrid-pi, high-frequency, small-signal model, also known as the Giacoletto model. Parameter notation varies among authors. $r_x = r_{bb'} = r_b =$ the ohmic base spreading resistance (includes resistance of BJT input lead to base semiconductor, also base semiconductor material, $15\ \Omega < r_x < 100\ \Omega$). $r_\pi = r_{b'e} =$ the small-signal base input resistance. $C_\pi = C_{b'e} = C_e =$ the small-signal capacitance of the forward-biased base-emitter junction. C_π is largely the base-emitter junction diffusion capacitance; it is generally about ten times the size of the base-collector junction capacitance, C_μ. $C_\mu = C_{b'c}$ is largely a depletion capacitance. $g_o = 1/r_o = g_{ce} =$ the small-signal output conductance, approximately equal to h_{oe} of the mid-frequency, common-emitter, h-parameter, small-signal model. $g_o \cong (h_{oe} - g_m h_{re})$, Millman and Halkias (1972). g_m is the BJT's transconductance.

circuited output conditions:

$$g_m \triangleq \left(\frac{\partial i_c}{\partial v_{b'e}}\right)_{v_{ce}=0} \cong \frac{h_{fe}}{r_\pi} = \frac{|I_{CQ}|}{V_T} \qquad\qquad 4.1$$

The g_m for BJTs is large compared to that for FETs; it increases with the quiescent collector current of the BJT. For example, if $I_{CQ} = 1$ mA, $g_m = 0.04$ S.

Another important BJT high-frequency parameter is f_T, the frequency at which the common-emitter, short-circuit output current gain reaches unity magnitude. In the circuit of Fig. 4.2(A), the output is short circuited, causing the small-signal voltage, $v_{ce} = 0$. The BJT common-emitter, h-parameter current gain is defined as

$$h_{fe} \triangleq \left(\frac{\partial i_c}{\partial i_b}\right)_{v_{ce}=0} \qquad\qquad 4.2$$

When i_b and i_c are sine waves, h_{fe} can be considered to be a complex function of frequency:

$$\mathbf{h}_{fe}(j\omega) = \left.\frac{\mathbf{I}_c}{\mathbf{I}_b}(j\omega)\right|_{V_{ce}=0} \qquad\qquad 4.3$$

Thus, by definition of f_T,

$$|\mathbf{h}_{fe}(j2\pi f_T)| = 1 \qquad\qquad 4.4$$

To develop an expression for f_T in terms of the hybrid-pi circuit parameters, we refer to Fig. 4.2(B). Note that because of the small-signal short circuit on the collector,

$$\mathbf{I}_c = g_m V_{b'e} - j\omega C_\mu V_{b'e} \cong g_m V_{b'e} \qquad\qquad 4.5$$

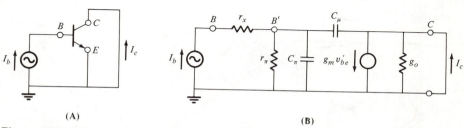

(A)   (B)

Figure 4.2 **(A)** Circuit for determining a BJT's short-circuit output, complex current gain, $\mathbf{h}_{fe}(j\omega)$. I_b is a sinusoidal small-signal current source. **(B)** Hybrid-pi HFSSM used to find $\mathbf{h}_{fe}(j\omega)$.

$V_{b'e}$ is found from Ohm's law: The input current, I_b, "sees" an impedance composed of r_π in parallel with the reactances of C_π and C_μ to ground. Thus we have

$$\mathbf{V}_{b'e} = \frac{I_b}{g_\pi + j\omega(C_\pi + C_\mu)} \qquad 4.6$$

Equation 4.6 can be substituted into Eq. 4.5, and the result used in Eq. 4.3 to obtain

$$\mathbf{h}_{fe}(j\omega) = \left.\frac{I_c}{I_b}\right|_{V_{ce}=0} = \frac{h_{fe}(0)}{1 + j\omega r_\pi(C_\pi + C_\mu)} \qquad 4.7$$

$h_{fe}(0)$ is the low- and mid-frequency common-emitter, h-parameter current gain, also known as transistor beta. Solution of Eq. 4.7 using the f_T definition given in Eq. 4.4 yields

$$f_T = \frac{g_m}{2\pi(C_\pi + C_\mu)} = \frac{h_{fe}(0)}{2\pi(C_\pi + C_\mu)r_\pi} = \frac{|I_{cQ}|}{2\pi(C_\pi + C_\mu)V_T} \text{ Hz} \qquad 4.8$$

Thus we see that the expression for the frequency for unity high-frequency, short-circuit, small-signal current gain magnitude is a function of both C_π and C_μ, as well as of the Q-point. Transistors designed to have small C_π and C_μ have high f_T's. This is an expensive feature of BJTs. The behavior of BJT $\mathbf{h}_{fe}(j\omega)$ is summarized in Fig. 4.3. Because $\mathbf{h}_{fe}(j\omega)$ has a single real pole at $\omega_\beta = 1/r_\pi(C_\pi + C_\mu)$ rad/s, the BJT has a gain-bandwidth product given by

$$\text{GBWP} = h_{fe}(0) \frac{1}{r_\pi(C_\pi + C_\mu)2\pi} \text{ Hz} \qquad 4.9$$

Not unexpectedly, GBWP = f_T Hz.

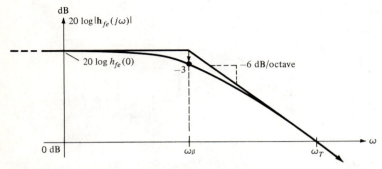

Figure 4.3   Frequency response of a BJT's $\mathbf{h}_{fe}(j\omega)$.

In closing, we remark that it is generally accepted that the BJT hybrid-pi HFSSM is valid at frequencies below $f_T/3$. Above this frequency, second- and third-order effects come into play, and detailed analysis becomes tedious, except by computer.

4.2 High-Frequency FET Models

The universal HFSSM for FETs is valid for all types of FETs: JFETs (p- and n-channel) and all types of MOSFETs. It is shown in Fig. 4.4. FET operation is assumed to be under saturated channel (normal linear) conditions.

In a typical small-signal JFET, the parameters range as follows: $0.1 \text{ mS} < g_m < 10 \text{ mS}$; $100 \text{ k}\Omega < r_d < 1 \text{ M}\Omega$; $0.1 \text{ pF} < C_{ds} < 1 \text{ pF}$; $1 \text{ pF} < (C_{gs}, C_{gd}) < 10 \text{ pF}$; r_{gs} (in parallel with C_{gs}, not shown) $> 10^8 \ \Omega$; and r_{gd} (in parallel with C_{gd}, not shown) $> 10^8 \ \Omega$.

Manufacturers generally do not give values for C_{gs}, C_{gd}, and C_{ds}. Instead, they specify C_{iss} and C_{rss}, the common-source, short-circuit input capacitance ($v_{ds} = 0$) and the reverse transfer capacitance ($v_{gs} = 0$), respectively. From the HFSSM of Fig. 4.4, we see that

$$C_{rss} \triangleq C_{ds} + C_{gd} \cong C_{gd} \qquad \qquad 4.10$$

Also,

$$C_{iss} \triangleq C_{gd} + C_{gs} \qquad \qquad 4.11$$

Therefore, from Eqs. 4.10 and 4.11, we have

$$C_{gd} \cong C_{rss} \qquad \qquad 4.12$$
$$C_{gs} \cong C_{iss} - C_{rss} \qquad \qquad 4.13$$
$$C_{ds} \cong C_{oss} - C_{rss} \qquad \qquad 4.14$$

When C_{oss}, the FET's output capacitance, is given, Eq. 4.14 can be used to approximate the small value of C_{ds}.

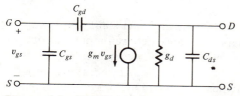

Figure 4.4   HFSSM for all discrete FETs.

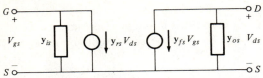

Figure 4.5   Discrete FET y-parameter HFSSM used at VHF and UHF.

At VHF and UHF, the lumped parameter FET HFSSM is no longer valid. Instead, it is possible to use a y-parameter model in which the four y-parameters are frequency dependent. Note that

$$y_{fs} = g_{fs} + jx_{fs} = \mathbf{y}_{fs}(j\omega) \qquad\qquad 4.15$$

and at low frequencies, $g_{fs} = g_m$, $y_{os} \to g_d$, $y_{rs} \to 0$, and $y_{is} \to \infty$.

FETs are characterized by a maximum operating frequency, f_{max}, at which the current into the gate node equals $|y_{fs}V_{gs}|$. It is easy to see from Fig. 4.4 that this occurs when

$$|I_g| = \frac{V_{gs}}{1/2\pi f_{max}(C_{gs} + C_{gd})} = |\mathbf{y}_{fs}(j2\pi f_{max}V_{gs}| \qquad\qquad 4.16$$

from which we can easily obtain

$$f_{max} \cong \frac{g_m}{2\pi C_{iss}} \text{ Hz} \qquad\qquad 4.17$$

f_{max} is a figure of merit for FETs similar to f_T used for BJTs. The practical maximum useful frequency of FET amplifiers is generally lower than f_{max}, however, because of the Miller effect and load capacitance effects.

Figure 4.5 illustrates the high-frequency y-parameter model useful at VHF and UHF. Note that all four y-parameters are, in general, frequency dependent.

4.3 The Miller Effect

The Miller effect is an effective increase in an inverting amplifier stage's input capacitance caused by negative feedback through a capacitance between input and output. It occurs in both FET and BJT single-stage amplifiers. The augmented C_{in} forms an RC low-pass filter with the Thevenin R_{th} driving the amplifier's input. If the amplifier is driven by a pure voltage source ($R_{th} = 0$), then there is no high-frequency attenuation caused by the Miller effect. Unfortunately, R_{th} is nonzero, so it is important to keep C_{in} as small as possible to avoid loss of high-frequency signal components.

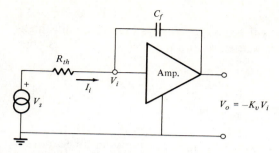

Figure 4.6 Negative feedback circuit illustrating the Miller effect. The amplifier is an ideal voltage amplifier with gain $-K_v$.

A generalized Miller effect circuit is shown in Fig. 4.6. If we look into the V_i node, we see an admittance

$$\mathbf{Y}_i = \frac{I_i}{V_i} = \frac{V_i - V_o}{V_i(1/j\omega C_f)} = \frac{j\omega C_f(V_i + V_i K_v)}{V_i} = j\omega C_f(1 + K_v) \tag{4.18}$$

It is apparent that the Thevenin circuit "sees" an enlarged equivalent input capacitance of $C_f(1 + K_v)$ to ground. This is the Miller effect. The equivalent low-pass filter formed by R_{th} and $C_f(1 + K_v)$ is shown in Fig. 4.7. The transfer function of this circuit is low-pass and is easily found to be

$$\frac{V_o}{V_s} = \frac{-K_v}{1 + j\omega R_{th} C_f(1 + K_v)} \tag{4.19}$$

The break frequency of the *Miller pole* is

$$f_o = \frac{1}{2\pi R_{th} C_f(1 + K_v)} \text{ Hz} \tag{4.20}$$

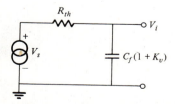

Figure 4.7 Low-pass filter formed at the input by the Miller effect in the circuit of Fig. 4.6.

Now let us examine the high-frequency behavior of the grounded-emitter BJT amplifier shown in Fig. 4.8(A). First we will do an exact analysis of the circuit's high-frequency response. Node equations are written on the $V_{b'e} = V_b$ and V_o nodes, then solved for V_o/V_s:

$$V_b(G_s + g_\pi + sC_\pi + sC_\mu) - V_o sC_\mu = V_s G_s \qquad 4.21$$

$$- V_b(sC_\mu - g_m) \qquad + V_o(sC_\mu + G_C) = 0 \qquad 4.22$$

After some algebra, we find the amplifier's high-frequency response in time-constant form:

$$A_v(s) = \frac{V_o}{V_s}$$

$$= \frac{-g_m R_C\left(\dfrac{r_\pi}{R_s + r_\pi}\right)\left(1 - s\dfrac{C_\mu}{g_m}\right)}{s^2 C_\mu C_\pi R_C\left(\dfrac{R_s r_\pi}{R_s + r_\pi}\right) + s\left[C_\pi\left(\dfrac{R_s r_\pi}{R_s + r_\pi}\right) + C_\mu R_C + C_\mu(1 + g_m R_C)\left(\dfrac{R_s r_\pi}{R_s + r_\pi}\right)\right] + 1}$$

$$4.23$$

Note that the collector-base junction capacitor, C_μ, gives rise to a very high frequency, right-half s-plane zero at $s_1 = +g_m/C_\mu$ rad/s. Indeed, $\mathbf{s}_1$ is generally at a frequency well above the BJT's ω_T, where the results predicted by the model of Fig. 4.8(B) are no longer valid. Common sense instructs us to disregard this zero, which we will do.

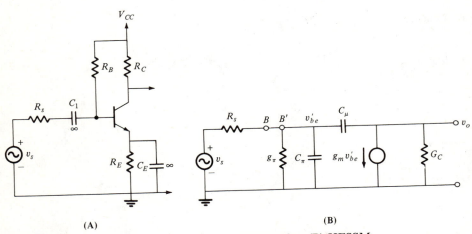

                              (A)   (B)

Figure 4.8 **(A)** Simple grounded-emitter BJT amplifier. **(B)** HFSSM of the amplifier in (A) using the hybrid-pi model. Here we assume $G_C \gg h_{oe}$, $r_x = 0$, $R_B \gg r_\pi$, and $v_{b'e} = v_b$.

The denominator of Eq. 4.23 is of the standard quadratic form $s^2/\omega_n^2 + s^2\xi/\omega_n + 1$, where the undamped natural frequency ω_n is

$$\omega_n = \sqrt{\frac{R_s + r_\pi}{R_s r_\pi R_C C_\mu C_\pi}}$$

4.24

The mid-frequency gain of the BJT amplifier is seen in the numerator of Eq. 4.23 and is

$$A_{v(mid)} = -\frac{g_m R_C r_\pi}{R_s + r_\pi}$$

4.25

The poles of $Av(s)$ can be real or complex-conjugate, depending on the value of the damping factor, ξ. ξ, in turn, depends on the size of C_μ and $g_m R_C$.

Next, we will examine the high-frequency response of the circuits of Fig. 4.8(A) and (B) using a simplification called the *unilateral approximation* (UA). It will be seen that the unilateral approximation makes the algebra of a pencil-and-paper analysis simpler at the expense of some accuracy. In writing the node equations for the circuit of Fig. 4.8(B) using the unilateral approximation, we assume that negligible current leaves the V_o node through C_μ. On the other hand, we do account for current leaving the V_b node through C_μ. In other words,

$$V_o G_C \gg j\omega C_\mu (V_o - V_b)$$

4.26

The simplified node equations are thus

$$V_b[G_s + g_\pi + s(C_\pi + C_\mu)] - V_o s C_\mu = V_s G_s$$

4.27

$$V_b g_m \qquad\qquad + V_o G_C = 0$$

4.28

Equation 4.28 is solved for V_b and the solution substituted into Eq. 4.27. The result is easily rearranged to give the amplifier's high-frequency gain under the unilateral approximation:

$$A_v(s) = \frac{V_o}{V_s} = \frac{-g_m R_C \left(\dfrac{r_\pi}{R_s + r_\pi}\right)}{1 + s[C_\pi + C_\mu(1 + g_m R_C)]\left(\dfrac{R_s r_\pi}{R_s + r_\pi}\right)}$$

4.29

This transfer function has only one real pole and no right-half s-plane zero. The Miller capacitance is clearly identifiable as

$$C_m = C_\mu(1 + g_m R_C)$$

4.30

The break frequency is simply

$$\omega_{ua} = \frac{1}{[C_\pi + C_\mu(1 + g_\mu R_C)]\left(\dfrac{R_s r_\pi}{R_s + r_\pi}\right)} \text{ rad/s}$$

4.31

Table 4.1 Comparison of analyses of the high-frequency response of the BJT amplifier

Parameter	Exact Method	Unilateral Approximation
Mid-frequency gain	-60 V/V	-60 V/V
ω_o (zero)	5.00×10^9 rad/s	————
ω_n	6.46×10^7 rad/s	————
ξ	4.95 (real roots, ω_1 and ω_2)	————
ω_1	6.60×10^6 rad/s	————
ω_2	6.32×10^8 rad/s	
ω_{ua}	————	7.51×10^6 rad/s
$\omega_{T/3}$	1.52×10^8 rad/s	1.52×10^8 rad/s

Let us now perform a numerical comparison of the UA results given in Eqs. 4.29 and 4.31, and the exact solution seen in Eqs. 4.23 through 4.25. Typical parameter values are selected: $g_m = 0.01$ S, $r_\pi = 1.5 \times 10^3$ Ω, $R_C = 10^4$ Ω, $R_s = 1000$ Ω, $C_\pi = 20$ pF, $C_\mu = 2$ pF, $r_x = 0$, and $R_b = \infty$. Table 4.1 compares the two methods of analysis.

Note that the break frequency found by the UA analysis, ω_{ua}, is 13.8% higher than the dominant pole found by the exact method, ω_1. The second high-frequency pole, ω_2, and the zero found by the exact method lie well above model validity limits set by $\omega_{T/3}$. They should be disregarded. Not surprisingly, both methods yield the same mid-frequency gain. The 13.8% error in using the unilateral approximation may or may not be acceptable; in any case, the reader should observe that the unilateral approximation is an approximation motivated by the need for algebraic simplification. If a circuit analysis software program is to be used, do not use the unilateral approximation; let the computer do the work.

The Miller effect also occurs in inverting FET amplifiers. Analysis of these amplifiers is quite similar to that for the BJT which we have just completed. An exact analysis gives a right-half s-plane zero at very high frequencies, a real, dominant pole, and a very high frequency real pole. Details of the analysis of the FET amplifier's Miller effect are left for the problems section.

4.4 Broadbanding Techniques

Many techniques have been developed to extend the high-frequency bandwidth of wideband transistor amplifier stages subject to the Miller effect. All such techniques seek to lower the input circuit RC time constant or compensate for it.

EXAMPLE 4.1

A Broadbanding Technique Used in DC Amplifiers

Consider the two-stage BJT amplifier shown in Fig. 4.9(A). We are concerned in this case with the Miller effect in Q_2. The Q_1 stage can be represented by a Thevenin equivalent (V_1, R_s). Figure 4.9(B) illustrates the HFSSM looking into Q_2's base. There is a loss in gain in coupling Q_1's collector to Q_2's base through R_1. However, the use of R_1 provides two benefits. The first is a desired shift in quiescent dc voltage level from Q_1's collector to Q_2's base. The second benefit is that the $R_1 C_1$ time constant can be adjusted to cancel approximately the effect of the input capacitance of Q_2, thus extending circuit bandwidth. Writing the voltage-divider expression for $V_{b(2)}/V_1$, we have

$$\frac{V_{b(2)}}{V_1} = \frac{\dfrac{1}{G_\pi + sC_T}}{\dfrac{1}{G_\pi + sC_T} + \dfrac{1}{G_1 + sC_1} + R_s} \qquad 4.32$$

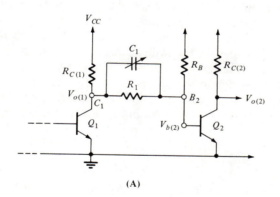

(A)

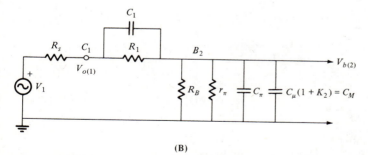

(B)

Figure 4.9 **(A)** Two-stage DC BJT amplifier with the Miller effect in Q_2. **(B)** HFSSM of the $Q_1 - Q_2$ interface; K_2 is the mid-frequency gain of Q_2, $V_{o(2)}/V_{b(2)}$.

where $G_\pi = (g_\pi + G_b)$ and $C_T = [C_\pi + C_\mu(1 + K_2)]$. If the attenuation given by Eq. 4.32 is put into time-constant form,

$$\frac{V_{b(2)}}{V_1} = \frac{\left(\dfrac{R_\pi}{R_\pi + R_1 + R_s}\right)(1 + sC_1R_1)}{s^2\left(\dfrac{C_1C_TR_1R_\pi R_s}{R_\pi + R_1 + R_s}\right) + s\,\dfrac{R_1C_1R_s + C_TR_\pi R_s + (C_1 + C_T)R_1R_\pi}{R_\pi + R_1 + R_s} + 1} \qquad 4.33$$

If we make $R_1C_1 = R_\pi C_T = RC$ and assume that $(R_1 + R_\pi) \gg R_s$, it is easy to show that Eq. 4.33 reduces to

$$\frac{V_{b(2)}}{V_1} = \frac{R_\pi}{R_\pi + R_1} \qquad 4.34$$

which is obviously frequency independent. ●

Another means of reducing high-frequency attenuation in coupling amplifier stages caused by the Miller effect is to follow every common-emitter BJT amplifier stage (or FET common-source amplifier) with a low output impedance emitter-follower stage (or FET source-follower). The follower presents a low capacitive load to the preceding stage and drives the next stage's input from a low source impedance, giving good high-frequency response despite the Miller capacitance. Figure 4.10 illustrates the alternating grounded-emitter/emitter-follower architecture. This type of circuit design is well suited for implementation in ICs.

Other broadbanding strategies use local (or implicit) negative feedback to trade off gain for bandwidth. Two such circuits are shown in Fig. 4.11.

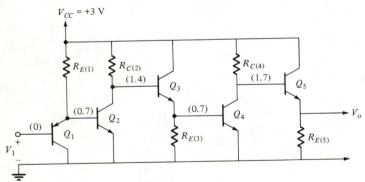

Figure 4.10 Use of emitter-followers to improve the gain-bandwidth product of a multistage BJT amplifier by reducing the effects of the Miller effect. Q_1, Q_3, and Q_5 are emitter-follower stages with low output impedance and high input impedance. Numbers in parentheses are quiescent dc voltages.

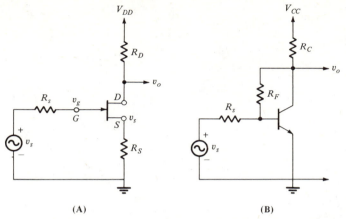

(A)   (B)

Figure 4.11 **(A)** Unbypassed source resistor acts to lower mid-band gain and raise the frequency of the dominant pole of the stage. The same strategy works with a BJT with an unbypassed emitter resistor. **(B)** Shunt feedback applied through R_f. Again, mid-band gain is exchanged for improved high-frequency response.

Finally, device orientation and circuit design can be used to reduce input capacitance and thus extend bandwidth. Examples of such broadbanded circuits are shown in Fig. 4.12.

EXAMPLE 4.2

Broadbanding: The Cascode Amplifier

To see how the two-transistor cascode circuit (Fig. 4.12[A]) improves bandwidth, we first construct a simplified HFSSM for this circuit (see Fig. 4.13[A]). To simplify this HFSSM, we note that the BJT's controlled current source, $g'_m V_{b'e}$, is in reality $g'_m V_2$ leaving the V_2 node. The V_2 node equation is unchanged if we replace the active source, $g'_m V_2$, with a passive conductance to ground, g'_m; this procedure follows the substitution theorem. The V_o node equation is preserved if the same controlled current source, $g'_m V_2$, enters the V_o node from ground. Now it is evident, from Ohm's law, that the output voltage is given by

$$V_o = \frac{V_2 g'_m}{G_C + sC_\mu} = \frac{V_2 g'_m R_C}{sR_C C_\mu + 1}$$
                                                                        4.35

The circuit containing the gate and drain nodes is easy to solve for V_2 if we consider the relative size of the conductances at the V_2 node. Using typical values, we

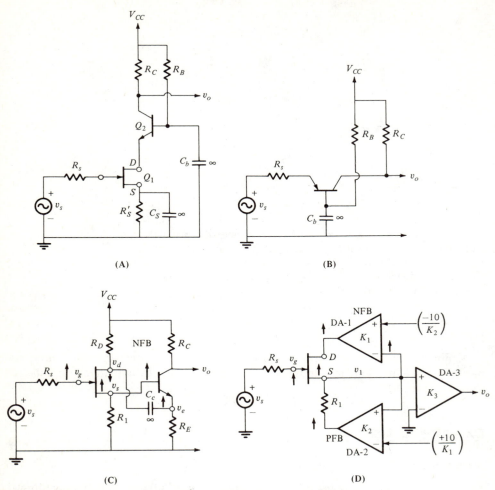

Figure 4.12 **(A)** FET-BJT cascode amplifier. (FET-FET and BJT-BJT architectures are also used.) **(B)** Grounded-base BJT amplifier. The FET equivalent is the grounded-gate. **(C)** Bootstrap circuit. **(D)** Second bootstrap circuit. The differential amplifiers are broadband IC differential amplifiers with low output impedance and differential gains K_i.

have $g_m' = 0.05 \text{ S} \gg g_\pi = 5 \times 10^{-4} \text{ S} \gg g_d = 2 \times 10^{-6} \text{ S}$. Thus the parallel combination of these conductances gives rise to a net resistance to ground from the V_2 node of 19.8 Ω. Because this is such a relatively low resistance value, it gives the capacitance, $(C_3 + C_\pi)$, negligible importance. (At what frequency does the reactance of $[C_3 + C_\pi]$ equal $1/[g_m' + g_\pi + g_d]$?) The low resistance also makes the unilateral approximation on the V_2 node fairly accurate (i.e., negligible current leaves the V_2 node

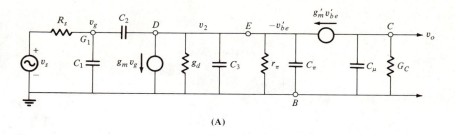

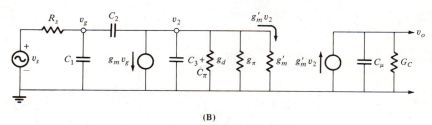

Figure 4.13 **(A)** High-frequency model for the FET-BJT cascode amplifier. $C_1 = C_{gs}$, $C_2 = G_{gd}$, $C_3 = C_{ds}$, $h_{oe} = 0$, $g'_m = h_{fe}/r = 0.05$ S, and $v_{b'e} = -v_2$. **(B)** HFSSM after simplification by means of a source split and use of the substitution theorem.

through C_2). The FET-BJT cascode amplifier node equation for the V_g node is

$$V_g[G_s + s(C_1 + C_2)] - V_2 s C_2 = V_s G_s \qquad 4.36$$

And using the unilateral approximation, we have, for the V_2 node,

$$V_2 = -\frac{V_g g_m}{g'_m + g_\pi} \qquad 4.37$$

Equations 4.36 and 4.37 are solved simultaneously for V_2/V_s, and the result is substituted into Eq. 4.35 to give the cascode amplifier's approximate high-frequency gain:

$$A_v(s) = \frac{V_o}{V_s} = \frac{-g_m R_C}{(1 + sR_C C_\mu)\left\{1 + sR_s\left[C_1 + C_2\left(1 + \dfrac{g_m}{g'_m + g_\pi}\right)\right]\right\}} \qquad 4.38$$

It is instructive to calculate typical numerical values for the high-frequency gain equation 4.38. Let $g'_m = 0.05$ S, $g_m = 4 \times 10^{-3}$ S, $g_\pi = 5 \times 10^{-4}$ S, $g_d = 2 \times 10^{-6}$ S, $R_C = 3300\ \Omega$, $C_1 = C_2 = 3$ pF, $C_\mu = 3$ pF, and $R_s = 1000\ \Omega$. The mid-frequency gain is

$$-g_m R_C = -(4 \times 10^{-3})(3300) = -13.20 \text{ V/V} \qquad 4.39$$

and the collector pole is at

$$f_c = \frac{1}{2\pi R_c C_\mu} = \frac{1}{2\pi(3300)(3 \times 10^{-12})} = 16.08 \text{ MHz} \qquad 4.40$$

The input or Miller pole is at

$$f_i = \frac{1}{2\pi R_s \left[C_1 + C_2 \left(1 + \dfrac{g_m}{g'_m + g_\pi} \right) \right]} = 25.52 \text{ MHz} \qquad 4.41$$

Note that the Miller capacitance,

$$C_m = C_2 \left(1 + \frac{g_m}{g'_m + g_\pi} \right) = 3.24 \text{ pF} \qquad 4.42$$

is very low (not much greater than C_2) because the FET gain, V_2/V_s, is close to unity. The secret of the cascode amplifier's extended high-frequency response is that the FET's input capacitance is only slightly larger than C_{iss}. Also notice that the mid-frequency gain of the cascode amplifier is the same as that of an FET grounded-source amplifier with an $R_d = R_C = 3300 \ \Omega$. Determination of a single FET grounded-source stage's high-frequency response is given in the problems at the end of the chapter. ●

EXAMPLE 4.3

Broadbanding: The Grounded-Base Amplifier

The BJT grounded-base amplifier shown in Fig. 4.12(B) uses a different circuit archi-tecture to gain high-frequency response. Its HFSSM is shown in Fig. 4.14(A). A source split and application of the substitution theorem again lead to a simplified HFSSM, shown in Fig. 4.14(B). By inspection of Fig. 4.14(B), we can write

$$V_o = V_e \frac{g_m RC}{1 + sR_C C_\mu} \qquad 4.43$$

V_e is given by a simple voltage-divider relation:

$$\frac{V_e}{V_s} = \frac{\dfrac{1}{g_\pi + g_m + sC_\pi}}{\dfrac{1}{g_\pi + g_m + sC_\pi} + B_s} \qquad 4.44$$

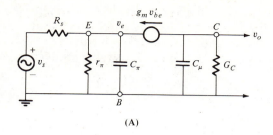

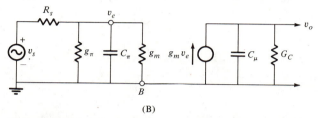

(B)

Figure 4.14 **(A)** High-frequency, hybrid-pi, BJT small-signal model used to describe a grounded-base amplifier. Assume $r_x = 0$ and $g_o = 0$. **(B)** The same HFSSM following a source split and application of the substitution theorem.

Substitution of Eq. 4.44 into Eq. 4.43 and rearrangement of terms gives the grounded-base BJT amplifier's transfer function:

$$A_v(s) = \frac{V_o}{V_s} = \frac{\dfrac{g_m R_C}{1 + R_s(g_m + g_\pi)}}{(1 + sC_\mu R_C)\left[1 + s\dfrac{R_s C_\pi}{1 + R_s(g_m + g_\pi)}\right]} \qquad 4.45$$

The first and dominant pole is the collector (output) pole, similar to that seen in the preceding analysis of the cascode amplifier. The input pole is at a very high frequency compared to the collector pole. Using the same numerical values for the BJT's parameters that we used in the cascode example, we find that

$$f_i = \frac{1}{\dfrac{2\pi C_\pi R_s}{1 + R_s(g_m + g_\pi)}} = \frac{1}{\dfrac{2\pi(100 \times 10^{-12})(10^3)}{1 + 10^3(0.05 + 5 \times 10^{-4})}} = 81.96 \text{ MHz} \qquad 4.46$$

The mid-frequency gain of the grounded-base amplifier is positive (noninverting) and is

$$A_{v(mid)} = \frac{g_m R_C}{1 + R_s(g_m + g_\pi)} = \frac{(0.05)(3300)}{1 + 10^3(0.0505)} = 3.20 \qquad 4.47$$

This is a relatively low voltage gain; it is due to the size of R_s (1 kΩ). Although increasing R_C will increase the gain, it is at the expense of bandwidth. A better solution is to try to reduce R_s in this case, to achieve a larger amplifier mid-band gain. Because $f_c \ll f_i$, it is easy to show that the appproximate gain-bandwidth product for the grounded-base amplifier is

$$\text{GBWP} = \frac{g_m/C_\mu}{2\pi} \text{ Hz} \qquad\qquad 4.48$$

R_s acts to trade off gain for bandwidth, a typical dilemma for the designer of broadband amplifiers.   ●

The bootstrap circuits shown in Figs. 4.12(C) and (D) use controlled sources to reduce the Miller capacitance of the input JFET. The reduction is accomplished by forcing the drain node to follow closely the small-signal voltage on the gate. Thus both ends of C_{gd} are at nearly the same small-signal voltage, and little input current flows through C_{gd}. In the differential amplifier bootstrap circuit of Fig. 4.12(D), the FET's source is also forced to follow v_g, so there is no small-signal current in C_{gs} either. Thus bootstrapping significantly reduces the circuit's total input capacitance and allows extended high-frequency operation with large input resistances R_s.

The bootstrap circuit of Fig. 4.12(C) is difficult to analyze with pencil and paper; its HFSSM has four nodes, even assuming C_c has zero reactance so that $V_d = V_e$. A computer circuit analysis is needed here.

The FET-IC bootstrap amplifier of Fig. 4.12(D) is much easier to analyze, however. Calculation of its high-frequency gain appears as a problem at the end of the chapter.

EXAMPLE 4.4   ●

Broadbanding: Shunt Peaking

The final broadbanding technique that we will consider in this chapter goes back historically to early vacuum tube days; it makes use of a small inductor in series with a resistive load. The technique is called *shunt peaking*. It can be used with grounded-emitter (grounded-source) or grounded-base (grounded-gate) amplifiers. We will illustrate it with a grounded-base amplifier, whose HFSSM is shown in Fig. 4.15. Note that $r_x = 0$ and $V_{b'e} = -V_e$, so we can use the substitution theorem and do a source split on the $g_m V_e$ current source. This leads to

$$V_o = V_e \frac{g_m}{sC_\mu + \dfrac{1}{R + sL}} \qquad\qquad 4.49$$

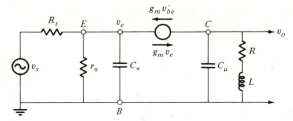

Figure 4.15 HFSSM for a shunt-peaked, grounded-base BJT amplifier.

which can be put in time-constant form:

$$V_o = V_e \frac{g_m R \left(1 + s \dfrac{L}{R} \right)}{s^2 L C_\mu + s C_\mu R + 1} \tag{4.50}$$

The purpose of using the inductance is to form a low-Q resonant circuit with C_μ (actually with C_μ plus the output loading capacitance) so that the -3 dB frequency is extended beyond $\omega_o = 1/RC_\mu$ rad/s, the break frequency with $L = 0$. One means of choosing L to extend the high-frequency bandwidth is to set the magnitude of the impedance seen by the controlled source, $g_m V_e$, to equal R at $\omega_o = 1/RC_\mu$ rad/s. Hence,

$$R^2 = \frac{R^2 + \omega_o^2 L^2}{(1 - \omega_o^2 L C_\mu)^2 + \omega_o^2 R^2 C_\mu^2} \tag{4.51}$$

If $\omega_o = 1/RC_\mu$ is substituted into Eq. 4.51 and L is solved for, we find that the expression

$$L = \frac{R^2 C_\mu}{2} \text{ H} \tag{4.52}$$

will cause the frequency response to be down 0 dB at ω_o, rather than -3 dB. It is left as an exercise to show that the new -3 dB frequency is at $\omega_o' = \sqrt{2}\omega_o$.

The shunt peaking broadbanding technique is most suitable for discrete-component high-frequency circuits; it also can be used with IC amplifiers operating in the HF–VHF range, where it is possible to integrate fractional microhenry inductors (not adjustable) on a chip. In the case of discrete-component circuits, C_μ is generally not precisely known, and capacitive loading of the next stage adds to its value; hence the inductors used in shunt peaking are usually made to be adjustable, with values in the low microhenry range. They are tuned individually to optimize the bandwidth of each stage. ●

4.5 High-Frequency Response of Multistage Amplifiers

In the preceding sections of this chapter, we have seen how single stages of amplification behave at high frequencies. To achieve requisite design goals for broadband high-frequency amplifiers, we generally must cascade two or more gain stages. In analyzing such multistage amplifiers, we assume either that the stages do not interact or that they do interact. Noninteraction requires that each stage not be affected by its neighbors; one way of obtaining this isolation is to make each stage have zero output impedance, that is, have an ideal voltage source at its output. The transfer function of an amplifier with noninteracting stages is simply the product of the transfer functions of its component stages. If the gain stages interact, the node equations describing overall behavior are coupled, and the solution of the overall transfer function requires their simultaneous solution, generally a tedious process best suited to a computer circuit analysis program. Often simplifications can be made, such as the unilateral approximation, or a source split using the substitution theorem, to aid analysis of amplifiers with interacting stages.

In the case of an amplifier made up from N identical noninteracting stages with simple real-pole, low-pass transfer functions (these could be op-amps), the overall transfer function is

$$A_v(s) = \prod_{i=1}^{N} \left(\frac{K_{vi}}{1 + s\tau_i} \right) = \frac{K_v^N}{(1 + s\tau)^N} \qquad\qquad 4.53$$

The half-power frequency of the overall amplifier is defined as ω_h such that

$$\frac{|A_v(j\omega_h)|}{|A_v(0)|} = \frac{\sqrt{2}}{2} \qquad\qquad 4.54$$

This implies that

$$[1 + (\omega_h\tau)^2]^N = 2 \qquad\qquad 4.55$$

Solving Eq. 4.55 for ω_h gives the well-known result

$$\omega_h = \omega_o\sqrt{2^{1/N} - 1} \qquad\qquad 4.56$$

Table 4.2   Values of the bandwidth reduction factor vs. the number of noninteracting identical gain stages

N	1	2	3	4	5
$\sqrt{2^{1/N} - 1}$	1.00	0.64	0.51	0.44	0.39

where $\omega_o = 1/\tau$. The $\sqrt{2^{1/N} - 1}$ factor in Eq. 4.56 is the bandwidth reduction factor; its values are given in Table 4.2.

EXAMPLE 4.5

Use of the Bandwidth Reduction Factor

As an example of the use of the bandwidth reduction factor, we will design a linear amplifier from three cascaded op-amp circuits, having an overall gain as high as possible and a -3 dB frequency of 1 MHz. The op-amps are to be used in the non-inverting configuration. Each op-amp has a unity-gain frequency (GBWP) of 25 MHz. From Eq. 4.56 and Table 4.2, we see that each closed-loop, noninverting op-amp circuit must have its -3 dB frequency at

$$\omega_o = \frac{\omega_h}{\sqrt{2^{1/3} - 1}} = \frac{2\pi \times 10^6}{0.51} = 1.23 \times 10^7 \text{ rad/s} \qquad 4.57$$

or

$$f_o = \frac{\omega_o}{2\pi} = 1.96 \times 10^6 \text{ Hz} \qquad 4.58$$

Hence the permissible mid-band gain per op-amp is

$$K_v = \frac{GBWP}{f_o} = \frac{2\pi \times 10^6}{1.96 \times 10^6} = 12.76 \qquad 4.59$$

and the overall amplifier's mid-band gain is

$$A_{v(\text{mid})} = K_v^3 = 2.076 \times 10^3 \qquad 4.60$$

SUMMARY

We have considered some of the basic factors in electronic circuit design that limit high-frequency response. Capacitive feedback in single inverting gain stages has been shown to increase the effective input capacitance of that stage, forming an RC low-pass filter with the Thevenin source resistance driving that stage. This enhancement of the input capacitance is called the Miller effect. Output capacitance to ground seen by amplifier stages also forms a low-pass filter. The output capacitance can include the Miller input capacitance of the next stage, as well as wiring capacitance.

Broadbanding strategies were introduced that seek to overcome the Miller effect and the effect of capacitive loading. Some of these strategies use circuit architecture to reduce the Miller capacitance; others use local negative feedback to trade off mid-band gain for high-frequency bandwidth.

The basic properties of cascaded noninteracting stages were examined. When the stages of a multistage amplifier interact, analysis is tedious and is best done on a computer.

PROBLEMS

4.1   Amplifier High-Frequency Response

A certain JFET is biased such that it has the following small-signal parameters: $g_{mo} = 6 \times 10^{-3}$ S, $g_d = 2 \times 10^{-5}$ S, $I_{DSS} = 10$ mA, $I_{DQ} = 5$ mA, $C_{gs} = C_{gd} = 2.5$ pF, and $C_{ds} = 0$.

(a) Use SSM analysis to find the amplifier's mid-frequency gain, A_o.

(b) Find algebraic and numerical expressions for the amplifier's mid- and high-frequency gain, $(V_o/V_1)(j\omega) = A_v(j\omega)$, in time-constant form. Make a detailed Bode magnitude plot for $A_v(j\omega)$; show asymptotes, asymptote slopes, dB corrections at break frequencies, the mid-frequency gain, the -3 dB frequency, and the f_T value. Use exact analysis.

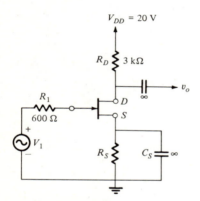

Figure P4.1(a) and (b)

(c) Repeat parts (a) and (b) for the same JFET used in the grounded-gate amplifier shown in Fig. P4.1(c).

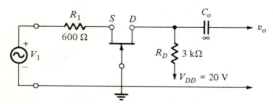

Figure P4.1(c)

(d) Repeat parts (a) and (b) for the source-follower amplifier shown in Fig. P4.1(d).

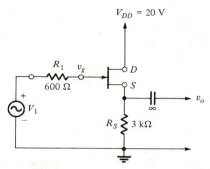

$V_{DD} = 20$ V

R_1 v_g
600 Ω

V_1

R_S 3 kΩ

v_o

Figure P4.1(d)

4.2   Amplifier High-Frequency Response

A grounded-emitter BJT amplifier has the simplified circuit shown. The BJT's hybrid-pi small-signal model has the following parameters: $g_o = r_x = 0$, $c_\pi = 30$ pF, $C_\mu = 3$ pF, $r_\pi = 2.5$ kΩ, $g_m = 0.05$ S, $R_1 = 300$ Ω, and $R_C = 4.7$ kΩ.

(a) Draw the complete HFSSM for the amplifier.
(b) Find a numerical expression for $A_v(j\omega) = (V_o/V_1)(j\omega)$ in time-constant form. Do not use any approximations.
(c) Give numerical values for the break frequencies, the mid-frequency gain, and the f_T for the amplifier.

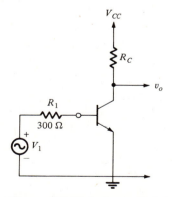

V_{CC}

R_C

v_o

R_1
300 Ω

V_1

Figure P4.2(a) through (c)

(d) Repeat parts (a) through (c) for the (simplified) grounded-base amplifier shown in Fig. P4.2(d)

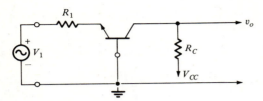

Figure P4.2(d)

(e) Repeat parts (a) through (c) for the simplified BJT-BJT cascode amplifier shown in Fig. P4.2(e). Q_1 and Q_2 are identical, and parameter values are as before.

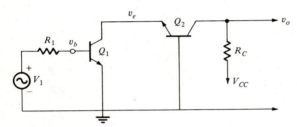

Figure P4.2(e)

4.3 Amplifier High-Frequency Response

A simplified circuit for an RC grounded-base/source-follower amplifier is shown in Fig. P4.3. (*Note:* This circuit has three nodes and can be most easily analyzed using an ECAP.) Assume the following at the operating points of the transistors:

$$\text{JFET:}\quad g_m = 3.5 \times 10^{-3}\ \text{S}$$
$$r_d = 40\ \text{k}\Omega$$
$$C_{gs} = C_{gd} = 3\ \text{pF}$$
$$C_{ds} = 0$$

$$\text{BJT:}\quad r_x = 0$$
$$g'_m = 4 \times 10^{-2}\ \text{S}$$
$$C_\mu = 2\ \text{pF}$$
$$C_\pi = 100\ \text{pF}$$
$$g_o = 10^{-5}\ \text{S}$$
$$r_\pi = 2.5\ \text{k}\Omega$$

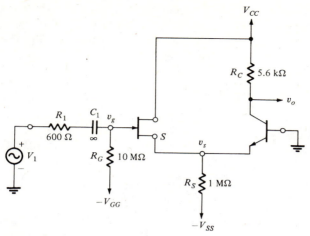

Figure P4.3

(a) Use small-signal analysis to find the amplifier's mid-frequency gain, A_o, and its -3 dB frequency.

(b) Find the amplifier's f_T.

4.4 A unity-gain, FET bootstrap amplifier uses two ideal differential amplifiers (not op-amps). The FET is characterized by nonzero parameters: g_d, g_m, C_{ds}, C_{gd}, C_{gs}, and an output capacitance, C_o, from the source node to ground.

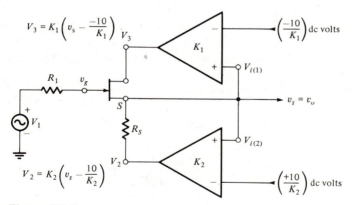

Figure P4.4

(a) Draw the HFSMM for the system.

(b) Find (V_o/V_g) in time-constant form at mid- and high frequencies. Note that K_1 and K_2 must have specific *real* numerical values to cancel currents in C_{gs}, C_{gd}, and C_{ds}.

(c) Find an expression for (V_o/V_1) in time-constant form. Find the system's poles and zeros for $R_1 = 1000\ \Omega$, $R_S = 10^4\ \Omega$, $C_{gs} = C_{gd} = 3\ \text{pF}$, $C_{ds} = 0.3\ \text{pF}$, and $C_o = 10\ \text{pF}$.

4.5 Let the differential amplifiers in Problem 4.4 be characterized by the open-circuit voltage gains

$$V_{oc(1)} = (V_{i(1)} - V'_{i(1)})(K_1/3 \times 10^{-9}s + 1)$$

$$V_{oc(2)} = (V_{i(2)} - V'_{i(2)})(K_2/3 \times 10^{-9}s + 1)$$

and a Thevenin equivalent output resistances of $100\ \Omega$. Use an ECAP to find the $-3\ \text{dB}$ frequency and f_T for the bootstrap amplifier of Problem 4.4.

4.6 For the JFET amplifier shown, assume $g_m = 3.5 \times 10^{-3}\ \text{S}$, $r_d = 5 \times 10^4\ \Omega$, $R_1 = 1000\ \Omega$, $R_d = 6.8\ \text{k}\Omega$, $C_{gs} = C_{gd} = 3\ \text{pF}$, and $C_{ds} \cong 0$.

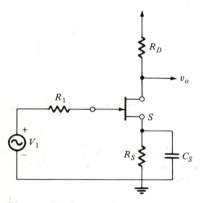

Figure P4.6

(a) Use exact analysis on the HFSSM to find the amplifier's mid-frequency gain, break frequencies (poles and zero), and f_T.

(b) Use the unilateral approximation to find the amplifier's mid-frequency gain, break frequency, and f_T.

4.7 A high-frequency bootstrap amplifier uses a large capacitor to couple the small-signal emitter voltage to the FET's drain (i.e., $v_e = v_d$) at mid- and high frequencies. Assume the following:

FET:		BJT:	
$g_m = 4 \times 10^{-3}\ \text{S}$		$r_x = 0 = g_o$	
$r_d = 3 \times 10^4\ \Omega$		$g_m = 4 \times 10^{-2}\ \text{S}$	
$C_{gs} = C_{gd} = 3\ \text{pF}$		$C_\pi = 40\ \text{pF}$	
$C_{ds} = 0$		$C_\mu = 3\ \text{pF}$	
		$r_\pi = 2.5 \times 10^3\ \Omega$	

$$R_1 = 10^3\ \Omega, \quad R_D = 2\ \text{k}\Omega, \quad R_E = 2\ \text{k}\Omega, \quad R_S = 2\ \text{k}\Omega, \quad R_C = 10\ \text{k}\Omega$$

Use an ECAP to plot the amplifier's frequency response. Give the mid-frequency gain, the -3 dB frequency, and the f_T for the amplifier. (Note that the HFSSM has four nodes.)

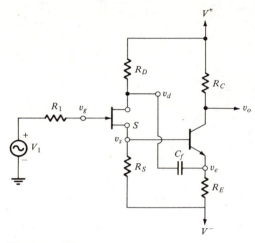

Figure P4.7

4.8   Amplifier High-Frequency Analysis

An op-amp is used with a JFET to reduce the Miller effect and to make a low-noise, video-frequency amplifier. Assume that at the JFET's operating point, $g_m = 3500\ \mu S$, $g_d = 0$, $C_{gs} = C_{gd} = 2.5$ pF, $C_{ds} = 0$, $R_F = 4.7$ kΩ, and the op-amp is ideal.

(a) Derive an algebraic expression for the mid- and high-frequency gain of the system in time-constant form.

(b) Calculate the mid-frequency gain, -3 dB frequency, and f_T for the system.

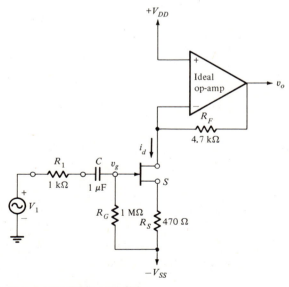

Figure P4.8(a) and (b)

(c) Repeat parts (a) and (b) for the FET circuit shown in Fig. P4.8(c). Assume the JFET has the same parameters.

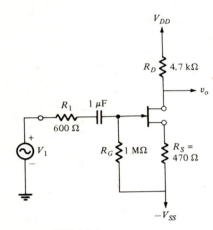

Figure P4.8(c)

4.9 A small inductor is used to extend the bandwidth of a grounded-source JFET amplifier; the design architecture is called shunt peaking. For the 2N4416 FET, $g_m = 4000\ \mu S$, $g_d = 10^{-4}\ S$, $C_{gd} = C_{gs} = 2\ pF$, and $C_{ds} = 0.5\ pF$.

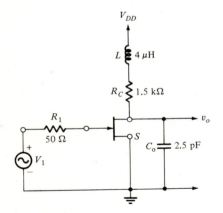

Figure P4.9

(a) Make a Bode magnitude plot for the amplifier extending from mid-band to well above f_T. Give the mid-band gain, peak gain, -3 dB frequency, and f_T for the amplifier.

(b) Repeat part (a) for $L = 0$.

4.10 A grounded-base/emitter-follower amplifier is shown in Fig. P4.10. Q_1 and Q_2 have identical hybrid-pi model parameters. Assume $R_1 = 600 \, \Omega$, $R_C = 2.2 \, \text{k}\Omega$, $R_E = 15 \, \text{k}\Omega$, $r_x = 0$, $g_\pi = 4 \times 10^{-4} \, \text{S}$, $g_o = 0$, $C_\mu = 3 \, \text{pF}$, $C_\pi = 75 \, \text{pF}$, and $g_m = 0.05 \, \text{S}$.

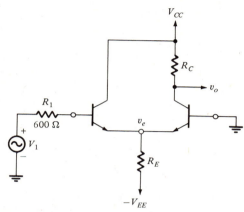

Figure P4.10

(a) Make a dimensioned Bode magnitude plot for this amplifier. Clearly show mid-band gain, $-3 \, \text{dB}$ frequency, and f_T.

(b) Using an ECAP, prepare a table of R_1 versus mid-band gain, $-3 \, \text{dB}$ frequency, and f_T for $R_1 = 0$, 52, 100, 300, 600, and 1200 Ω.

4.11 Plot graphs of the $-3 \, \text{dB}$ frequency of f_T versus R_1 for $0 < R_1 < 1200 \, \Omega$ for the JFET grounded-source amplifier in Fig. P4.11. Assume $g_m = 4 \times 10^{-3} \, \text{S}$, $g_d = 2 \times 10^{-5} \, \text{S}$, $C_{gs} = C_{gd} = 3 \, \text{pF}$, $C_{ds} = 0.5 \, \text{pF}$, $C_o = 6 \, \text{pF}$, and $R_D = 6.8 \, \text{k}\Omega$.

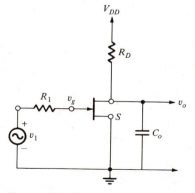

Figure P4.11

4.12 Plot graphs of the -3 dB frequency and f_T versus R_1 for $0 < R_1 < 1200\ \Omega$ for the BJT grounded-emitter amplifier. Assume $g_m = 0.04$ S, $g_o = 0$, $r_x = 0$, $r_\pi = 2.5$ kΩ, $R_C = 10$ kΩ, $R_B = 470$ kΩ, $C_\pi = 63$ pF, $C_\mu = 3.5$ pF, $C_o = 5$ pF, and $C_1 = \infty$.

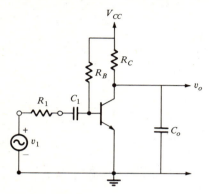

Figure P4.12

4.13 Amplifier High-Frequency Response

The circuit shown in Fig. P4.13. is a high-frequency, **BJT-BJT bootstrap amplifier**. Q_1 and Q_2 are identical and have nominal h_{fe}'s $= 49$. Resistor values were chosen to give $I_{B(1)Q} = 5\ \mu$A, $V_{C(1)Q} = 7$ V, $V_{E(2)Q} = 3$ V, $V_{C(2)Q} = 6$ V, and $V_{B(2)Q} = 3.65$ V. Use small-signal analysis in solving this problem. Let the hybrid-pi small-signal model for Q_1 have $r_x = 0$, $r_{\pi(1)} = 5.2$ kΩ, $g_{m(1)} = 0.00942$ S, $C_\pi = 35$ pF, $C_\mu = 3$ pF, and $g_{o(1)} = 0$. The hybrid-pi small-signal model for Q_2 has $r_x = 0$, $r_{\pi(2)} = 0.104$ kΩ, $g_{m(2)} = 0.47$ S, $C_\pi = 100$ pF, $C_\mu = 3$ pF, and $g_{o(1)} = 0$.

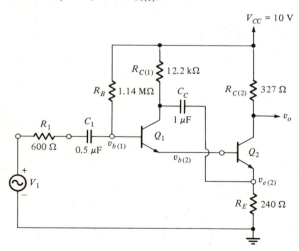

Figure P4.13

(a) Draw the complete HFSSM for the bootstrap amplifier.

(b) Use an ECAP to plot the bootstrap amplifier's frequency response; clearly show the amplifier's mid-frequency gain, -3 dB frequency, and f_T.

4.14 A p-channel JFET and on npn BJT are used to make a "follower" buffer amplifier shown in Figure P4.14. Assume the following:

FET: $g_m = 3500 \ \mu S$
$$g_d = 2 \times 10^{-5} \ S$$
$$C_{gd} = C_{gs} = 3 \ pF$$
$$C_{ds} = 0$$

BJT hybrid-pi parameters: $g'_m = 0.04 \ S$
$$r_x = 0$$
$$r_\pi = 2.5 \ k\Omega$$
$$C_\pi = 40 \ pF$$
$$C_\mu = 3 \ pF$$
$$R_L = 3.9 \ k\Omega$$
$$R_1 = 100 \ \Omega$$

Make Bode magnitude and phase plots for the amplifier covering mid- and high frequencies (above f_T). Give mid-band gain, -3 dB frequency and phase, and f_T and phase at f_T.

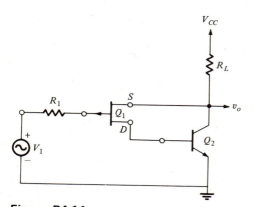

Figure P4.14

4.15 Repeat Problem 4.14 for the circuit shown here. The same transistors and resistor values are used.

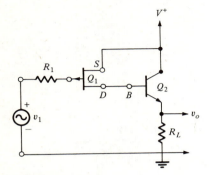

Figure P4.15

Chapter   *5*

Differential Amplifiers

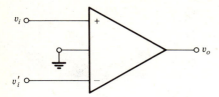

Figure 5.1   Ideal differential amplifier.

The differential amplifier (DA), also known as the difference amplifier, is one of the most widely used analog electronic circuit designs, in both integrated and discrete circuits. Differential amplifiers have many forms; they are used as the input stages in nearly all modern operational and instrumentation amplifiers, voltage comparators, analog multipliers, and other analog ICs. They are also used in oscilloscope vertical and horizontal amplifiers.

An ideal differential amplifier is shown in Fig. 5.1. Its output is given by

$$V_o = A_d(V_i - V_i')$$ 5.1

where A_d is the differential voltage gain of the ideal differential amplifier. In practice, the actual DA output is more closely described by the following equation for sinusoidal inputs:

$$V_o = \mathbf{A}V_i - \mathbf{A}'V_i'$$ 5.2

where $\mathbf{A}$ and $\mathbf{A}'$ are complex gains and $\mathbf{A} - \mathbf{A}' = \mathbf{e}$, a small vector.

A comprehensive analysis of DA behavior follows. It is based on Middlebrook's (1963) detailed study on this topic and uses his notation.

5.1 Analysis of Differential Amplifiers

The most general form of differential amplifier has a differential (push-pull) output as well as a differential input. It is shown in Fig. 5.2. In circuits with bilateral symmetry, such as differential amplifiers, it is convenient to define two forms of input

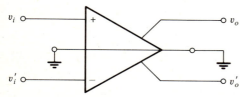

Figure 5.2   Differential amplifier with differential outputs.

and output signals: the common-mode (average) and difference-mode forms. They are defined in Eqs. 5.3 through 5.6:

$$V_{ic} = \frac{V_i + V'_1}{2} \qquad \text{(common-mode input)} \tag{5.3}$$

$$V_{oc} = \frac{V_o + V'_o}{2} \qquad \text{(common-mode output)} \tag{5.4}$$

$$V_{id} = \frac{V_i - V'_i}{2} \qquad \text{(difference-mode input)} \tag{5.5}$$

$$V_{od} = \frac{V_o - V'_o}{2} \qquad \text{(difference-mode output)} \tag{5.6}$$

The input/output relationships for the amplifier of Fig. 5.2 are given by the basic Middlebrook equations, presented here in frequency response format:

$$V_{od} = \mathbf{A}_{dd}V_{id} + \mathbf{A}_{dc}V_{ic} \tag{5.7}$$
$$V_{oc} = \mathbf{A}_{cd}V_{id} + \mathbf{A}_{cc}V_{ic} \tag{5.8}$$

It is easy to show that the single-ended outputs are given by

$$V_o = (\mathbf{A}_{dd} + \mathbf{A}_{cd})V_{id} + (\mathbf{A}_{dc} + \mathbf{A}_{cc})V_{ic} \tag{5.9}$$
$$V'_o = (\mathbf{A}_{cd} - \mathbf{A}_{dd})V_{id} + (\mathbf{A}_{cc} + \mathbf{A}_{dc})V_{ic} \tag{5.10}$$

Also, the single-ended noninverting output can be written in terms of the actual inputs to the amplifier:

$$V_o = \frac{(\mathbf{A}_{dd} + \mathbf{A}_{cd} + \mathbf{A}_{dc} + \mathbf{A}_{cc})V_i}{2} + \frac{(\mathbf{A}_{cc} + \mathbf{A}_{dc} - \mathbf{A}_{dd} - \mathbf{A}_{cd})V'_i}{2} \tag{5.11}$$

Notice that Eq. 5.11 is of the form of the basic gain equation, Eq. 5.2, and it is evident that

$$\mathbf{A} = \frac{\mathbf{A}_{dd} + \mathbf{A}_{cd} + \mathbf{A}_{dc} + \mathbf{A}_{cc}}{2} \tag{5.12}$$

$$\mathbf{A}' = \frac{\mathbf{A}_{dd} + \mathbf{A}_{cd} - \mathbf{A}_{cc} - \mathbf{A}_{dc}}{2} \tag{5.13}$$

In a DA circuit that has perfectly matched symmetrical components, it is obvious that the cross-gains, $\mathbf{A}_{dc}$ and $\mathbf{A}_{cd}$, equal zero. In practice, $|\mathbf{A}_{dc}|$ and $|\mathbf{A}_{cd}|$ are very small at dc, low, and mid-frequencies, and tend to increase at high frequencies. Also, $|\mathbf{A}_{dd}| \gg |\mathbf{A}_{cc}|$ over a wide range of frequencies.

5.2 Common-Mode Rejection Ratio

The common-mode rejection ratio (CMRR) is a figure of merit for differential amplifiers; it is usually given in decibels. It describes how well a differential amplifier's behavior approximates that of an ideal differential amplifier, having $A_{cd} = A_{dc} = A_{cc} = 0$. We define the CMRR as

$$\text{CMRR} = \frac{V_{ic} \text{ to give a certain output}}{V_{id} \text{ to give the same output}} \qquad 5.14$$

Obviously, the CMRR should be as large as possible.

For oscilloscope vertical amplifiers and other differential amplifiers with push-pull outputs,

$$V_{od} = A_{dd}V_{id} + A_{dc}V_{ic} \qquad 5.7$$

Using Eqs. 5.7 and 5.14, we can write an expression for the magnitude of the differential output CMRR, CMRR_d:

$$\text{CMRR}_d = \left| \frac{A_{dd}}{A_{dc}} \right| \qquad 5.15$$

When the differential amplifier has a single-ended output, which is the most common case, we have

$$V_o = (A_{dd} + A_{cd})V_{id} + (A_{cc} + A_{dc})V_{ic} \qquad 5.9$$

Using the definition 5.14 with Eq. 5.9, we find

$$\text{CMRR}_s = \left| \frac{A_{dd} + A_{cd}}{A_{cc} + A_{dc}} \right| \qquad 5.16$$

If the amplifier is symmetrical, this reduces to

$$\text{CMRR}_s = \left| \frac{A_{dd}}{A_{cc}} \right| \qquad 5.17$$

Note that all CMRR expressions are functions of frequency. In practice, CMRR tends to decrease with increasing frequency; more will be said on this phenomenon later. In modern differential amplifiers, it is not uncommon to find CMMRs of 120 dB (10^6) at 60 Hz.

5.3 Analysis of Simple DA Stages

Analysis of differential amplifiers using small-signal models for transistors is easily carried out using the bisection theorem, described in Chapter 2, Sec. 2.3. This analysis can give parameters such as A_{dd} and A_{cc} (hence CMRR_s), and input and output impedances.

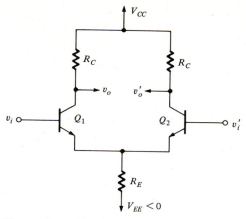

Figure 5.3   Simple BJT differential amplifier.

From the small-signal analysis of DA circuits, we can see what circuit parameters contribute to inherently large CMRRs and input impedances, desirable features in differential amplifiers. A simple BJT differential amplifier is shown in Fig. 5.3. We assume bilateral symmetry in the circuit. The MFSSM for this differential amplifier is illustrated in Fig. 5.4. To use the bisection theorem on this circuit, we first assume a purely difference-mode input ($V_{ic} = 0$). This assumption permits us to ground all nodes on the axis of symmetry, because the nodes have zero small-signal potential with respect to ground. Thus the small-signal model with difference-mode excitation

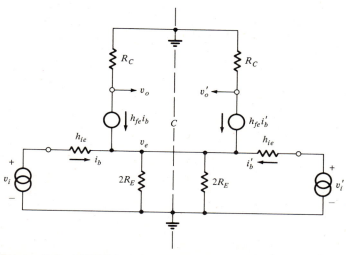

Figure 5.4   MFSSM of the differential amplifier of Fig. 5.3 showing the axis of symmetry.

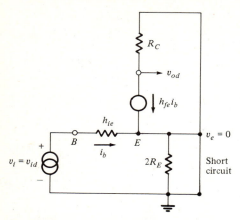

Figure 5.5   Reduction of the small-signal model of Fig. 5.4 for difference-mode input signals.

of Fig. 5.4 can be reduced to that of Fig. 5.5. By inspection of Fig. 5.5, we can write

$$\frac{V_{od}}{V_{id}} = A_{dd} = \frac{-h_{fe}R_C}{h_{ie}} \qquad\qquad 5.18$$

Next we consider the BJT differential amplifier with only a common-mode input ($V_{id} = 0$). In this case, we can cut all the links in the circuit of Fig. 5.4 that cross the axis of symmetry. This is possible because there is zero net current in these links under common-mode excitation. The reduced circuit obtained is shown in Fig.

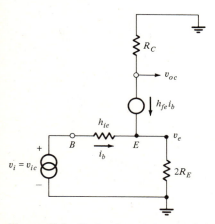

Figure 5.6   Reduction of the small-signal model of Fig. 5.4 for common-mode signals.

5.6. Here we note that $V_{oc} = -i_b h_{fe} R_C$ and $i_b = V_{ic}/[h_{ie} + 2R_E(1 + h_{fe})]$, so

$$\frac{V_{oc}}{V_{ic}} = A_{cc} = \frac{-h_{fe} R_C}{h_{ie} + 2R_E(1 + h_{fe})} \qquad 5.19$$

From Eqs. 5.17 through 5.19, we can write the magnitude of the CMRR_s as

$$\text{CMRR}_s = 1 + \frac{2R_E(1 + h_{fe})}{h_{ie}} = 1 + \frac{2R_E(1 + h_{fe})|I_{BQ}|}{V_T} \qquad 5.20$$

The last term in Eq. 5.20 follows from the fact that the BJT's h_{ie} is inversely proportional to the quiescent (average) base bias current, I_{BQ}. In order that CMRR_s be a maximum, it is clear that the designer should choose matched transistors with high h_{fe}'s, make R_E as large as possible, and set the Q-point for as high an I_{BQ} as reasonable.

A practical limitation to the size of R_E is placed by Ohm's law, that is, by the total dc voltage drop across R_E divided by the total quiescent emitter current that must flow through R_E. It turns out that the best way to obtain a large effective value of R_E is by using one or more transistors as an active, high-impedance current source (or sink) in place of a passive R_E.

As an example, let us assume that dc biasing conditions require that each transistor have a quiescent emitter current of 0.5 mA, and that their emitters be at a dc quiescent voltage of -0.7 V. Also, let the negative supply voltage, V_{EE}, be -15 V. Hence, by Ohm's law $R_E = 14.3/0.001 = 14.3$ kΩ. This is a low resistance compared to what can be achieved with a suitable active device current source.

EXAMPLE 5.1

Use of a BJT as an Active Current Sink

As a first example of an active current source, consider the BJT circuit of Fig. 5.7. The collector of the BJT is attached to the emitters of the DA BJTs in Fig. 5.3, instead of R_E. The emitter resistor of Q_3 is returned to the negative dc supply, V_{EE}.

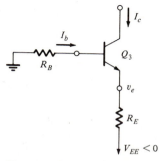

Figure 5.7 Simple BJT current sink for the emitter currents of Q_1 and Q_2 of the differential amplifier in Fig. 5.3.

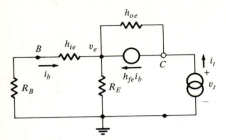

Figure 5.8   MFSSM for the current sink shown in Fig. 5.7. A small-signal test source, v_t, is used to find the resistance looking into the collector.

To find the small-signal equivalent resistance looking into Q_3's collector, R_{eq}, we consider the MFSSM of the circuit of Fig. 5.7 and use a small-signal test source, v_t, as shown in Fig. 5.8. R_{eq} will be simply the ratio of v_t to i_t; hence we will find it in this circuit.

Note that $i_b = -v_e G_1$, where $G_1 = 1/(h_{ie} + R_B)$. The node equation on v_e yields

$$v_e[G_E + h_{oe} + G_1(1 + h_{fe})] = v_t h_{oe} \qquad\qquad 5.21$$

but

$$i_t = v_e(G_E + G_1) \qquad\qquad 5.22$$

so

$$i_t = \frac{v_t h_{oe}(G_E + G_1)}{G_E + h_{oe} + G_1(1 + h_{fe})} \qquad\qquad 5.23$$

Hence R_{eq} can be shown to equal

$$R_{eq} = \frac{v_t}{i_t} = \frac{1 + \dfrac{h_{fe}R_E}{R_E + h_{ie} + R_B}}{h_{oe}} \qquad\qquad 5.24$$

Inspection of Eq. 5.24 shows that if $R_E = 0$, then $R_{eq} = 1/h_{oe}$, which is usually around several hundred kilohms for a small-signal BJT. If we insert typical values for the parameters in Fig. 5.8—$h_{fe} = 100$, $R_E = 10\ \text{k}\Omega$, $h_{ie} = 3\ \text{k}\Omega$, and $R_B = 230\ \text{k}\Omega$—we find that R_{eq} is increased by a factor of 4.12 over $1/h_{oe}$. Thus it is possible to obtain an R_{eq} value of over 1 MΩ with this simple active current source, versus 14.3 kΩ for a passive resistor.   ●

In many integrated and discrete circuit DA designs, effective high-impedance temperature-stabilized current sources (or sinks) are made from two or three active devices. Examples of these DA *long tails* are shown in Fig. 5.9. In the analysis of the current source circuits shown in this figure, we are interested in finding expressions for the dc current I_{EC} sunk by the collector (or drain) of Q_3, as well as for the

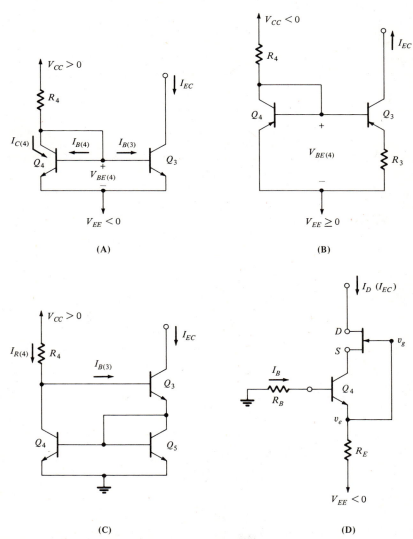

Figure 5.9   Current sinks and sources used for the *long tail* in the design of IC and discrete differential amplifiers. **(A)** Basic two-BJT current sink. **(B)** Widlar current source. **(C)** Wilson current sink. **(D)** JFET-BJT cascode long tail.

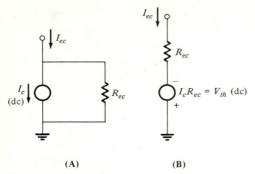

(A)   (B)

Figure 5.10   Norton (**A**) and Thevenin (**B**) models for current sources (or sinks) used with differential amplifiers. Note that in SSM analysis, the dc current source or voltage source drops out, leaving only R_{ec}.

small-signal resistance R_{eq} seen looking into this collector. With I_{EC} and R_{eq}, it is possible to simplify each active current source by representing it as either a Norton or a Thevenin equivalent circuit, as shown in Fig. 5.10.

To find I_{EC} for the circuit of Fig. 5.9(A), we assume $Q_3 = Q_4$ and a large h_{fe}. Because of transistor matching and the fact that $V_{BE(3)} = V_{BE(4)}$, $I_{C(3)} = I_{C(4)} = I_{EC}$, we have, from Ohm's law,

$$I_{C(4)} = I_{R(4)} = \frac{V_{CC} - V_{BE(4)}}{R_4} = I_{EC} \qquad 5.25$$

I_{EC} may also be found for the Widlar current source of Fig. 5.9(B). Applying Kirchhoff's voltage law around the base loop, we have, assuming high h_{fe}'s,

$$V_{BE(3)} + I_{C(3)}R_3 = V_{BE(4)} \qquad 5.26$$

Also, from the Ebers-Moll equations (Eqs. 1.16A and B), we have, for a forward-biased base-emitter junction,

$$I_C \cong \alpha_F I_{ES} e^{V_{BE}/V_T} \qquad 5.27$$

Hence Eq. 5.26 may be written using Eq. 5.27 as

$$V_T \ln\left(\frac{I_{C(3)}}{\alpha_F I_{ES(3)}}\right) + I_{C(3)}R_3 = V_T \ln\left(\frac{I_{C(4)}}{\alpha_F I_{ES(4)}}\right) \qquad 5.28$$

Because the transistors are matched, $\alpha_F I_{ES(3)} = \alpha_F I_{ES(4)}$, and we have

$$\ln\left(\frac{I_{C(4)}}{I_{C(3)}}\right) = \frac{I_{C(3)}R_3}{V_T} \qquad 5.29$$

Equation 5.29 is transcendental and may be solved numerically by Newton-Raphson or other trial-and-error methods to give the $I_{C(3)} = I_{EC}$ desired. Note that I_{EC} will be less than $I_{C(4)}$.

Finding I_{EC} for the Wilson current source is more complex. Gray and Meyer (1984) show that I_{EC} for the Wilson current source is given by

$$I_{EC} = \frac{I_{R(4)}(h_{fe}^2 + 2h_{fe})}{h_{fe}^2 + 2h_{fe} + 2} \cong I_{R(4)} \cong \frac{V_{CC} - 2V_{BE(5)}}{R_4} \qquad 5.30$$

Note that $V_{BE(5)}$ can be approximated by $V_T \ln(I_{EC}/I_S)$. Hence I_{EC} for the Wilson source can be found by solving a transcendental equation using Eq. 5.30.

The circuit of Fig. 5.9(D) also defies an easy analysis for the dc I_{EC}. We note that the V_{GS} of the FET equals $-V_{CE}$ of the BJT, and $I_{EC} = I_D = I_C$. It appears that device characteristics are critical in this circuit; I_B must be used to adjust the operating points of Q_3 and Q_4 so that both devices are in their constant-current regions. Although the dc current I_{EC} in this circuit can be found using a graphical analysis on the devices' characteristic curves, it is possible to find the small-signal Thevenin resistance, R_{eq}, looking into the drain of Q_3 with a test source. The MFSSM for this circuit is shown in Fig. 5.11. Node equations are written on v_e and v_s. Note $ib = -v_e G_1$ and $v_{gs} = v_e - v_s$.

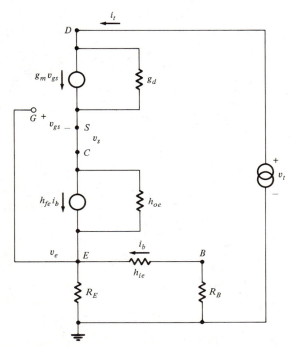

Figure 5.11   MFSSM for the JFET-BJT cascode current sink of Fig. 5.9(D).

$$v_e[G_E + h_{oe} + G_1(h_{fe} + 1)] - v_s h_{oe} = 0 \qquad\qquad 5.31$$

$$-v_e(h_{fe}G_1 + h_{oe} + g_m) \qquad\qquad + v_s(g_m + g_d + h_{oe}) = g_d v_t \qquad 5.32$$

Also, we note that

$$i_t = v_e(G_1 + G_E) \qquad\qquad 5.33$$

Solving Eqs. 5.31 and 5.32 for v_e and substituting into Eq. 5.33, we can write, after some algebraic manipulation,

$$R_{eq} = \frac{v_t}{i_t} = \frac{1 + \mu + h_{oe}r_d}{h_{oe}} + \left(\frac{R_1 R_E}{R_1 + R_E}\right)\left[1 + \frac{h_{fe}(1 + \mu)}{R_1 h_{oe}}\right] \qquad 5.34$$

If typical values are assumed for the parameters in Eq. 5.34—$\mu = 99$, $h_{fe} = 100$, $R_E = 10 \text{ k}\Omega$, $R_1 = 1 \text{ M}\Omega$, $h_{oe} = 5 \times 10^{-6}$ S, and $r_d = 5 \times 10^4 \ \Omega$—then R_{eq} is found to be about 40 MΩ, a reasonable value for a dynamic current source. If this value of R_{eq} is substituted into Eq. 5.20 with $h_{fe} = 100$ and $h_{ie} = 4 \times 10^3 \ \Omega$, the CMRR is seen to be 2×10^6, or 126 dB, an excellent value.

5.4 Use of Common-Mode Negative Feedback to Increase Differential Amplifier CMRR

In common-mode negative feedback (CMNF), the common-mode output of a differential amplifier is amplified by a factor, B_c, then subtracted from the common-mode input signal. A signal flow graph based on the Middlebrook equations 5.7 and 5.8 and the difference-mode and common-mode defining relations 5.3 through 5.6 is shown in Fig. 5.12. Reduction of this signal flow graph yields the following

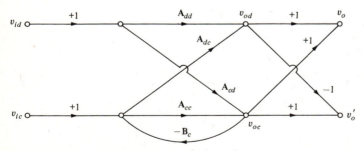

Figure 5.12   Signal flow graph representation of a differential amplifier with common-mode negative feedback.

relations for the Middlebrook gain terms with CMNF:

$$A'_{dd} = \frac{v_{od}}{v_{id}} = A_{dd} + \frac{A_{cd}A_{dc}(-\beta_c)}{1 + A_{cc}\beta_c} \cong A_{dd} \qquad 5.35$$

$$A'_{cd} = \frac{v_{oc}}{v_{id}} = \frac{A_{cd}}{1 + A_{cc}\beta_c} \qquad 5.36$$

$$A'_{dc} = \frac{v_{od}}{v_{ic}} = \frac{A_{dc}}{1 + A_{cc}\beta_c} \qquad 5.37$$

$$A'_{cc} = \frac{v_{oc}}{v_{ic}} = \frac{A_{cc}}{1 + A_{cc}\beta_c} \qquad 5.38$$

From Eqs. 5.15 and 5.16, we observe that a differential amplifier with CMNF has a CMRR that is increased by a factor of $(1 + A_{cc}\beta_c)$, as shown in the following two equations:

$$\text{CMRR}_d = \frac{A'_{dd}}{A'_{dc}} = \frac{(1 + A_{cc}\beta_c)A_{dd}}{A_{dc}} \qquad 5.39$$

$$\text{CMRR}_s = \frac{A'_{dd} + A'_{cd}}{A'_{cc} + A'_{dc}}$$

$$= \frac{A_{dd}(1 + A_{cc}\beta_c) + A_{cd}}{A_{cc} + A_{dc}} \qquad 5.40$$

Implementation of CMNF requires access to v_{oc}. If the differential amplifier has a push-pull output (v_0 and v'_o), v_{oc} can be derived easily by a simple summer using two matched resistors. The v_{oc} obtained is amplified and returned to the input as a common-mode feedback voltage.

EXAMPLE 5.2

Implementation of Common-Mode Negative Feedback

Figure 5.13 illustrates one means of implementing CMNF. An external FET head-stage is used to drive an IC differential amplifier. The differential amplifier is described by the basic Middlebrook equations, where A_{dc} and A_{cd} are assumed to be zero because of symmetry. Thus we have

$$v_o = A_{dd}v_{id} + A_{cc}v_{ic} \qquad 5.41$$

and

$$v'_o = -A_{dd}v_{id} + A_{cc}v_{ic} \qquad 5.42$$

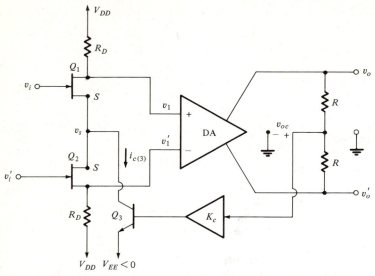

Figure 5.13 Differential amplifier with an external low-noise JFET headstage with common-mode negative feedback.

v_{oc} is given by Eq. 5.4 and is simply

$$v_{oc} = A_{cc}v_{ic} \tag{5.43}$$

It is easy to see that the small-signal current into Q_3's collector is given by

$$i_{c(3)} = \frac{K_c h_{fe} v_{oc}}{h_{ie}} \tag{5.44}$$

assuming $h_{oe(3)} = 0$. Figure 5.14 shows the MFSSM for the differential headstage. For easier analysis, this small-signal model can be reconfigured as shown in Fig. 5.15 using a source split that does not alter the system's node equations. Note that v_{gs} in Fig. 5.14 is simply $-v_s$. Writing node equations for v_s and v_1, we have

$$v_s(2g_d + 2g_m) - v_1(2g_d) = \frac{-h_{fe}K_c v_{oc}}{h_{ie}} \tag{5.45}$$

$$-v_s(g_m + g_d) + v_1(G_D + g_d) = 0 \tag{5.46}$$

Solving Eqs. 5.45 and 5.46, we find

$$v_1 = \left(\frac{-0.5h_{fe}R_D K_c}{h_{ie}}\right)v_{oc} \tag{5.47}$$

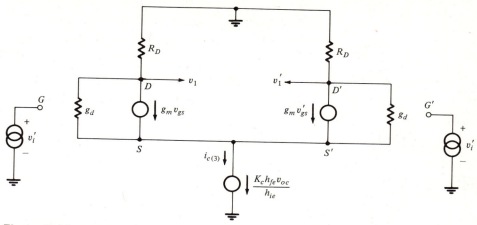

Figure 5.14   MFSSM for the differential headstage of Fig. 5.13.

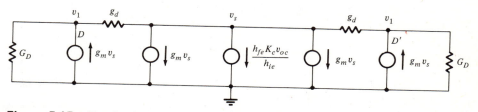

Figure 5.15   Circuit of Fig. 5.14 after a $g_m v_{gs}$ source split. v_i and $v_i' = 0$.

Since we are interested in the common-mode input component from the feedback, use of Eq. 5.3 gives

$$v_{1(cf)} = \left(\frac{-0.5 h_{fe} R_D K_c}{h_{ie}}\right) v_{oc} \tag{5.48}$$

$v_{1(cf)}$ is added to $v_{1(cs)}$ derived from the input common-mode component, v_{ic}. Again, from Fig. 5.13, we may draw a MFSSM for the headstage differential amplifier, this time assuming that there is no feedback to Q_3 and that the sources of the FETs "see" a resistance $1/h_{oe(3)}$ to ground looking into Q_3's collector. Application of the bisection theorem for common-mode inputs yields the split circuit of Fig. 5.16 and the relation

$$v_{1(c)} = \frac{v_{ic}\left(\dfrac{-g_m h_{oe(3)}}{2}\right)}{\dfrac{(g_d + G_D) h_{oe(3)}}{2} + (g_d + g_m) G_D} \tag{5.49}$$

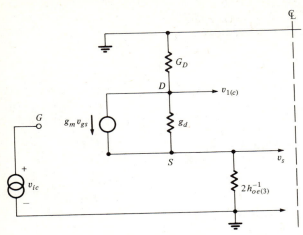

Figure 5.16 JFET differential amplifier headstage small-signal model for common-mode excitation, following application of the bisection theorem.

which is the same as

$$v_{1(c)} = \frac{v_{ic}(-\mu R_D)}{r_d + R_D + \dfrac{(\mu + 1)2}{h_{oe(3)}}} \qquad\qquad 5.50$$

where $\mu = g_m r_d$ for the matched JFETs.

The results derived in the preceding discussion are written in signal flow graph form in Fig. 5.17. Now the single-ended output CMRR can be found using the

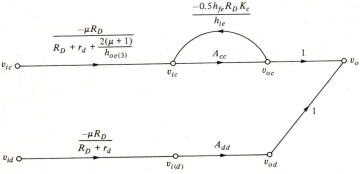

Figure 5.17 Signal flow graph representation of the differential amplifier of Fig. 5.13 with common-mode negative feedback.

definition Eq. 5.14:

$$\text{CMRR}_s = \left(\frac{A_{dd}}{A_{cc}}\right)\left(1 + \frac{A_{cc}h_{fe}R_D K_c}{2h_{ie}}\right)\left[1 + \frac{2(\mu + 1)}{R_D + r_d h_{oe(3)}}\right]$$
5.51

We see that using CMNF with the external headstage significantly increases the amplifier's CMRR_s. Substitution of typical numerical values into Eq. 5.51—$h_{fe} = 100$, $R_D = 10 \text{ k}\Omega$, $K_c = 200$, $h_{ie} = 2 \text{ k}\Omega$, $1/h_{oe(3)} = 100 \text{ kS}$, $\mu = 199$, $r_d = 50 \text{ k}\Omega$, $A_{cc} = 0.1$, and $A_{dd} = 100$—gives

$$\text{CMRR}_s = \left(\frac{100}{0.1}\right)(5001)(688) = 3.34 \times 10^6$$
5.52

in which the CMRR_s is seen to increase by a factor of 5001 as a result of the CMNF.

Other examples of the use of CMNF to raise differential amplifier CMRR are given in this chapter's problem section. ●

5.5 Effect of Source Impedance Asymmetry on CMRR

In this section, we will show that source impedance unbalance in a DA system can significantly alter the CMRR of that system. A differential amplifier with finite source and input resistances is shown in Fig. 5.18.

First we will consider the differential amplifier's input resistances. Manufacturers of differential amplifiers specify values of R_{in} for common-mode and difference-mode inputs. For a common-mode input, $R_{in} = R_{ic}$ is given by the ratio of V_{ic} to I_{ic}. In this case, Fig. 5.18 may be shown in a more simplified form (see Fig. 5.19). Note that one-half the total common-mode input current flows in each resistor.

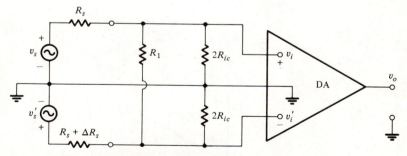

Figure 5.18 Differential amplifier with finite input and source resistances. R_{ic} is the amplifier's common-mode input resistance to ground. R_1 is the equivalent shunting resistor seen for difference-mode inputs (see Eq. 5.55).

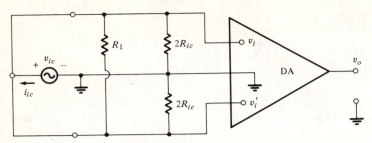

Figure 5.19   Equivalent DA input circuit for pure common-mode inputs. No current flows in R_1.

For pure difference-mode excitation, the difference-mode input resistance is given by v_{id}/i_{id}. This condition is shown in Fig. 5.20. Writing a node equation for the positive input terminal, we find

$$i_{id} = \frac{v_{id}}{2R_{ic}} + \frac{2v_{id}}{R_1} \qquad\qquad 5.53$$

This expression can be rearranged to find the difference-mode input conductance, G_{id}:

$$\frac{i_{id}}{v_{id}} = \frac{1}{R_{id}} = \frac{1}{2R_{ic}} + \frac{2}{R_1} \qquad\qquad 5.54$$

Solving for the required shunting resistance, we have

$$R_1 = \frac{4R_{ic}R_{id}}{2R_{ic} - R_{id}} \qquad\qquad 5.55$$

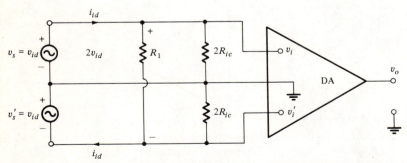

Figure 5.20   Equivalent DA input circuit for pure difference-mode inputs.

which is an expression for the shunting resistor R_1 in terms of the specified common-mode and difference-mode input resistances. In many differential amplifiers, $R_{ic} = R_{id}$, so $R_1 = 4R_{id}$.

We will now develop an expression for the CMRR of the unbalanced-input differential amplifier of Fig. 5.18 in terms of the external source resistors, the input resistors, and the amplifier's Middlebrook gain coefficients. For a single-ended DA output,

$$v_o = (A_{dd} + A_{cd})v_{id} + (A_{cc} + A_{dc})v_{ic} \qquad 5.9$$

First we assume common-mode excitation in v_s and v_s'. That is, $v_s = v_s' = v_{sc}$. Because of unbalance in the source resistances, a new difference-mode component will appear at (v_i, v_i') and contribute to a changed CMRR. The source unbalance will have similar effects for difference-mode excitation at (v_s, v_s'). That is, a new common-mode component will appear at (v_i, v_i'), as well as an altered difference-mode component.

Analyzing the circuit of Fig. 5.18 for common-mode input in v_s will result in a considerable simplification if we treat the shunting resistor as an open circuit. This is because for common-mode inputs and small unbalances in R_s, negligible current will flow in R_1.

It is easy to show, using superposition and the definitions in Eqs. 5.3 and 5.5, that a common-mode excitation, v_{sc}, yields an unwanted difference-mode component at the difference amplifier's input terminals:

$$\frac{v_{id}}{v_{sc}} = \frac{R_{ic}\Delta R_s}{(2R_{ic} + R_s)^2} \qquad 5.56$$

and, similarly, a common-mode component:

$$\frac{v_{ic}}{v_{sc}} = \frac{2R_{ic}}{2R_{ic} + R_s} + \frac{R_{ic}\Delta R_s}{(2R_{ic} + R_s)^2} \qquad 5.57$$

The ΔR_s term in Eq. 5.57 is generally several orders of magnitude less than the first term and may be neglected.

Although R_1 carries current for difference-mode excitation, its exclusion from the circuit of Fig. 5.18 is justified because of the algebraic simplification, and also substitution of typical numerical values for the exact solution shows its omission to have a negligible effect on the computation of the system's CMRR. If we set $R_1 = \infty$ and let $v_s = -v_s' = v_{sd}$, then it is easy to show, using superposition and the definitions of Eqs. 5.3 and 5.5, that difference-mode excitation yields

$$\frac{v_{id}}{v_{sd}} = \frac{2R_{ic}}{2R_{ic} + R_s} - \frac{R_{ic}\Delta R_s}{(2R_{ic} + R_s)^2} \qquad 5.58$$

$$\frac{v_{ic}}{v_{sd}} = \frac{R_{ic}\Delta R_s}{(2R_{ic} + R_s)^2} \qquad 5.59$$

If Eqs. 5.58 and 5.59 are substituted into Eq. 5.9, we obtain

$$v_o = \frac{v_{sc}(A_{dd} + A_{cd})R_{ic}\Delta R_s}{(2R_{ic} + R_s)^2} + \frac{v_{sc}(A_{cc} + A_{dc})2R_{ic}}{2R_{ic} + R_s} \qquad 5.60$$

Also, consideration of the difference-mode input to the system gives the result

$$v_o = \frac{v_{sd}(A_{dd} + A_{cd})2R_{ic}}{2R_{ic} + R_s} + \frac{v_{sd}(A_{cc} + A_{dc})R_{ic}\Delta R_s}{(2R_{ic} + R_s)^2} \qquad 5.61$$

Now, if we go to the basic definition of CMRR given by Eq. 5.14 and take the ratio of v_{sc} (to give $v_o = 1$ V) to v_{sd} (to give $v_o = 1$ V), we can write an expression for the system's CMRR. Note that we have dropped the numerator's ΔR_s term because it is many orders of magnitude less than the amplifier's common-mode rejection ratio, $CMRR_a$.

$$CMRR_{sys} = \frac{CMRR_a}{\dfrac{\Delta R_s(CMRR_a)}{2(2R_{ic} + R_s)} + 1} \qquad 5.62$$

Relation 5.62 for the overall CMRR of a differential amplifier driven through un-balanced source resistors is plotted in Fig. 5.21.

Inspection of Eq. 5.62 shows that the magnitude of the system's CMRR goes to infinity for

$$CMRR_a\left(\frac{0.5\Delta R_s}{2R_{ic} + R_s}\right) + 1 = 0 \qquad 5.63$$

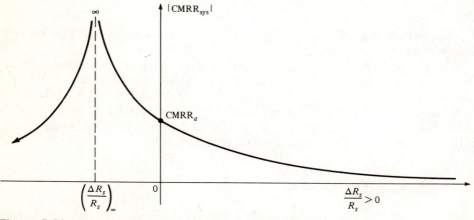

Figure 5.21   System CMRR magnitude vs. fractional unbalance in source resistance, $\Delta R_s/R_s$.

Solving for ΔR_s to give infinite CMRR_{sys}, we find

$$\frac{\Delta R_s}{R_s} = -\frac{4R_{ic}/(R_s + 2)}{\text{CMRR}_a} \qquad\qquad 5.64$$

For a numerical example, we choose $R_{ic} = 100 \text{ M}\Omega$, $R_s = 100 \text{ k}\Omega$, and $\text{CMRR}_a = 10^5$. Hence $\Delta R_s = -4000 \ \Omega$ will give an infinite CMRR_{sys}. This is only a 4% decrease in R_s. Note that if CMRR_a is decreased, a proportionally larger change in R_s will be required to achieve an infinite CMRR_{sys}.

Another, more general way of looking at a differential amplifier with unbalanced source or input resistors is now considered. The general system is shown in Fig. 5.22.

Using an approach similar to that used in the analysis of the DA circuit of Fig. 5.18, we derive a general expression for the CMRR_{sys}:

$$\text{CMRR}_{\text{sys}} = \frac{\text{CMRR}_a + \dfrac{R_1 R_s' - R_1' R_s}{2R_1 R_1' + R_1 R_s' + R_1' R_s}}{\text{CMRR}_a\left(\dfrac{R_1 R_s' - R_1' R_s}{2R_1 R_s + R_1 R_s' + R_1' R_s}\right) + 1} \qquad\qquad 5.65$$

Inspection of this expression shows that when the system is balanced, that is, when $R_1 R_s' = R_1' R_s$, the system CMRR, CMRR_{sys}, equals that of the amplifier alone, CMRR_a. Further, it is possible to solve for an input resistor setting R_1' that will give *infinite* CMRR_{sys}. Setting the denominator of Eq. 5.65 equal to zero, we find, assuming $\text{CMRR}_a \gg 1$,

$$R_1' = \frac{R_1 R_s'(\text{CMRR}_a)}{R_s(\text{CMRR}_a) - 2R_1} \qquad\qquad 5.66$$

or, equivalently,

$$R_1 = \frac{R_1' R_s(\text{CMRR}_a)}{R_s'(\text{CMRR}_a) - 2R_1'} \qquad\qquad 5.67$$

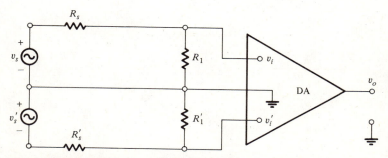

Figure 5.22 DA system with both unbalanced source and input resistances.

Let us assume that $R_s = R_s' = 10\text{ k}\Omega$, $R_1 = 10\text{ M}\Omega$, and $\text{CMRR}_a = 10^4$. Substitution of these values into Eq. 5.66 gives $R_1' = 1.250 \times 10^7\ \Omega$. Evaluation of Eq. 5.67 with these values ($R_1' = 10\text{ M}\Omega$) gives $R_1 = 8.333 \times 10^6\ \Omega$ for infinite CMRR_{sys}.

Although manipulation of the ratio R_1/R_1' can in theory achieve a maximization of CMRR_{sys}, it is not practical to build variable resistors or potentiometers with resistances in excess of 10 MΩ. Therefore, realization of a theoretically infinite CMRR_{sys} may be better done by other balancing means if R_1's in excess of $10^7\ \Omega$ are required. For example, if we solve for the value of R_s to set the denominator of Eq. 5.65 equal to zero, we find that infinite CMRR_{sys} will occur when

$$R_s = \frac{R_s'R_1(\text{CMRR}_a) + 2R_1R_1'}{R_1(\text{CMRR}_a)} \qquad 5.68$$

Substitution into Eq. 5.68 of the numerical values used in evaluating relations 5.66 and 5.67 shows that $R_s = 12\text{ k}\Omega$ will give an infinite CMRR_{sys}. Thus the addition of a 2 kΩ resistor in series with the 10 kΩ R_s will obtain the desired result.

It should be pointed out that R_s, R_s', or CMRR_a may be imprecisely known, leading to a trial-and-error process in adjusting R_1 or R_1', or in augmenting R_s or R_s', to maximize the CMRR_{sys}.

5.6 High-Frequency Behavior of Differential Amplifiers

Differences in the frequency responses $A_{dd}(j\omega)$ and $A_{cc}(j\omega)$ can lead to deterioration of CMRR at high frequencies. To examine how this effect occurs, consider the simple FET differential amplifier stage shown in Fig. 5.23. Note that a small shunt capacitance, C_s, from the source node to small-signal ground exists in a practical circuit. C_s is generally in parallel with a large R_s, or long tail.

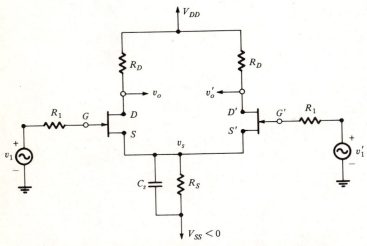

Figure 5.23 Symmetrical JFET differential amplifier stage.

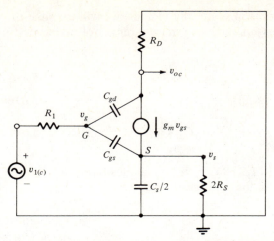

Figure 5.24 HFSSM of the JFET differential amplifier in Fig. 5.23 valid for common-mode inputs. We assume $C_{gs} = C_{gd}$, $g_d = 0$, $C_{ds} = 0$, and $2R_S \gg R_D$.

For common-mode excitation, we apply the bisection theorem and redraw the small-signal model for the half-circuit, shown in Fig. 5.24. The difference-mode small-signal model for the FET differential amplifier is shown in Fig. 5.25. The small capacitances C_{gd}, C_{gs}, and C_s are included. A detailed description of the common-mode circuit's transfer function requires the solution of the following three simultaneous

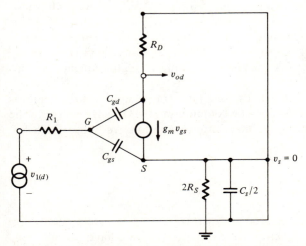

Figure 5.25 HFSSM of the JFET differential amplifier of Fig. 5.23 valid for difference-mode inputs. The same assumptions are made as for the common-mode small-signal model.

node equations. Because such analysis is algebraically tedious, we will examine the gain of the common-mode circuit, $A_{cc}(j\omega)$, in three separate ranges of frequency to see how the CMRR varies with frequency.

$$v_g(G_1 + sC_{gs} + sC_{gd}) - v_{oc}(sC_{gd}) - v_s(sC_{gs}) = v_{1(c)}G_1 \qquad \text{5.69A}$$

$$v_g(g_m - sC_{gd}) + v_{oc}(G_d + sC_{gd}) - v_s(g_m) = 0 \qquad \text{5.69B}$$

$$v_g(g_m + sC_{gs}) + v_{oc}(0) + v_s\left(g_m + \frac{G_S}{2} + \frac{sC_s}{2} + sC_{gs}\right) = 0 \qquad \text{5.69C}$$

In the low-frequency range (dc to $1/2\pi R_s C_s$), all capacitance terms drop out of Eqs. 5.69A through C, and the solution for A_{cc} is easily found to be

$$\frac{v_{oc}}{v_{1(c)}} = -\frac{g_m R_d}{2R_S g_m + 1} \qquad \text{5.70}$$

At frequencies around and above $f_b = 1/2\pi R_s C_s$, the $C_s/2$ term enters the common-mode circuit's node equations. For typical values of $C_s = 3$ pF and $R_s = 10$ MΩ, f_b is 5.3 kHz. This f_b is below frequencies where C_{gs} and C_{gd} dominate the circuit's high-frequency response. If we find $A_{cc}(s)$ with C_s terms in Eqs. 5.69A through C, leaving out C_{gs} and C_{gd} terms, we obtain a lead-lag type of transfer function:

$$A_{cc}(s) = \frac{v_{oc}}{v_{1(c)}} = -\left(\frac{g_m R_d}{2g_m R_S + 1}\right)\frac{1 + sC_s R_S}{1 + s\dfrac{C_s R_S}{2g_m R_S + 1}} \qquad \text{5.71}$$

If the pole frequency of the gain expression Eq. 5.71 is evaluated using typical numerical values for the parameters—$g_m = 4 \times 10^{-3}$ S, $C_s = 3$ pF, and $R_S = 10$ MΩ— we find that it is well above the frequency where C_{gd} and C_{gs} express their influences on $A_{cc}(j\omega)$. In fact, it is above the range of frequencies where the simple lumped-parameter HFSSM for the JFET is valid, about 424 MHz. Thus the presence of C_{gs} and C_{gd} prevents the magnitude of $A_{cc}(j\omega)$ from reaching its high-frequency asymptotic value of $g_m R_d$.

At very high frequencies, C_s bypasses the FET's source node, and we can assume that $v_s = 0$. Under these conditions, the common-mode HFSSM reduces to the difference-mode HFSSM. Using the unilateral approximation, we find at high frequencies that

$$A_{cc}(s) \cong A_{dd}(s) = -\frac{-g_m R_d}{1 + sR_1[C_{gs} + C_{gd}(1 + g_m R_d)]} \qquad \text{5.72}$$

Consideration of $A_{cc}(j\omega)$ and $A_{dd}(j\omega)$ as described before leads to a description of the CMRR as a function of frequency. At low frequencies, we have

$$\text{CMRR}(0) = \frac{A_{dd}(0)}{A_{cc}(0)} = 1 + 2g_m R_S \qquad \text{5.73}$$

Unfortunately, at $f_b > 1/2\pi R_S C_s$, the magnitude of $\mathbf{A}_{cc}(j\omega)$ is increasing at $+6$ dB/octave, causing $|\mathbf{CMRR}(j\omega)|$ to decrease at -6 dB/octave. $|\mathbf{A}_{dd}(j\omega)|$ begins to decrease around the Miller break frequency, f_m, given by the unilateral approximation as

$$f_m = \frac{1}{2\pi R_1 [C_{gs} + C_{gd}(1 + g_m R_d)]} \text{ Hz} \qquad 5.74$$

Thus, above f_m, the **CMRR**$(j\omega)$ magnitude rolls off at -12 dB/octave. At limiting high frequencies, the magnitude of $A_{dd} = A_{cc}$ and the CMRR $\rightarrow 1.0$, or 0 dB. The presence of C_s is seen to cause deterioration in a differential amplifier's CMRR at relatively low frequencies. Design of the active long tail circuit must thus involve not only making the resistance R_S large but also keeping C_s as small as possible if a good, broadband CMRR is to be obtained.

Another remedy to the problem of CMRR deterioration at high frequencies is to use DA architecture that has inherently low difference-mode input capacitance, such as a cascode differential amplifier, shown in Fig. 5.26. The cascode DA design will cause the differential amplifier's CMRR to remain high over a broader bandwidth than for a conventional DA design, such as that of Fig. 5.23, where there is a pronounced difference-mode Miller effect.

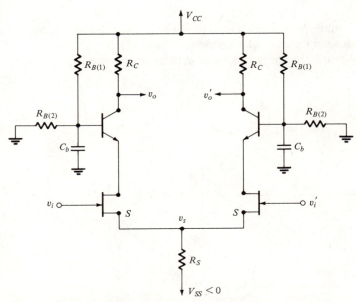

Figure 5.26   JFET-BJT cascode DA headstage. Note that the bases of the BJTs are bypassed to ground.

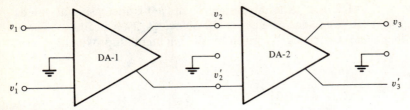

Figure 5.27   Two cascaded DA stages.

5.7 Cascaded Differential Amplifiers and CMRR

Often, in order to realize high gain-bandwidth product differential amplifiers with high gains, it is necessary to cascade two or more amplification stages. In this section, we examine the effects of cascading two imperfect differential amplifiers, as shown in Fig. 5.27.

The Middlebrook equations for the amplifiers are written as

$$v_{2(d)} = A_{dd(1)}v_{1(d)} + A_{dc(1)}v_{1(c)} \tag{5.75}$$

$$v_{2(c)} = A_{cd(1)}v_{1(d)} + A_{cc(1)}v_{1(c)} \tag{5.76}$$

$$v_{3(d)} = A_{dd(2)}v_{2(d)} + A_{dc(2)}v_{2(c)} \tag{5.77}$$

$$v_{3(c)} = A_{cd(2)}v_{2(d)} + A_{cc(2)}v_{2(c)} \tag{5.78}$$

From the Middlebrook equations, we construct a signal flow graph giving the overall characteristics of the cascaded differential amplifiers, shown in Fig. 5.28.

Assuming all cross-terms are nonzero, we can write an expression for the single-ended output, v_3:

$$v_3 = (A_{dd(1)}A_{dd(2)} + A_{cd(1)}A_{dc(2)} + A_{cd(1)}A_{cc(2)} + A_{dd(1)}A_{cd(2)})v_{1(d)} \tag{5.79}$$
$$+ (A_{cc(1)}A_{cc(2)} + A_{cc(1)}A_{dc(2)} + A_{dc(1)}A_{cd(2)} + A_{dc(1)}A_{dd(2)})v_{1(c)}$$

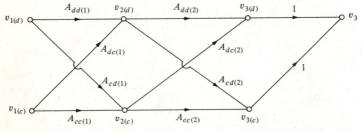

Figure 5.28   Signal flow graph representation of the Middlebrook equations applied to the differential amplifiers of Fig. 5.27.

The CMRR of the single-ended output, cascaded differential amplifier is found using relation 5.14 on Eq. 5.79:

$$\mathrm{CMRR}_{\mathrm{sys}} = \frac{A_{dd(1)}A_{dd(2)} + A_{cd(1)}A_{dc(2)} + A_{cd(1)}A_{cc(2)} + A_{dd(1)}A_{cd(2)}}{A_{cc(1)}A_{cc(2)} + A_{cc(1)}A_{dc(2)} + A_{dc(1)}A_{cd(2)} + A_{dd(2)}A_{dc(1)}} \qquad 5.80$$

If all cross-terms are assumed to be zero, then the system CMRR reduces to

$$\mathrm{CMRR}_{\mathrm{sys}} = \frac{A_{dd(1)}A_{dd(2)}}{A_{cc(1)}A_{cc(2)}} = (\mathrm{CMRR}_{s(1)})(\mathrm{CMRR}_{s(2)}) \qquad 5.81$$

However, even for small cross-terms, there may be a significant change in the $\mathrm{CMRR}_{\mathrm{sys}}$ found by Eq. 5.80. To illustrate this point, we assume that $A_{dd(1)} = 1000$, $A_{cc(1)} = 0.1$, $A_{dc(1)} = A_{cd(1)} = 0.01$, $A_{dd(2)} = 100$, $A_{cc(2)} = 0.01$, and $A_{dd(2)} = A_{cd(2)} = 0.01$. The CMRR_s of DA-1 is found from Eq. 5.16 to be 9.091×10^3, and that for DA-2 is 5.001×10^3. Their product is 4.546×10^7, or 153 dB. Using Eq. 5.80, we find that the $\mathrm{CMRR}_{\mathrm{sys}}$ of the cascaded differential amplifiers is 9.979×10^4, or 100 dB, well below the product of the CMRRs of the two separate differential amplifers but more than the $\mathrm{CMRR}_{\mathrm{sys}}$ of either individual amplifier. Thus we can infer that cascading DA stages will, in general, improve the overall differential amplifier CMRR over that of the individual amplifiers. The maximum improvement will generally be less than the product of the individual amplifier's CMRR_s's due to the presence of cross-terms (A_{cd} and A_{dc}) resulting from circuit asymmetries.

5.8 Input Impedance of Differential Amplifiers

Effective low-noise, high-frequency DA design generally requires the use of circuit geometries that give high input impedance to both common-mode and difference-mode components of the input signals. The equivalent, shunt input capacitance to ground seen by these input sources should be as small as possible, and the resistance to ground and between the input terminals should be large.

To illustrate the calculation of input impedance, we first consider the simple, MFSSM of the BJT differential amplifier shown in Fig. 5.4. For pure difference-mode inputs, inspection of Fig. 5.5 shows that v_{id} sees an input resistance of

$$R_{id} = h_{ie} \qquad 5.82$$

Inspection of the small-signal model for common-mode inputs, Fig. 5.6, reveals

$$R_{ic} = h_{ie} + 2R_E(1 + h_{fe}) \qquad 5.83$$

It is obvious from Eqs. 5.82 and 5.83 that the difference-mode input resistance is orders of magnitude lower than the common-mode input resistance. This relationship can lead to poor frequency response to the desired difference-mode signal. Fortunately, there are several ways to mitigate this problem. One obvious way is to use

JFETs in the differential amplifier, as shown in Fig. 5.23. As long as the gate-drain junctions are reverse biased, the dc and mid-frequency input impedance will be high. At high frequencies, however, the input impedance to difference-mode inputs will fall because of the Miller effect. To overcome the Miller effect for difference-mode inputs, a cascode DA geometry is effective, as shown in Fig. 5.26.

If BJTs are to be used, it is possible to realize high input impedance for both difference-mode and common-mode inputs through the use of Darlington modules in the differential amplifier, as shown in Fig. 5.29. It is left as an exercise to show that the low- and mid-frequency difference-mode input resistance for a Darlington differential amplifier is given by

$$R_{id} = h_{ie(1)} + h_{ie(2)}(1 + h_{fe(1)}) + R_e(1 + h_{fe(1)})(1 + h_{fe(2)}) \qquad 5.84$$

R_{id} is typically of the order of low megohms for typical high-beta, small-signal transistors and associated resistors. On the other hand, R_{ic} is enormous!

Some designers have employed positive feedback to raise the difference-mode input impedance of differential amplifiers. Although effective in some cases at low frequencies, positive feedback generally reduces bandwidth for increased gain and in some cases can lead to instability at high frequencies. The input source resistance interacts with the differential amplifier's input impedance and can adversely affect DA behavior if it is too large and positive feedback is used.

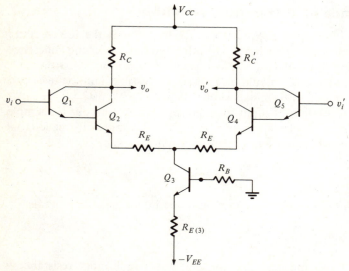

Figure 5.29 BJT Darlington differential amplifier. For symmetry, $Q_1 = Q_5$, $Q_2 = Q_4$, and $R_C = R_C'$. Q_3 provides a low-conductance current sink long tail for dc biasing of the differential amplifier.

SUMMARY

In this chapter, we have stressed the analysis, design, and application of differential amplifier stages because of the importance of this amplifier in the design of modern analog IC electronic systems. Although we have used small-signal linear analysis of differential amplifiers, the reader should be aware that an alternate approach to their analysis which stresses DA nonlinearity can also be employed. See, for example, Section 3.4 in Gray and Meyer (1984) for a detailed treatment of large-signal analysis of BJT differential amplifiers.

Of interest is the investigation of the behavior of the Middlebrook gains and the CMRR when asymmetries are assumed to exist in the otherwise symmetrical architecture of differential amplifiers. For example, transistor Q has a transconductance g_m; transistor Q' has a transconductance $g'_m = g_m + \delta g_m$. Note that the bisection theorem cannot be applied to asymmetrical circuits; hence any algebraic study of the effects of parameter asymmetries must be done with the full small-signal model of the differential amplifier. Analysis of asymmetric circuit behavior can be algebraically complex. Another means of studying the effects of asymmetries on DA performance is to use computer simulations in which a worst-case scenario is considered along with the matched circuit's behavior. The role of the common-emitter resistance (long tail) and common-mode negative feedback in reducing the effects of certain unbalance errors is an interesting topic for future investigations.

PROBLEMS

5.1   A differential amplifier is described by the equation

$$v_o = A_C v_{1(c)} + A_D v_{1(d)}$$

The measurements are as shown in Fig. P5.1. Find values for A_D, A_C, and the CMRR in decibels.

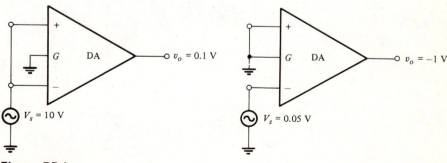

Figure P5.1

5.2   Mid-frequency Analysis of a BJT Using the Bisection Theorem
Assume $Q_1 = Q_2$ and $h_{re} = h_{oe} = 0$. Find expressions for A_{dd}, A_{cc}, and the CMRR.

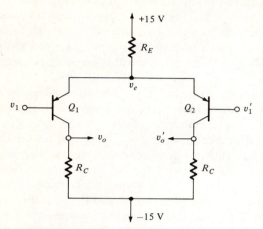

Figure P5.2

5.3   A BJT differential amplifier uses a third BJT as a high-impedance long tail or emitter resistance. Assume $Q_1 = Q_2$, $h_{ie} = 2.5$ kΩ, $h_{fe(1)} = 199$, $h_{oe(1)} = 10^{-5}$ S, and $h_{re(1)} = 0$; also assume $h_{ie(3)} = 1.5$ kΩ, $h_{fe(3)} = 149$, $h_{oe(3)} = 10^{-5}$ S, and $h_{re(3)} = 0$.

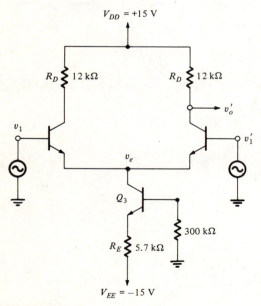

Figure P5.3

(a) Find a numerical value for the small-signal resistance seen looking into Q_3's collector.

(b) Find numerical values for A_{dd}, A_{cc}, and the single-ended CMRR.

5.4 A JFET differential amplifier uses a BJT to create a high-impedance long tail or source resistance. Assume $Q_1 = Q_2$, $g_m = 0.002$ S, and $r_d = 250$ kΩ; also assume $h_{fe(3)} = 150$, $h_{oe(3)} = 10^{-6}$ S, $h_{ie(3)} = 1.5$ kΩ, and $h_{re(3)} = 0$. Find algebraic expressions and numerical values for A_{dd}, A_{cd}, A_{dc}, A_{cc}, and the single-ended output CMRR.

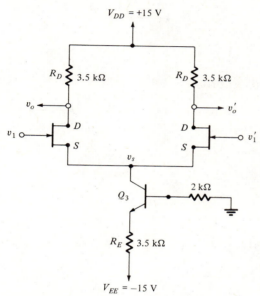

Figure P5.4

5.5 A simple two-DA stage op-amp design is shown in Fig. P5.5. Assume $Q_1 = Q_2$, and $Q_3 = Q_4$ (matched JFET pairs). Let $g_{m(1)} = 4000$ S, $R_{S(1)} = 1.2$ MΩ, $g_{m(3)} = 2500$ S, $r_{d(3)} = 62$ kΩ, $r_{d(1)} = 50$ kΩ, $R_{D(1)} = 5$ kΩ, $R_{S(2)} = 1$ MΩ, $R_{D(3)} = 10$ kΩ, $g_{m(5)} = 2000$ S, $r_{d(5)} = 40$ kΩ, $R_{D(5)} = 10$ kΩ, $g_{m(6)} = 8000$ S, $r_{d(6)} = 50$ kΩ, and $R_{S(6)} = 5$ kΩ.

(a) Find the numerical value for $v_o/v_{1(d)}$ (gain for a single-ended output from a pure difference-mode input).

(b) Find the numerical value for $v_o/v_{1(c)}$ (gain for a single-ended output from a pure common-mode input).

(c) Find the amplifiers' CMRR.

5.6 The differential amplifier shown in Fig. P5.6 uses common-mode negative feedback. Assume $Q_1 = Q_2$. Let $h_{fe(1)} = 200$, $h_{oe(1)} = h_{re(1)} = 0$, $h_{ie(1)} = 2.5$ kΩ, $R_1 = 700$ kΩ, $R_C = 4$ kΩ, $h_{oe(3)} = 10^{-5}$ S, $h_{fe(3)} = 200$, $h_{ie(3)} = 1.2$ kΩ, and $h_{re(3)} = 0$. Numbers in parentheses are quiescent dc bias voltages; ZD is a 14.3 V, low-current zener diode. Assume its small-signal resistance is $r_z = 0$.

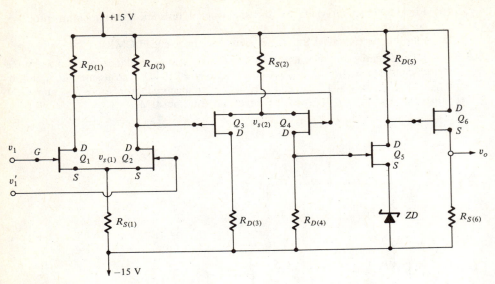

Figure P5.5

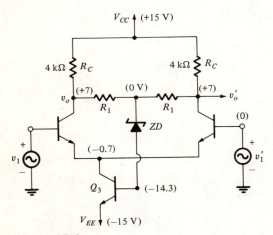

Figure P5.6

(a) Use the bisection theorem to draw a MFSSM for the circuit, given pure difference-mode excitation ($v'_1 = -v_1$).

(b) Repeat part (a) for pure common-mode excitation ($v_1 = v'_1$).

(c) Develop an algebraic expression for $v_o/v_{1(d)}$. Evaluate numerically. Repeat for $v_o/v_{1(c)}$.

(d) Evaluate the circuit's CMRR.

(e) Repeat parts (c) and (d) for the case where the zener diode is removed and Q_3's base is tied directly to small-signal ground.

5.7   The circuit shown in Fig. P5.7 is a grounded-base version of a differential amplifier. Both BJTs are matched and have $h_{oe} = h_{re} = 0$. Using the bisection and reduction theorems, derive expressions for A_{dd}, A_{cc}, and the single-ended CMRR.

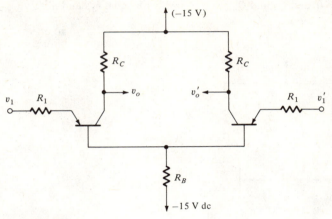

Figure P5.7

5.8   A FET/BJT cascode differential amplifier is shown in Fig. P5.8, in which

$$Q_1 = Q_2, \text{ with } h_{oe} = h_{re} = 0, \ h_{fe} = 200, \ h_{ie} = 3 \times 10^3 \ \Omega$$

$$Q_3 = Q_4, \text{ with } g_m = 5 \times 10^{-3} \ \text{S}, \ r_d = 10^5 \ \Omega$$

$$R_C = 10^4 \ \Omega, \ R_S = 45 \ \text{M}\Omega, \ R_1 \| R_2 = 10^5 \ \Omega$$

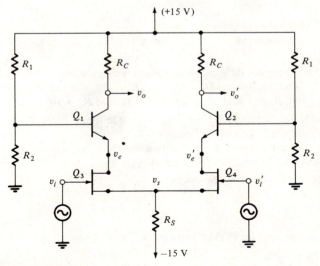

Figure P5.8

(a) Use the bisection theorem to draw the MFSSMs valid for difference-mode and common-mode excitation.

(b) Find algebraic and numerical expressions for A_{dd}, A_{cc}, and the single-ended CMRR.

5.9 An FET-BJT cascode differential amplifier is shown in which Q_1 and Q_2 are matched, with h_{oe} and $h_{re} = 0$, and Q_3 and Q_4 are matched, with equal r_d's and μ's.

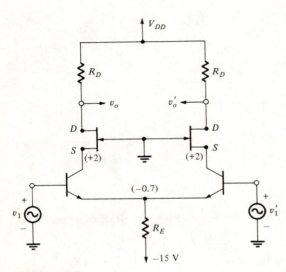

Figure P5.9

(a) Draw the MFSSM for the complete circuit.

(b) Find algebraic expressions for A_{dd}, A_{cc}, and the CMRR using the bisection theorem.

5.10 A scheme proposed for raising the difference-mode input resistance of a BJT differential amplifier is shown in Fig. P5.10. Assume $Q_1 = Q_2$, with $h_{oe} = h_{re} = 0$, and also $R_B \gg R_C$. Use $h_{ie} = 2$ kΩ, $h_{fe} = 100$, $R_E = 1$ MΩ, $R_B = 1.01 \times 10^6$ Ω, and $R_C = 10^4$ Ω.

(a) Draw the MFSSM for the amplifier for common-mode input. Find $R_{in} = v_{1(c)}/i_1$ for common-mode input (algebraic and numerical).

(b) Draw the MFSSM for the amplifier for difference-mode input. Find $R_{in} = v_{1(d)}/i_1$ for difference-mode input (algebraic and numerical).

5.11 A differential amplifier is described by the relation

$$v_o = 100v_{1(d)} + 0.01v_{1(c)}$$

It is connected as shown in Fig. P5.11. Resistors R_1 and R_s are matched exactly.

(a) Find $v_o = f(v_s, v_s') = A_D v_{sd} + A_C v_{sc}$.

(b) Find the system's CMRR and compare to the CMRR of the amplifier.

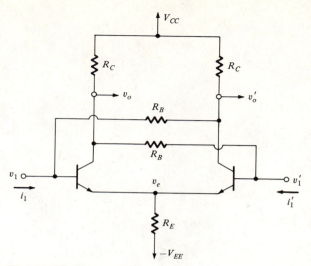

Figure P5.10

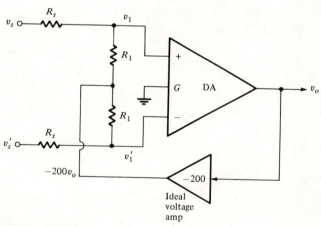

Figure P5.11

5.12 A differential amplifier is given common-mode negative feedback as shown in Fig. P5.12. Assume the following:

$$A_{cd} = A_{dc} = 0$$

$$v_o = A_{dd}v_{1(d)} + A_{cc}v_{1(c)}$$

$$v'_o = -A_{dd}v_{1(d)} + A_{cc}v_{1(c)}$$

$$v_{oc} = A_{cc}v_{1(c)}$$

$$\mathrm{CMRR}_a \triangleq \frac{A_{dd}}{A_{cc}}$$

$$\alpha = \frac{R_1}{R_1 + R_s}$$

$$\beta = \frac{R_s}{R_1 + R_s}$$

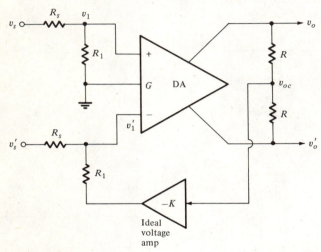

Figure P5.12

(a) Find expressions for $v_{1(c)}$ and $v_{1(d)}$ in terms of v_{sc} and v_{sd}, K, and other systems parameters.

(b) Write an expression for v_o in terms of v_{sc} and v_{sd}.

(c) Find the system's single-ended CMRR.

(d) Does a K value exist that will cause the single-ended CMRR to go to infinity? If so, find it.

5.13 A JFET amplifier uses identical transistors Q_1 and Q_2 having transconductance g_m and drain resistance r_d. Q_3 has transconductance g'_m and drain conductance $g'_d = 0$.

(a) Draw the MFSSM for the amplifier as a symmetrical circuit. Note that Q_3, R_D, and R'_D lie on the axis of symmetry and can be split according to the protocols shown in Chapter 2, Fig. 2.9. Assume v_g of Q_3 is at small-signal ground.

(b) Find algebraic expressions for A_{dd} and A_{cc}.

5.14 In the BJT differential amplifier shown in Fig. P5.14, $Q_1 = Q_2$, with $h_{re} = h_{oe} = 0$ and $h_{fe(1)}$ and $h_{ie(1)} > 0$. Also, $Q_2 = Q_4$, with $h_{oe} = h_{re} = 0$ and $h_{fe(2)}$ and $h_{ie(2)} > 0$.

(a) Use MFSSM analysis to find an expression for the R_{in} that v_1 sees when the differential amplifier has difference-mode excitation ($v'_1 = -v_1$).

(b) Find an expression for the differential amplifier's CMRR.

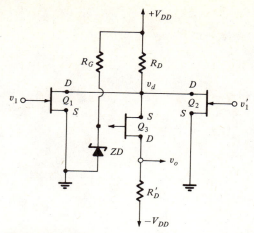

Figure P5.13

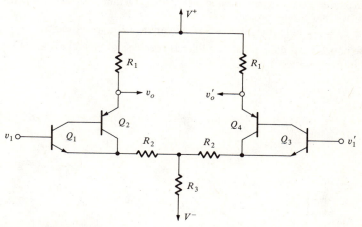

Figure P5.14

5.15 Darlington stages are used in a BJT differential amplifier to increase the gain and raise the difference-mode input resistance, as shown in Figure 5.29. Assume $Q_1 = Q_5$, with $h_{oe} = h_{re} = 0$ and $h_{fe(1)}$ and $h_{ie(1)} > 0$. Also, $Q_2 = Q_4$, with $h_{oe} = h_{re} = 0$ and $h_{fe(2)}$ and $h_{ie(2)} > 0$. Verify relation 5.28 for R_{id} in the text using the bisection theorem and h-parameter MFSSMs for the BJTs as described here.

5.16 In the simple BJT differential amplifier shown in Fig. P5.16, assume $Q_1 = Q_2$, $h_{oe} = h_{re} = 0$, $h_{fe} = 99$, $h_{ie} = 2.5$ kΩ, $R_E = 15$ kΩ, and $R_C = 9.1$ kΩ.

(a) Find an expression for the mid-frequency small-signal input resistance at either input node for purely difference-mode excitation.

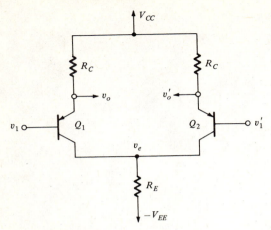

Figure P5.16

(b) Find an expression for the mid-frequency small-signal input resistance at either input node for purely common-mode excitation.

(c) Assume $v_1' = 0$. Find an expression for the input resistance v_1 sees.

5.17 The CMRR of a certain differential amplifier is specified to be 130 dB at 60 Hz. Find A_D and A_C in the equation

$$v_o = A_C v_{1(c)} + A_D v_{1(d)}$$

Assume $v_o = 10$ V rms at 60 Hz when the amplifier is connected as shown in Fig. P5-17.

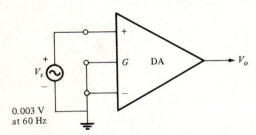

Figure P5.17

5.18 Two matched JFETs are used to make a differential amplifier. Cross-neutralization capacitors C_n are used to cancel the Miller effect at high frequencies. Make Bode magnitude plots over mid- and high frequencies for $\mathbf{A}_{dd}(j\omega)$ and $\mathbf{A}_{cc}(j\omega)$ for $C_n = 0$, and also for $C_n = 3$ pF. Clearly show the mid-frequency gain, the -3 dB frequency, and f_T for A_{dd}. Show the mid-frequency gain, the gain at the -3 dB frequency, and the gain at f_T for A_{cc}. Assume $Q_1 = Q_2$, $g_m = 3500$ μS, $r_d = 40$ kΩ, $C_{gs} = C_{gd} = 3$ pF, $C_{ds} = 0$, $R_S = 1$ MΩ, $R_1 = 0.5$ kΩ, and $R_D = 5.6$ kΩ.

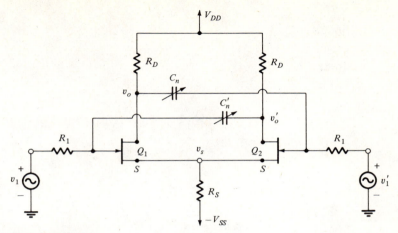

Figure P5.18

5.19 An FET differential amplifier is operated at high frequencies. Assume $Q_1 = Q_2$, $g_m = 4000\ \mu S$, $g_d = 0$, $C_{gs} = C_{gd} = 3$ pF, $C_{ds} = 0$, $R_D = 10$ kΩ, $R_S = 2$ MΩ, and $R_1 = 100$ Ω. Make Bode magnitude plots for $A_{dd}(j\omega)$ and $A_{cc}(j\omega)$ over the mid- and high-frequency ranges. Show the mid-frequency gain, the -3 dB frequency, and f_T for A_{dd}. Give the mid-frequency gain, the gain at the -3 dB frequency, and the gain at f_T for A_{cc}.

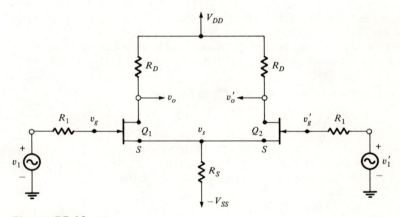

Figure P5.19

5.20 A Widlar current sink is to be used as a long tail for a differential amplifier. We wish to find the small-signal Norton conductance, G_{in}, looking into the collector of Q_2. Assume the following:

Q_1: $r_x = g_o = 0$
$r_\pi = 2.5 \times 10^3\ \Omega$
$g_m = 0.045$ S

Q_2: $r_x = 0$

$g_o = 1 \times 10^{-5}$ S

$g_m = 0.06$ S

$r_\pi = 2 \times 10^3$ Ω

$R_3 = R_4 = 1$ kΩ

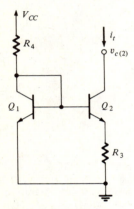

Figure P5.20

(a) Draw the MFSSM for the circuit. Use the mid-frequency hybrid-pi model for the BJTs (see Chapter 4, Fig. 4.1) in which $C_\pi = C_\mu = 0$.

(b) Find the resistance the base of Q_2 sees looking into the base and collector of Q_1.

(c) Use a test current source, i_t, to find the voltage v_c at Q_2's collector. Note $G_{in} = i_t/vc$.

5.21 A differential op-amp is described by the gain equation

$$v_o = A_D v_{id} + A_C v_{ic}$$

The op-amp's $\text{CMRR}_a = A_D/A_C$. The op-amp is connected as a differential amplifier using precision matched resistors; $R_1 = R_1'$ and $R_f = R_f'$. We wish to examine the effects of resistor mismatch on the CMRR of the system. Here

$$\text{CMRR}_{sys} = \frac{v_{sc} \text{ to give } v_o = 1 \text{ V}}{v_{sd} \text{ to give } v_o = 1 \text{ V}}$$

Let

$$\alpha = \frac{R_f}{R_f + R_1}$$

$$\alpha' = \frac{R_f'}{R_f' + R_1'}$$

$$\beta = \frac{R_1}{R_f + R_1}$$

$$\beta' = \frac{R'_1}{R'_f + R'_1}$$

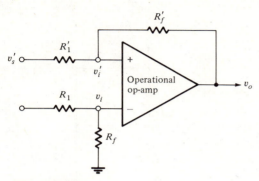

Figure P5.21

(a) Find expressions for v_{ic} and v_{id} for the cases v_{sc} ($v_{sd} = 0$) and v_{sd} ($v_{sc} = 0$). Use these expressions to find v_o, and substitute v_o into the relation for CMRR$_{sys}$ just given. Assume $A_D/A_C \gg (\alpha - \alpha')/(\alpha + \alpha')$.

(b) Now assume that $R_1 = R'_1$ and $R'_f = R_f + \Delta R$. Let $R_f = 10^5 \, \Omega$, $R_1 = 10^4 \, \Omega$, and CMRR$_a = 80$ dB. Find ΔR to make CMRR$_{sys}$ infinite.

5.22 A BJT is used to apply common-mode negative feedback around a two-state FET differential amplifier. See Fig. P5.22. Use MFSSM analysis in which $Q_1 = Q_2$, $g_{d(1)} = 0$, and $g_{m(1)} = 4 \times 10^{-3}$ S. Also, $Q_3 = Q_4$, $g_{d(3)} = 0$, and $g_{m(3)} = 4 \times 10^{-3}$ S. Q_5 is represented by a mid-frequency hybrid-pi model in which $r_x = 0$, $g_{m(5)} = 0.04$ S, $g_o = 2 \times 10^{-5}$ S, and $r_\pi = 2$ kΩ. Note that Q_5, the zener diode, and the ideal dc current source, $2I_{SQ}$, are on the axis of symmetry. The zener's small-signal resistance, r_z, equals 0.5 kΩ, and $R_{d(1)} = R_{d(2)} = 5$ kΩ.

(a) Draw the MFSSM for the differential amplifier for difference-mode excitation.

(b) Find an algebraic expression and the numerical value for $A_D = v_o/v_{1(d)}$.

(c) Draw the MFSSM for the differential amplifier for common-mode excitation. Note that the components on the axis of symmetry must be split.

(d) Find an algebraic expression and the numerical value for $A_D = v_o/v_{1(c)}$.

(e) Find a numerical value for CMRR$_{sys}$ with common-mode negative feedback, and also for the case where $g_{m(5)} = 0$ (CMNF = 0).

5.23 A BJT differential amplifier uses two matched 2N2369 transistors. See Fig. P5.23. The equivalent circuit of the long tail includes a parallel 2 MΩ resistor, 5 pF capacitor, and 2.0125 mA dc current source for dc biasing.

(a) Use an ECAP to plot the common-mode dc input/output curve for the differential amplifier.

(b) Plot the common-mode frequency response of the amplifier over the range 10^4 Hz $< f < 10^8$ Hz. How is the common-mode frequency response affected by the shunt capacitance to ground in the long tail?

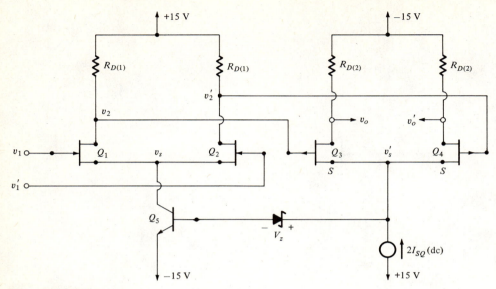

Figure P5.22

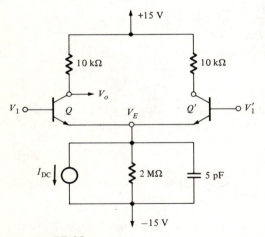

Figure P5.23

(c) Repeat parts (a) and (b) for pure difference-mode excitation. Plot the dc input/output curve for $-0.2 \text{ V} < V_{1(d)} < 0.2 \text{ V}$. Plot the difference-mode frequency response for $10^5 \text{ Hz} < f < 10^9 \text{ Hz}$.

(d) Use the data from the common-mode and difference-mode frequency response curves to plot the differential amplifier's CMRR versus frequency from 10 kHz to 100 MHz.

Chapter 6

Introduction to Feedback

In an electronic circuit, feedback is defined as a process whereby a signal proportional to the circuit's output is combined with a signal proportional to the input to form an *error signal* which affects the output. Nearly all electronic circuits contain feedback, either implicitly, as the result of circuit geometry (e.g., from an unbypassed BJT emitter resistor), or explicitly, by design. In this chapter, we will focus our attention on explicit, or externally applied feedback. We will show that it adds a number of desirable properties to electronic amplifier performance, and a few undesirable ones as well. One of the consequences of applying feedback is a change in the closed-loop amplifier's frequency response. This change can be beneficial if the feedback is correctly designed, or lead to problems, including instability, if poorly engineered. Chapter 7 treats the frequency domain effects of feedback in detail, including instability, oscillations, and oscillator design.

We will use op-amps to illustrate many of the concepts of feedback circuits, but by no means do we wish to imply that op-amps are the only electronic systems using feedback. Much of the treatment of feedback amplifiers in this chapter and in Chapter 7 is based on fundamental concepts and techniques from the area of control systems analysis. We use the root-locus method to describe and predict quantitatively and qualitatively electronic feedback circuit behavior in the frequency domain.

6.1 Classification of Electronic Feedback Systems

We begin this section with a general treatment of feedback in electronic amplifiers. Figure 6.1(A) shows a simple single-loop feedback system represented by a Mason signal flow graph. The same system is shown in conventional block diagram format in Fig. 6.1(B).

In an electronic circuit system, α, β, and K_v are, in general, transfer functions, usually functions of frequency. The summation of the signals αX and βY can be accomplished in practice by several electronic means. Resistive summing may be used; the signals may be subtracted by a differential amplifier; one signal may be applied to the gate of a JFET amplifier while the other is put into the JFET's source across an unbypassed source resistor; or transformer windings can be used to algebraically add the feedback signal to the input. In an electronic system, the input and output (X and Y) are usually either a voltage or a current but can also be a parameter such

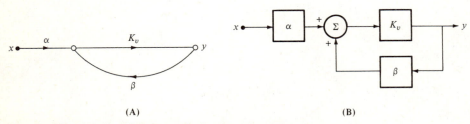

(A)   (B)

Figure 6.1   **(A)** Signal flow graph diagram of a simple feedback system. **(B)** Simple feedback system in block diagram form.

as frequency, phase, power, or the like. For example, if X is a voltage and Y is a current β must have the dimensions of resistance and α must be dimensionless in order that like units (volts) be summed at the summer. Note that K_v will have the dimension of siemens in this case.

With reference to Fig. 6.1(A), a reduction of the elementary signal flow graph yields the closed-loop transfer function

$$\frac{Y}{X} = \frac{\alpha K_v}{1 - \beta K_v} = \frac{\alpha K_v}{1 - A_L} \qquad\qquad 6.1$$

This simple result allows us to define the following important quantities used in analyzing and describing electronic feedback systems:

Forward gain = open-loop gain = αK_v

Loop gain = $\beta K_v = A_L$

Return difference = RD = $1 - A_L$

dB of feedback = $-20 \log|(1 - \mathbf{A}_L(j\omega)|$

A *negative feedback system* has a minus sign associated with either K_v or β or the summer, or with all three elements of the signal flow graph. Control systems generally use negative feedback, and it is customary to assume a subtraction of βY at the summer; K_v and β are generally taken as positive quantities at dc or at mid-frequencies. This is by no means the case in electronic feedback systems. $\mathbf{K}_v(j\omega)$ and $\boldsymbol{\beta}(j\omega)$ may or may not have a sign inversion at dc or at mid-frequencies, depending on the circuit design. The summer will not have a sign inversion if a simple resistive summer is used; it will invert if a differential amplifier is used to add in the feedback.

To see if an electronic system is using negative feedback, we examine the loop gain. $\mathbf{A}_L(j0)$ will have a minus sign if the amplifier is a direct-coupled negative feedback system. If the feedback amplifier is reactively coupled, then $\mathbf{A}_L(j\omega)$ will have a minus sign (its phase will be $-180°$) at mid-frequencies if the system uses negative feedback.

In summary, an electronic amplifier may have either negative or positive feedback depending on the sign of $\mathbf{A}_L(j\omega)$. Feedback amplifier output is generally a voltage, although current output amplifiers are encountered. Hence we can have

Negative voltage feedback

Positive voltage feedback

Negative current feedback

Positive current feedback

We will treat the properties of negative and positive voltage feedback in most detail because they are most important. Negative and positive current feedback are used in the design of some oscillators and will be treated in Chapter 7.

Some authors of electronic circuit textbooks classify electronic feedback circuits according to the configuration of the circuit through which the feedback is applied. Although these distinctions are interesting, they do not contribute significantly to our understanding of the effects of feedback, and so we do not treat them here.

6.2 Effects of Negative Voltage Feedback

Negative voltage feedback (NVF) has gained the reputation as a panacea for a variety of electronic problems. NVF can create problems, as well. We will use simple illustrations to show the most important properties resulting from NVF.

6.2.1 Reduction of Output Resistance

A schematic of a simple inverting voltage amplifier is shown in Fig. 6.2. R_o is the amplifier's output resistance without feedback (generally a few hundred ohms). A resistive summer is used to add the input, V_s, and the fed-back output at the summing junction node, V_i. Note that in practice, $R_1 \gg (R_s, R_f) \gg R_o$.

To find the feedback amplifier's transfer function, we write a node equation on V_i:

$$V_i(G_s + G_f) - V_o G_f = G_s v_s \qquad\qquad 6.2$$

Obviously,

$$V_o = -K_v V_i \qquad\qquad 6.3$$

Hence,

$$V_i = \frac{-V_o}{K_v} \qquad\qquad 6.4$$

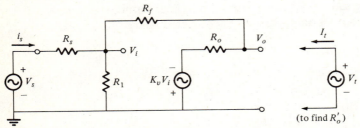

Figure 6.2 Simple inverting negative feedback amplifier. This schematic also represents an op-amp inverter.

so

$$-\left(\frac{V_o}{K_v}\right)(G_s + G_f) - V_o G_f = G_s V_s \qquad\qquad 6.5$$

Hence the closed-loop amplifier's transfer function is

$$\frac{V_o}{V_s} = -\frac{K_v G_s}{G_s + G_f + K_v G_f} \qquad\qquad 6.6$$

Equation 6.6 may be easily written in the standard form of Eq. 6.1:

$$\frac{V_o}{V_s} = A_v = -\frac{\left(\dfrac{R_f}{R_f + R_s}\right)K_v}{\left[1 + K_v\left(\dfrac{R_s}{R_f + R_s}\right)\right]} \qquad\qquad 6.7$$

In Eq. 6.7, we observe by comparison with Eq. 6.1 that

$$\alpha = \frac{R_f}{R_f + R_s} \qquad\qquad 6.8$$

$$\beta = \frac{R_s}{R_f + R_s} \qquad\qquad 6.9$$

$$A_L = -K_v\left(\frac{R_s}{R_f + R_s}\right) \qquad \text{(a NVF system)} \qquad\qquad 6.10$$

If K_v is very large, as in the case of an op-amp, Eq. 6.7 reduces to the well-known relation

$$\frac{V_o}{V_s} = -\frac{R_f}{R_s} \qquad\qquad 6.11$$

This is the gain of an ideal op-amp inverter circuit.

To examine the output resistance of the NVF amplifier of Fig. 6.2, we set $V_s = 0$ and place an active test source, V_t, across the output terminals. The output resistance, R_o', is given by

$$R_o' = \frac{V_t}{I_t} \qquad\qquad 6.12$$

By a simple voltage divider,

$$V_i = V_t\left(\frac{R_s}{R_f + R_s}\right) \qquad\qquad 6.13$$

We neglect the small output current through R_f and R_s and write

$$I_t = \frac{V_t - (-K_v V_i)}{R_o} \qquad\qquad 6.14$$

Substituting Eq. 6.13 into Eq. 6.14, we get

$$I_t = \frac{V_t + K_v\left(\dfrac{V_t R_s}{R_f + R_s}\right)}{R_o} \qquad\qquad 6.15$$

Equation 6.15 is used to solve Eq. 6.12 for the output resistance with NVF:

$$R_o' = \frac{V_t}{I_t} = \frac{R_o}{\left[1 + K_v\left(\dfrac{R_s}{R_s + R_f}\right)\right]} = \frac{R_o}{[1 - A_L]} \qquad\qquad 6.16$$

In general, it is easy to show that NVF reduces the output impedance by the return difference. Thus,

$$\mathbf{Z}_o'(j\omega) = \frac{\mathbf{Z}_o(j\omega)}{\mathbf{RD}(j\omega)} \qquad\qquad 6.17$$

In an op-amp circuit where there may be over 100 dB of NVF at low frequencies, R_o' values of less than 0.001 Ω are commonly seen. It can be shown that $|\mathbf{Z}_o'(j\omega)|$ increases with increasing ω, a significant result for audio amplifiers and regulated dc power supplies using NVF.

6.2.2 Input Impedance

Consider R_{in}' "seen" by v_s in Fig. 6.2 as

$$R_{in}' = \frac{V_s}{I_s} \qquad\qquad 6.18$$

We will include the amplifier's input resistance, R_1, and write a node equation on V_i:

$$V_i(G_1 + G_s + G_f) - V_s G_s - v_o G_f = 0 \qquad\qquad 6.19$$

Substituting Eq. 6.3 for v_o into Eq. 6.19 and rearranging, we get

$$V_i = \frac{V_s G_s}{G_1 + G_s + G_f(1 + K_v)} \qquad\qquad 6.20$$

Ohm's law tells us that

$$I_s = (V_s - V_i)G_s \qquad\qquad 6.21$$

When Eq. 6.20 is substituted into Eq. 6.21, and the i_s obtained is put into Eq. 6.18, we find the input resistance with NVF:

$$R'_{in} = \frac{V_s}{I_s} = R_s\left[1 + \frac{G_s}{G_1 + G_f(1 + K_v)}\right] \qquad\qquad 6.22$$

And because $G_1 \ll G_f$, this reduces to

$$R'_{in} \cong R_s\left(1 + \frac{R_f}{R_s K_v}\right) \cong R_s \qquad\qquad 6.23$$

for large K_v. Hence the V_i node appears as a *virtual ground*, and R_s carries i_s to that virtual ground, causing V_s to see $R'_{in} \cong R_s$.

An alternate NVF amplifier circuit is shown in Fig. 6.3. A differential amplifier receives a single-ended input, v_s, to its noninverting input terminal; NVF is applied through a voltage divider (R_a and R_b) to the inverting input. This amplifier's closed-loop gain can be easily found. The output is simply

$$V_o = K_v(V_i - V'_i) = K_v(V_s - V'_i) \qquad\qquad 6.24$$

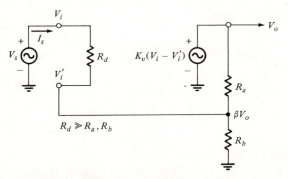

Figure 6.3   Differential amplifier with NVF and a noninverting input.

The node equation on V_i' is written:

$$V_i'(G_d + G_b + G_a) - V_s G_d = V_o G_a \qquad 6.25$$

Solving Eq. 6.24 for V_i' and substituting in Eq. 6.25 yields

$$\frac{V_o}{V_s} = A_v = \frac{(G_b + G_a)K_v}{G_a K_v + G_a + G_b + G_d} \qquad 6.26$$

which, as K_v becomes very large, reduces to the well-known expression for the gain of a noninverting ideal op-amp:

$$\frac{V_o}{V_s} = A_v = 1 + \frac{R_a}{R_b} = \frac{1}{\beta} \qquad 6.27$$

If there were no feedback ($R_b = 0$, $\beta = 0$) in the circuit of Fig. 6.3, the input resistance seen by V_s would be just R_d. With NVF, however, R_{in}' is more difficult to find and is given in general by Eq. 6.18.
To find I_s, we note that

$$I_s = (V_s - V_i')G_d \qquad 6.28$$

V_i' is found by writing a node equation:

$$V_i'(G_d + G_b + G_a) - V_s G_d = V_o G_a = (V_s - V_i')K_v G_a \qquad 6.29$$

Equation 6.29 is solved for V_i':

$$V_i' = \frac{V_s(G_d + K_v G_a)}{[G_a(1 + K_v) + G_b + G_d]} \qquad 6.30$$

The expression for V_i' is substituted into Eq. 6.28, which is then put into Eq. 6.18 for R_{in}'. After some algebra, assuming that $K_v G_a \gg G_d$, we find

$$R_{in}' = R_d\left(1 + \frac{K_v R_b}{R_b + R_a}\right) = R_d(1 - A_L) = R_d(\text{RD}) \qquad 6.31$$

where it is obvious that $\beta = R_b/(R_b + R_a)$, and the loop gain is

$$A_L = -K_v\beta \qquad 6.32$$

Note that the effective input resistance is very large in this example; R_{in} is increased by a factor of the return difference by NVF. The situation is not that simple, however. Refer to Chapter 5, Sec. 5.8 and Fig. 5.18. Because the differential amplifier has a symmetrical input stage, its input resistance is more accurately modeled by the three resistors: R_d between the input nodes and the resistors $2R_{ic}$ from each input

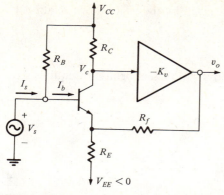

Figure 6.4 BJT headstage with NVF. $-K_v$ is a gain block derived from one or more subsequent BJT stages, or from an IC amplifier.

node to ground. Although NVF was seen to give a large R'_{in} in the preceding case, V_s actually sees the R'_{in} given by Eq. 6.31 in parallel with a resistor $2R_{ic}$ to ground. The $2R_{ic}$ resistor is not changed by NVF in this case. ($2R_{ic}$ is generally very large [about $10^9 \, \Omega$] in modern op-amps.)

EXAMPLE 6.1

Effect of Negative Voltage Feedback on R_{in}

As an example of how NVF can affect input resistance, consider the circuit of Fig. 6.4. First note that R_b will be in parallel with whatever small-signal resistance is seen looking into the BJT's base, regardless of the amount of NVF. The MFSSM for the circuit of Fig. 6.4 is shown in Fig. 6.5.

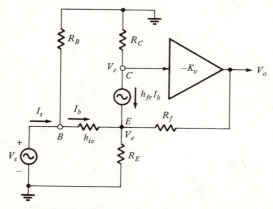

Figure 6.5 MFSSM of the feedback circuit of Fig. 6.4.

For zero feedback ($G_f = 0$, $\beta = 0$), R_{in} is given by the well-known relation

$$R_{in} = R_B \,\|\, [h_{ie} + R_E(1 + h_{fe})] \qquad\qquad 6.33$$

When NVF is applied through a finite R_f, analysis of system behavior requires writing a node equation for V_e:

$$V_e(G_E + g_{ie} + G_f) - V_oG_f - V_sg_{ie} - h_{fe}I_b = 0 \qquad\qquad 6.34$$

V_o is given by

$$V_o = -h_{fe}I_b(-K_v)R_C \qquad\qquad 6.35$$

and I_b may be written as

$$I_b = (V_s - V_e)g_{ie} \qquad\qquad 6.36$$

where it is understood that $g_{ie} = 1/h_{ie}$. When Eq. 6.36 is substituted into Eqs. 6.34 and 6.35, and Eq. 6.35 is in turn substituted into Eq. 6.34, we can solve for V_e in terms of V_s:

$$\frac{V_e}{V_s} = \frac{g_{ie}[1 + h_{fe}(1 + R_CK_vG_f)]}{\{g_{ie}[1 + h_{fe}(1 + R_CK_vG_f)] + G_E + G_f\}} \qquad\qquad 6.37$$

Now the closed-loop gain for this system is found by substituting Eq. 6.37 for V_e into Eq. 6.36, and then Eq. 6.36 for I_b into Eq. 6.35:

$$\frac{V_o}{V_s} = A_v = \frac{g_{ie}h_{fe}R_CK_v}{1 + \dfrac{g_{ie}}{G_E + G_f}[1 + h_{fe}(1 + R_CK_vG_f)]} \qquad\qquad 6.38$$

If K_v is very large, Eq. 6.38 reduces to

$$\frac{V_o}{V_s} = 1 + \frac{R_f}{R_E} \qquad\qquad 6.39$$

The input resistance is now easily found from Eqs. 6.36 and 6.37:

$$R'_{in} = R_B \left\|\, \frac{V_s}{I_b} \right. \qquad\qquad 6.40$$

$$R'_{in} = R_B \left\|\, \left\langle h_{ie}\left\{1 + \frac{g_{ie}}{G_E + G_f}[1 + h_{fe}(1 + R_CK_vG_f)]\right\}\right\rangle \right. \qquad\qquad 6.41$$

That is, looking into the BJT's base, V_s sees h_{ie} multiplied by the RD factor. If K_v is very large, this reduces to

$$R'_{in} = R_B \left\| \left\{ h_{ie} \left[1 + \left(\frac{R_E}{R_E + R_f} \right) \left(\frac{h_{fe}R_C}{h_{ie}} \right) K_v \right] \right\} \right. \qquad 6.42$$

in which the feedback attenuation is easily seen to be

$$\beta = \frac{R_E}{R_E + R_f} \qquad 6.43$$

and the loop gain is approximately

$$A_L = -\left(\frac{R_E}{R_E + R_f} \right) \left(\frac{h_{fe}R_C}{h_{ie}} \right) K_v \qquad 6.44$$

Note that $h_{fe}R_C/h_{ie}$ is the gain magnitude of the BJT amplifier ($|V_c/V_s|$) with $R_E = 0$. ●

In summary, we have seen that NVF can either lower the effective input resistance of an amplifier or raise it; the result depends on the circuit geometry. One should consider each case individually when investigating this effect.

6.2.3 Reduction of Total Harmonic Distortion

NVF has the very useful property of reducing the total harmonic distortion (THD) in power amplifiers and other signal-conditioning systems. To illustrate how this happens, we represent an amplifier with distortion by the signal flow graph model in Fig. 6.6(A). First, we let $\beta = 0$ and pick a V_s such that

$$V_o = V_s \alpha K_v + V_d = V_{os} + V_{od}(V_{os}) \qquad 6.45$$

Next, we allow $\beta > 0$ and readjust V_s to a larger value, V'_s, such that $V'_{os} = V_{os}$. These changes yield the same THD, so $V'_{od}(V'_{os}) = V_{od}(V_{os})$. It is easy to see that the larger V'_s required with NVF is

$$V'_s = V_s(1 + \beta K_v) = V_s(1 - A_L) = V_s(RD) \qquad 6.46$$

From the signal flow graph of Fig. 6.6(A), the THD with NVF is

$$\text{THD}' = V'_{od} = \frac{V_d(v_{os})}{(1 + \beta K_v)} = \frac{V_d(V_{os})}{[RD]} \qquad 6.47$$

Thus we see that for the same output signal component, V_{os}, the THD of the amplifier with NVF is reduced by a factor of $1/RD$. Of course, we pay for the reduction

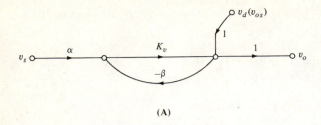

(A)

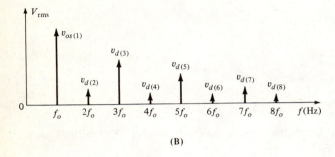

(B)

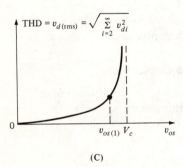

(C)

Figure 6.6 **(A)** Signal flow graph representation of an amplifier with NVF and distortion. v_d is the rms THD introduced by nonlinearities in the output stage. **(B)** Output root spectrum of a distorting amplifier with a pure sinusoidal input voltage with frequency, f_o. Note that an infinite sequence of harmonic components is generated as the result of the output distortion. **(C)** Typical relation between the rms THD (v_d) and the rms signal component at the output, v_{os}. Note that v_d increases abruptly as $v_{os} \rightarrow v_c$, the output level at which hard clipping occurs in the output power stage, creating a severely distorted sinusoidal output.

of THD by having less overall signal voltage gain or, equivalently, requiring a larger input, V'_s, to get the same output signal.

An important design principle for power amplifiers emerges from this analysis:

Always design the amplifier so that the output stage clips or reaches saturation first.

This design principle insures that the distortion products enter the loop at the output node and therefore are attenuated by 1/RD. The input stage and intermediate stages should be designed to be as distortion-free (linear) as possible, even through the use of local NVF loops around each of the stages to reduce their (local) THD. Of course, the input stage normally works at low signal levels, so its distortion will be minimal compared to that for the output stage.

6.2.4. Improvement in Gain Sensitivity

NVF can be shown to improve gain sensitivity, the variation in closed-loop gain resulting from changes in amplifier circuit component values. The closed-loop gain of the NVF amplifier of Fig. 6.6(A) (with $V_d = 0$) is

$$\frac{V_o}{V_s} = A_v = \frac{\alpha K_v}{(1 + \beta K_v)} \qquad 6.48$$

The sensitivity of A_v with respect to the gain K_v is defined as

$$S_{K_v}^{A_v} \triangleq \frac{\Delta A_v/A_v}{\Delta K_v/K_v} \qquad 6.49$$

To find an analytical expression for $S_{K_v}^{A_v}$, we need to find an expression for ΔA_v. Accordingly, we note that

$$\Delta A_v = \left(\frac{\partial A_v}{\partial K_v}\right)\Delta K_v \qquad 6.50$$

where

$$\frac{\partial A_v}{\partial K_v} = \alpha \frac{[(1 + \beta K_v) - \beta K_v]}{(1 + \beta K_v)(1 + \beta K_v)} = \frac{\alpha K_v}{(1 + \beta K_v)}\frac{1/K_v}{[RD]} \qquad 6.51$$

so

$$\Delta A_v = A_v \frac{1/K_v}{[RD]} \Delta K_v \qquad 6.52$$

and

$$S_{K_v}^{A_v} = \frac{[A_v/K_v(\text{RD})](\Delta K_v/A_v)}{\Delta K_v/K_v} = \frac{1}{[\text{RD}]}$$ 6.53

From Eq. 6.53, we see that $S_{K_v}^{A_v}$ is equal to the reciprocal of the return difference. In the case of op-amps, $1/\text{RD}$ is very small, generally around 10^{-5}. Thus an op-amp circuit with NVF is generally quite insensitive to variations in its dc open-loop gain, K_v.

In many NVF circuits, we are interested in knowing how a sensitivity varies as a function of frequency. Complex algebra must be used in the calculation of such sensitivities, because the closed-loop gain is complex. Here we express the closed-loop gain in polar form for convenience:

$$\mathbf{A}_v(j\omega) = |\mathbf{A}_v(j\omega)|e^{j\theta(\omega)}$$ 6.54

The gain sensitivity, given by Eq. 6.49, can also be written as

$$\mathbf{S}_x^{A_v} = \frac{d\mathbf{A}_v/\mathbf{A}_v}{dx/x} = \frac{d(\ln \mathbf{A}_v)}{d(\ln x)}$$ 6.55

where x is a parameter in the NVF circuit. From Eq. 6.54, we can write

$$\ln \mathbf{A}_v(j\omega) = \ln|\mathbf{A}_v(j\omega)| + j\theta(\omega)$$ 6.56

Substitution of Eq. 6.56 into Eq. 6.55 yields

$$\mathbf{S}_x^{A_v} = \frac{d(\ln|\mathbf{A}_v(j\omega)|)}{dx/x} + j\frac{d\theta(\omega)}{dx/x}$$ 6.57

The real part of $\mathbf{S}_x^{A_v}$ is called the magnitude sensitivity; the imaginary part $\mathbf{S}_x^{A_v}$ is the phase sensitivity. Generally, x will be a real number, which simplifies the calculations of Eq. 6.57. Sensitivity analysis using Eq. 6.57 can be utilized to evaluate active filter designs.

6.2.5. Gain-Bandwidth Product

A trade-off between low- or mid-frequency gain and bandwidth is a general property of amplifiers with NVF. Stated mathematically, closed-loop gain times bandwidth equals (approximately, in most cases) a constant. This aspect of NVF amplifier behavior is widely used to extend the high- (and sometimes the low-) frequency response of various gain stages and IC amplifiers. We will illustrate this principle using two conventional op-amp circuits.

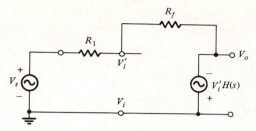

Figure 6.7 Simple op-amp inverter circuit.

Figure 6.7 is the schematic of a simplified op-amp inverter circuit in which the op-amp is frequency compensated so that its open-loop gain is given by

$$\frac{V_o}{V_i - V_i'} = H_a(s) = \frac{K_v}{1 + \tau_a s} \tag{6.58}$$

To find the gain-bandwidth product for this circuit, we need to find the circuit's closed-loop gain, $A_v(s) = V_o/V_s$. A node equation is written on the V_i' node:

$$V_i'(G_1 + G_f) - V_o G_f = V_s G_s \tag{6.59}$$

Noting that the V_i input is grounded, we can write V_i' from Eq. 6.58 and substitute into Eq. 6.59 to obtain

$$V_o\left[-\left(\frac{\tau_a s + 1}{K_v}\right)(G_1 + G_f) - G_f\right] = V_s G_s \tag{6.60}$$

Equation 6.60 is solved for the closed-loop gain in time-constant form:

$$\frac{V_o}{V_s} = A_v(s) = \frac{\dfrac{-K_v R_f}{[R_f + R_1(1 + K_v)]}}{\left[s\,\dfrac{\tau_a(R_f + R_1)}{R_f + R_1(1 + K_v)} + 1\right]} \tag{6.61}$$

The low- and mid-frequency gain is

$$|A_v(0)| = \frac{K_v R_f}{[R_f + R_1(1 + K_v)]} \tag{6.62}$$

and the frequency in hertz, f_o, where $|A_v(j2\pi f_o)|$ is down 3 dB from its low-frequency value $(0.707|A_v(0)|)$, is given by

$$f_o = \frac{1}{2\pi}\frac{R_f + R_1(1 + K_v)}{\tau_a(R_f + R_1)} \tag{6.63}$$

Because the closed-loop amplifier's bandwidth extends to dc, f_o is, in this case, its bandwidth. Thus the closed-loop gain-bandwidth product for the simple op-amp inverter is

$$
\text{GBWP}_c = \frac{K_v R_f}{[R_f + R_1(1 + K_v)]} \frac{[R_f + R_1(1 + K_v)]}{\tau_a(R_f + R_1)2\pi} \text{ Hz}
$$

$$
= \frac{K_v}{2\pi\tau_a} \frac{R_f}{(R_f + R_1)} = (\text{GBWP}_a)\alpha \tag{6.64}
$$

Inspection of Eq. 6.64 shows that the closed-loop amplifier's GBWP_c can be expressed as the product of the gain-bandwidth product of the open-loop amplifier GBWP_a, and α, the feed-forward attenuation introduced in Fig. 6.1 and Eq. 6.8. As an example, if the op-amp inverter circuit of Fig. 6.7 is given a gain of -1, $|A_v(0)| = 1$, $\alpha = R/2R = \frac{1}{2}$, and the op-amp itself has a specified gain-bandwidth product (GBWP_a) of 12 MHz, then the -3 dB frequency (BW_1) for the unity-gain system will be, from Eq. 6.64,

$$
\text{BW}_1 = \frac{(\text{GBWP}_a)\alpha}{|A_v(0)|} = \frac{(12)(1/2)}{|-1|} = 6 \text{ MHz} \tag{6.65}
$$

As $|A_v(0)|$ is increased, $\alpha \rightarrow 1$, and GBWP_c of the closed-loop system approaches that of the op-amp, GBWP_a.

EXAMPLE 6.2

Amplifier with a Constant Gain-Bandwidth Product

Now consider a noninverting op-amp circuit, shown in Fig. 6.8. Here, as in the previous example, Eq. 6.58 describes the op-amp's open-loop gain as a function of frequency. The closed-loop gain is obtained from

$$
V_o = (V_s - \beta V_o) \frac{K_v}{\tau_a s + 1} \tag{6.66}
$$

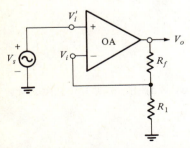

Figure 6.8   Noninverting op-amp circuit.

where

$$\beta = \frac{R_1}{R_1 + R_f}$$

6.67

Solving for $A_v(s)$, we find

$$\frac{V_o}{V_s} = A_v(s) = \frac{\dfrac{K_v}{(1 + \beta K_v)}}{\left[s\left(\dfrac{\tau_a}{1 + \beta K_v} \right) + 1 \right]}$$

6.68

The gain-bandwidth product for this NVF amplifier circuit is written by inspection of Eq. 6.68:

$$\text{GBWP}_c = \frac{K_v}{1 + \beta K_v} \frac{1 + \beta K_v}{2\pi\tau_a} = \text{GBWP}_a$$

6.69

In the example just presented, the closed-loop GBWP_c is exactly equal to the op-amp's GBWP_a, regardless of the closed-loop gain. This is found to be a special case of the constancy of the ideal gain-bandwidth product. In general, there is a reduction of mid-band gain along with a broadening of bandwidth as the amount of NVF is increased. As will be seen, if the open-loop amplifier has two or more poles, the gain-bandwidth trade-off soon breaks down at high loop gain values.

EXAMPLE 6.3

Feedback Amplifier with a Constant Bandwidth

As a further example of the gain-bandwidth relations in operational amplifiers, we consider a unique system designed and patented by Comlinear Corporation to avoid the gain-bandwidth trade-off in the design of video-frequency op-amps. The Comlinear op-amp architecture is shown in Fig. 6.9. Of special interest are the broadband, unity-gain buffer amplifier between the noninverting and the inverting op-amp inputs and the use of a current-controlled voltage source (CCVS) to provide open-loop gain. We observe that at the V_i' node,

$$V_i' = V_i$$

6.70

Also,

$$I_d = I_1 - I_f$$

6.71

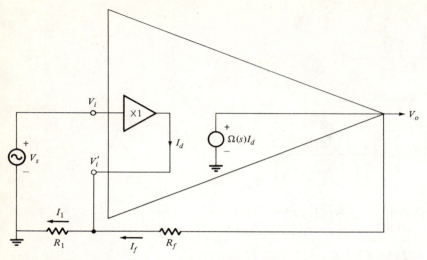

Figure 6.9 Comlinear op-amp connected as a noninverting voltage amplifier. $\Omega(s)$ is a current-controlled voltage source (transimpedance).

and, by Ohm's law,

$$I_f = (V_o - V_i)G_f \qquad\qquad 6.72$$

and

$$I_1 = V_iG_1 \qquad\qquad 6.73$$

so

$$V_o = \Omega I_d = \Omega(I_1 - I_f) = \Omega[V_iG_f - G_f(V_o - V_i)] \qquad\qquad 6.74$$

Noting that $V_i = V_i' = V_s$, we solve Eq. 6.74 for the noninverting gain, A_v:

$$A_v = \frac{V_o}{V_s} = \frac{\Omega(G_1 + G_f)}{(1 + \Omega G_f)} \qquad\qquad 6.75$$

It is seen that $A_v \to (1 + R_f/R_1)$ as $\Omega \to \infty$ in Eq. 6.75. Now let the transresistance, Ω, have a single real-pole transfer function of the form

$$\Omega(s) = \frac{\Omega_0}{1 + \tau s} \; \Omega \qquad\qquad 6.76$$

If we substitute Eq. 6.76 into Eq. 6.75 and rearrange terms to put the gain in time-constant form, we get

$$A_v(s) = \frac{V_o}{V_s} = \frac{\dfrac{\Omega_0(G_1 + G_f)}{(1 + \Omega_0 G_f)}}{\left(s\dfrac{\tau}{1 + \Omega_0 G_f} + 1\right)} \qquad\qquad 6.77$$

The corner frequency of the closed-loop, noninverting Comlinear op-amp circuit is just

$$\omega_0 = \frac{(1 + G_f\Omega_0)}{\tau} \text{ rad/s} \qquad\qquad 6.78$$

The value of ω_0 depends on the dc transresistance, Ω_0, and the feedback conductance, G_f. It is independent of G_1. Hence G_f can be made large and fixed, and the dc gain, $A_v(0)$, can be raised with G_1 without reducing ω_0, eliminating the gain-bandwidth trade-off we have seen in conventional op-amp circuits. (The foregoing analysis assumed that the unity-gain buffer amplifier has a bandwidth far in excess of ω_0.) ●

6.3 Negative Current Feedback

In negative current feedback (NCF) amplifiers, a signal is fed back proportional to the output current through the load. A two-op-amp circuit using NCF is shown in Fig. 6.10. Assume op-amp 2 is ideal. It is easy to show that the voltage at its output,

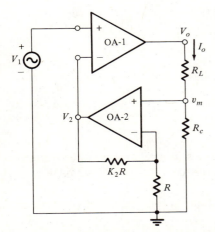

Figure 6.10   NCF system. Op-amp 1 provides the controlled current, I_o; op-amp 2 is used in a noninverting voltage amplifier configuration.

V_2, is given by

$$V_2 = I_o R_c (1 + K_2) \qquad 6.79$$

Assume op-amp 1 has a finite scalar differential gain, $K_{v(1)}$. An expression for I_o can be written using Eq. 6.79 and Ohm's law:

$$I_o = \frac{V_o}{R_L + R_c} = \frac{K_{v(1)}[V_1 - I_o R_c (1 + K_2)]}{(R_L + R_c)} \qquad 6.80$$

Solving for the system's transconductance, G_m, we find

$$G_m = \frac{I_o}{V_1} = \frac{K_{v(1)}}{R_L + R_c (1 + K_{v(1)} + K_{v(1)} K_2)} \qquad 6.81$$

If $K_{v(1)}$ is very large,

$$G_m = \frac{1}{R_c (1 + K_2)} \qquad 6.82$$

To find the effect of NCF on the effective source impedance seen by $R_L + R_c$, we treat the output nodes, V_o and V_m, as a Thevenin equivalent port. From Eq. 6.81, we find the short-circuit output current, I_{osc}:

$$I_{osc}\Big|_{R_L = 0} = \frac{K_{v(1)} V_1}{R_c (1 + K_{v(1)} + K_{v(1)} K_2)} \qquad 6.83$$

The open-circuit voltage across the output port is just

$$V_{ooc}\Big|_{R_L = \infty, I_o = 0} = V_1 K_{v(1)} \qquad 6.84$$

The Thevenin output resistance that RL sees is thus

$$R_{out} = \frac{V_{ooc}}{I_{osc}} = R_c + R_c K_{v(1)} (1 + K_2) \qquad 6.85$$

If $R_c = 1\ \Omega$, $K_{v(1)} = 10^6$, and $K_2 = 99$, then RL sees a very large source resistance (about $10^8\ \Omega$), indicating a good, constant current source. Thus we see that NCF raises output impedance, whereas NVF lowers it.

How does NCF affect gain-bandwidth relations in this circuit? We assume the gain of op-amp 1 in Fig. 6.10 is given by

$$\frac{V_o}{V_1 - V_2} = K_{v(1)}(s) = \frac{K_{v(1)o}}{1 + \tau s} \qquad 6.86$$

and that op-amp 2 has infinite (ideal) bandwidth. Substitution of this gain for $K_{v(1)}$ in Eq. 6.81 and expression in time-constant form yields the frequency-dependent transconductance:

$$G_m(s) = \frac{I_o}{v_1} = \frac{\left(\dfrac{K_{v(1)o}}{R_L + R_c[1 + K_{v(1)o}(1 + K_2)]}\right)}{\left\{s\dfrac{\tau(R_L + R_c)}{R_L + R_c[1 + K_{v(1)o}(1 + K_2)]} + 1\right\}} \qquad 6.87$$

The dc transconductance, $G_m(0)$, is independent of R_L for large $K_{v(1)o}$ and may be raised by making G_c large; it is unaffected by increases in $K_{v(1)o}$ and K_2 as long as $K_{v(1)o}K_2 \gg 1$. The break frequency of $G_m(j\omega)$ is approximately

$$\omega_0 \cong \frac{R_c K_{v(1)o}(1 + K_2)}{\tau_1(R_L + R_c)} \text{ r/s} \qquad 6.88$$

and obviously depends on the load, R_L, as well as on the current-measuring resistor, R_c. Note that increasing the loop gain of the NCF circuit by increasing K_2 or $K_{v(1)o}$ will increase this system's closed-loop bandwidth, ω_0.

EXAMPLE 6.4

Amplifier with Negative Current Feedback

Another example of an NCF amplifier is shown in Fig. 6.11. This system is a high-current driver for an electron beam deflection coil. Analysis is begun by noting that the complementary symmetry power amplifier has a dc small-signal gain of v_o/v_2, which is slightly lower than unity. Hence the gain from the compensated power op-amp's (OA-1) inverting input to the v_o node can be written as

$$\frac{V_o}{V_i'} = \frac{-K_{v(1)}}{1 + \tau_1 s} \qquad 6.89$$

The controlled output current, I_o, is given by Ohm's law:

$$I_o = \frac{V_o}{sL + R_L + R_c} = \frac{\dfrac{V_o}{R_L + R_c}}{\left(s\dfrac{L}{R_L + R_c} + 1\right)} \qquad 6.90$$

The feedback voltage proportional to I_o is

$$V_c = I_o R_c \qquad 6.91$$

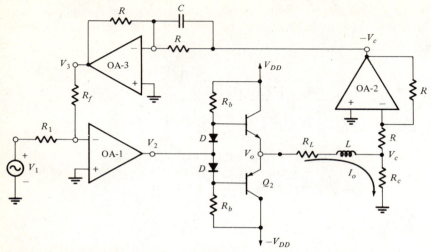

Figure 6.11 Power amplifier with NCF. Q_1 and Q_2 are power transistors; diodes, D, are for class B biasing. Because $R \gg R_c$, the V_c node can be considered to be a voltage source. Op-amp 1 is a high-voltage, high-slew-rate op-amp used to drive the $Q_1 - Q_2$ power stage. Op-amps 2 and 3 are conventional high-performance op-amps; op-amp 3 is used to put a zero in the loop gain.

Op-amp 2 is a unity-gain inverter; its bandwidth is considered infinite compared to other system break frequencies. Op-amp 3 is used to insert a zero in the loop gain for purposes of high-frequency compensation. Therefore,

$$\frac{V_3}{-V_c} = \frac{-R}{\left(\dfrac{1}{G + sC}\right)} = -(1 + sRC) \qquad 6.92$$

Equations 6.89, through 6.92 and the node equation on V_1' can be combined in a signal flow graph, shown in Fig. 6.12. Mason's rule is used to find the closed-loop system's transconductance, $G_m(s)$:

$$G_m(s) = \frac{I_o}{V_1} = \frac{-\alpha\left(\dfrac{K_{v(1)}}{\tau_1 s + 1}\right)\left(\dfrac{1/(R_L + R_c)}{\tau_L s + 1}\right)}{\left\{1 - \left[\left(\dfrac{-K_{v(1)}}{\tau_1 s + 1}\right)\left(\dfrac{1/(R_L + R_c)}{\tau_L s + 1}\right)R_c(-1)[-(1 + sRC)]\beta\right]\right\}} \qquad 6.93$$

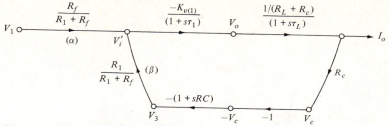

Figure 6.12   Signal flow graph representation of the NCF power amplifier circuit of Fig. 6.11.

We can put Eq. 6.93 into standard time-constant form, noting that $\tau_L = L/(R_L + R_c)$:

$$G_m(s) = \frac{I_o}{V_1} = \frac{-\dfrac{\alpha K_{v(1)}/(R_L + R_c)}{[1 + K_{v(1)}\beta R_c/(R_L + R_c)]}}{\left\{ s^2 \dfrac{\tau_1 \tau_L}{1 + K_{v(1)}\beta R_c/(R_L + R_c)} + s \dfrac{[\tau_1 + \tau_L + K_{v(1)}\beta R C R_c/(R_L + R_c)]}{1 + K_{v(1)}\beta R_c/(R_L + R_c)} + 1 \right\}}$$

$$6.94$$

From this rather algebraically intense expression, we note that as $K_{v(1)} \to \infty$, the dc transconductance (numerator of Eq. 6.94) approaches

$$G_m(0) = -G_c \frac{\alpha}{\beta}$$

$$6.95$$

(Do not worry about the minus sign in Eqs. 6.94 and 6.95; it simply means that I_o flows into the V_o node when $V_1 > 0$.) The system's break frequency, ω_n, is

$$\omega_n = \sqrt{\frac{1 + K_{v(1)}\beta R_c/(R_L + R_c)}{\tau_1 \tau_L}} \text{ rad/s}$$

$$6.96$$

and obviously increases with $K_{v(1)}\beta$. Opening the loop by setting $R_c = 0$ reduces the system to two slow real poles at $\omega_1 = 1/\tau_1$ and $\omega_2 = 1/\tau_L$.

Because of potential instability of the closed-loop NCF system, design of this system requires a wise choice of dc loop gain:

$$A_L(0) = -\frac{K_{v(1)}\beta R_c}{(R_L + R_c)}$$

$$6.97$$

Pole positions for the closed-loop system may be found from root-locus techniques, discussed in detail in Chapter 7. The root-locus diagram for this amplifier is shown in Fig 6.13. It gives the loci of all possible closed-loop pole positions as $A_L(0)$ is varied from zero to minus infinity.   ●

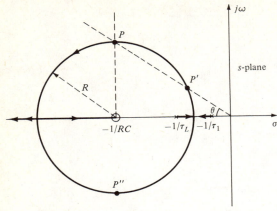

Figure 6.13   Root-locus diagram for the NCF power amplifier of Figs. 6.11 and 6.12. The closed-loop system's poles (poles of $G_m(s)$) start at $-1/\tau_1$ and $-1/\tau_L$ when $A_L(0) = 0$, and move symmetrically around the real axis in the s-plane in the direction of the arrows on the locus branches as $A_L(0) \to -\infty$. A specific value of K_v can be found so that the closed-loop system's poles lie at P and P'', giving the desired closed-loop system transient response.

In summary, we see that NCF raises the Z_{out} of amplifiers markedly, making them appear like current sources to the loads they drive. NCF acts like NVF in that it extends system bandwidth.

6.4 Positive Voltage Feedback

Positive voltage feedback (PVF) generally has the opposite effect of NVF on amplifiers, because the return difference,

$$\textbf{RD} = [1 - \textbf{A}_L^{'}] \qquad\qquad 6.98$$

has a positive A_L term:

$$A_L = +\beta K_v \qquad\qquad 6.99$$

And as $0 \to \beta \to K_v^{-1}$, $1 \to \text{RD} \to 0$. Zero RD generally results in amplifier instability, as we will see in Chapter 7.

Where, then, is PVF used if its use tends to reverse the beneficial effects of NVF and leads to stability problems? One important use of PVF is in capacitance neutralization, a technique used to reduce the input capacitance (C_i) of amplifiers. The reduction of C_i raises the amplifier high-frequency input impedance. Another use of PVF is in several classes of nonlinear switching circuits to introduce instability and increase switching speed purposely. Although not covered in this text, flip-flops, one-shot multivibrators, and astable multivibrators generally use positive feedback in their designs. Capacitance neutralization is studied in the next example.

EXAMPLE 6.5

Amplifier with Capacitance Neutralization

Figure 6.14 shows a noninverting op-amp circuit that uses both NVF and PVF. If $C_N = 0$ (no PVF), then the circuit's gain is given by

$$\frac{V_o}{V_s} = A_v(s) = \frac{1}{(sR_sC_1 + 1)} \frac{\dfrac{K_v}{(1 + \beta K_v)}}{\left[s\dfrac{\tau_a}{(1 + \beta K_v)} + 1 \right]} \qquad 6.100$$

Note that a second real pole occurs as the result of the low-pass filter formed by the source resistance, R_s, and the input capacitance, C_i.

Now consider the V_i node with PVF through CN. The node equation is

$$V_i(sC_i + sO_N + G_s) - V_osC_N = V_sG_s \qquad 6.101$$

Equation 6.101 is solved for V_i and substituted into

$$V_o = (V_i - V_i')\frac{K_v}{(\tau_a s + 1)} \qquad 6.102$$

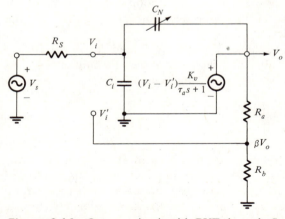

Figure 6.14 Op-amp circuit with PVF through C_N and NVF through the voltage divider, R_a and R_b. C_i is the input capacitance of the amplifier's noninverting input. R_S is the Thevenin source resistance associated with V_s.

to obtain, after some algebra, the closed-loop gain

$$A_v(s) = \frac{V_o}{V_s} = \frac{\dfrac{K_v}{(1 + \beta K_v)}}{\left\{ s^2 \dfrac{\tau_a \tau_s}{(1 + \beta K_v)} + s \dfrac{[\tau_a + \tau_s - C_N R_s K_v + \beta K_v \tau_s]}{(1 + \beta K_v)} + 1 \right\}} \qquad 6.103$$

As $K_v \to \infty$, Eq. 6.103 reduces to

$$A_v(s) = \frac{V_o}{V_s} \cong \frac{(1/\beta)}{\left\{ s^2 \dfrac{\tau_a \tau_s}{\beta K_v} + s \dfrac{[\beta \tau_s - C_N R_s]}{\beta} + 1 \right\}} \qquad 6.104$$

The gain expression 6.104 has the general quadratic low-pass format

$$A_v(s) = \frac{A_{vo}}{\left\{ \dfrac{s^2}{\omega_n^2} + s \dfrac{2\xi}{\omega_n} + 1 \right\}} \qquad 6.105$$

where

A_{vo} = the closed-loop dc gain
ω_n = the break frequency in rad/s
ξ = the damping factor

It is evident that when $CN = \beta \tau_s / Rs$, then $\xi = 0$ and the closed-loop system's poles lie on the $j\omega$ axis. This configuration means that the closed-loop system is an oscillator (unstable). Obviously, the value of C_N must lie between zero and $\beta \tau_s / R_s$ F in order for the amplifier to be useful (stable).

Let us examine the input admittance, Y_{in}, seen by V_s and R_s. A fixed C_i to ground appears in parallel with the dynamic capacitance seen looking into C_N. At low frequencies, it is not difficult to show that Y_{in} is given by

$$\mathbf{Y}_{in} = sC_i + s\left(\frac{-R_a}{R_b}\right)C_N \text{ S} \qquad 6.106$$

When $C_N = C_i R_b / R_a$, then $\mathbf{Y}_{in} = 0$ and neutralization is ideal. At this C_N value, however, the $(\beta \tau_s - C_N R_s)/\beta$ term in Eq. 6.104 is zero, implying amplifier instability! Clearly it is not possible to obtain complete neutralization (i.e., a 100% cancellation of input capacitance) with PVF because of stability problems. A feedback technique known as bootstrapping has proven to be far more successful in reducing amplifier input capacitance. Bootstrapping is illustrated in Chapter 4, Fig. 4.12(C) and (D).

●

SUMMARY

Most electronic amplifier circuits use negative voltage feedback, which has the following properties:

1. Reduces mid-band gain.
2. Extends bandwidth.
3. Decreases total harmonic distortion at a given power output level.
4. Decreases R_{out}.
5. Can either decrease or increase R_{in}, depending on the circuit.
6. Decreases gain sensitivity to certain circuit parameters.
7. Can lead to instability if applied in excess.
8. Has little effect on the output signal-to-noise ratio.

Negative current feedback has all of the properties just listed except item 4; R_{out} is made very large with NCF.

Positive voltage feedback has many undesirable properties when used on linear amplifiers; generally, it reverses the properties of NVF as well as presents problems with amplifier stability.

PROBLEMS

6.1 Negative feedback is combined with the input signal, V_1, at the source of a JFET. Use mid-frequency small-signal analysis to find the mixing gains K_1 and K_2 in the relation

$$V_2 = K_1 V_1 + K_2 V_o$$

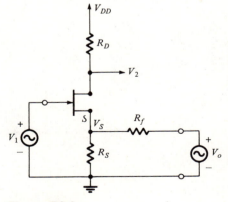

Figure P6.1

6.2 Repeat Problem 6.1 for the BJT circuit shown in Fig. P6.2. Let $h_{re} = h_{oe} = 0$ in the BJT's common-emitter, h-parameter MFSSM.

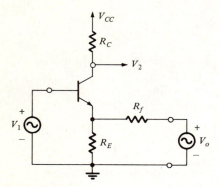

Figure P6.2

6.3 A certain power op-amp (POA) has an open-loop gain of $V_o/V_i' = -5 \times 10^4$. It is connected as a gain of -250 amplifier as shown in Fig. P6.3. When the sinusoidal fundamental component of the output voltage is 28 V peak-to-peak, the THD is measured to be 0.5% of the rms fundamental output voltage, V_{os}. Find the percent THD when the circuit is given unity (inverting) gain and V_{os} is again 28 V peak-to-peak.

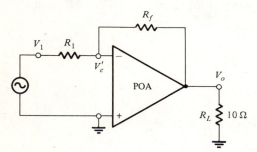

Figure P6.3

6.4 Negative current feedback is used to make a voltage-controlled current sink. Assume:

FET: (μ, r_d)

Op-amp: $V_2 = K_v(V_1 - V_s)$

(a) Find the system transconductance, $G_m = I_2/V_1$.
(b) Find the VCCS's (Norton) output resistance.

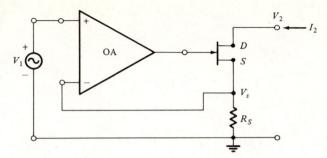

Figure P6.4

6.5 The op-amp circuit shown in Fig. P6.5 is a form of *current mirror*. Assume the following:

BJT: $h_{oe} = h_{re} = 0$; use MFSSM

Op-amp: $V_1 = \dfrac{K_v}{\tau s + 1}(V_i - V_i')$

Find an expression for $(I_2/I_1)(s)$ in time-constant form. (*Note:* I_2 is the output short-circuit current.)

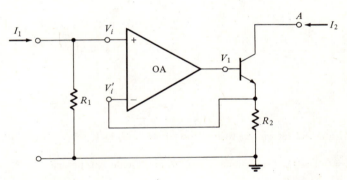

Figure P6.5

6.6 Two op-amps are used to make constant-current supply for an impedance Z_L. Op-amp 1 is nonideal with open-loop gain $V_2/V_i = +K_{v(1)}$. Op-amp 2 is ideal. See Fig. P6.6.
  (a) Derive an algebraic expression for the system's transconductance, $G_m = I_L/v_1$. Show what happens as $K_{v(1)} \to \infty$.
  (b) Find an expression for the output resistance Z_L sees looking into the output of op-amp 1 under closed-loop conditions (let $K_{v(1)}$ be finite).

6.7 A power VMOSFET is used with a differential amplifier to make a voltage-regulated dc power supply. See Fig. P6.7. Assume $V_z = 6.8$ V (dc voltage source), $g_m = 2 \times 10^{-3}$ S, $r_d = 2 \times 10^4 \ \Omega$, $K_v = 10^4$, $R_1 = 10^4 \ \Omega$, $R_f = 1.2 \times 10^4 \ \Omega$, and $R_s = 100 \ \Omega$.

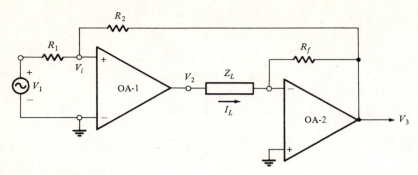

Figure P6.6

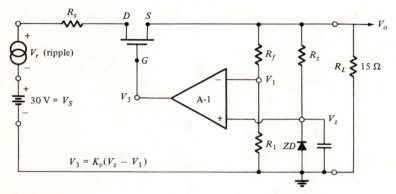

Figure P6.7

(a) Use the MFSSM of the circuit to find the ripple gain, V_o/V_r, of the regulator. Use $R_L = 15\ \Omega$.

(b) Assume $V_r = 0$. Use the small-signal model to find the dc V_o (open-circuit, $R_L = \infty$).

(c) Find R_{out} that R_L sees.

6.8 An ideal op-amp is used to make a VCCS.

(a) Find an expression for $I_L = f(V_1)$.

(b) What conditions must exist on resistors R_1, R_2, R_3, and R_4 for I_L to be independent of Z_L?

6.9 The sensitivity function for the corner (-3 dB) frequency, ω_0, in the op-amp circuit shown in Fig. P6.9 is given by

$$S_{K_v}^{\omega_0} = \frac{\Delta\omega_0/\omega_0}{\Delta K_v/K_v} = \frac{d\omega_0}{dK_v}\frac{K_v}{\omega_0}$$

Assume $V_o = -K_v V_i'$. Find an expression for $S_{K_v}^{\omega_0}$ in terms of circuit parameters.

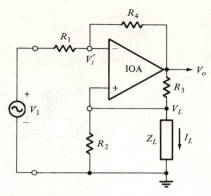

Figure P6.8

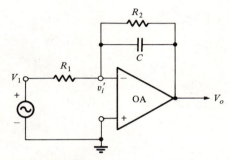

Figure P6.9

6.10 Positive current feedback and negative voltage feedback are used with a power op-amp having a differential voltage gain

$$V_o = \frac{K_v}{\tau s + 1} (V_i - V'_i)$$

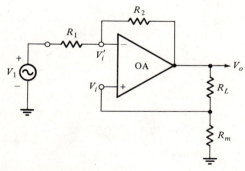

Figure P6.10

Assume $(R_1 + R_2) \gg (R_L + R_m)$, $\alpha = R_2/(R_1 + R_2)$, $\beta = R_1/(R_1 + R_2)$, and $\gamma = R_m/(R_L + R_m)$.

(a) Assume the op-amp has $R_{out} = 0$. Find an expression for V_o/V_1 in time-constant form. Comment on how PCF affects the dc gain, corner frequency, and gain-band-width product.

(b) Now let the op-amp have a finite R_{out}. Use a test source, V_t, in place of R_L to find an expression for the Thevenin output resistance R_L sees.

6.11 Find an algebraic expression for the voltage gain, $K_v = V_o/V_1$, of the JFET negative voltage feedback amplifier. Assume $Q_1 = Q_2$, $g_d = 0$, and $g_m \neq 0$. Give K_v for KA $\to \infty$.

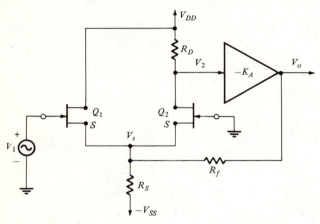

Figure P6.11

6.12 A power BJT is used to make a series-regulated, constant-voltage, dc power supply. Let $h_{fe} = 10$, $h_{oe} = 1 \times 10^{-5}$ S, $h_{re} = 0$, $h_{ie} = 1.5$ kΩ, DA gain $A_D = 1 \times 10^4$, $R_s = 100$ Ω, $R_L = 15$ Ω, $(R_F + R_1) \gg R_L$, $\alpha K_v \gg 1$, and $\alpha = 0.667$.

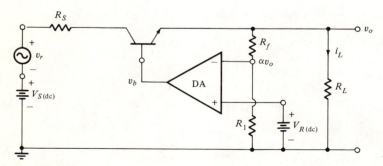

Figure P6.12

(a) Find an algebraic expression for, and evaluate numerically, the regulator's ripple gain, v_o/v_r, at mid-frequencies.

(b) Assume $h_{oe} = v_r = 0$. Find an algebraic expression for, and evaluate numerically, the regulator's Thevenin output resistance, R_{out}, that R_L sees.

(c) Find an expression for, and evaluate numerically, the regulator's regulation, $\Delta V_o/\Delta R_L$.

6.13 A differential amplifier and a power BJT are used to make a constant-current regulator. Negative current feedback is used. Assume the differential amplifier's $K_d = 1 \times 10^5$, $h_{oe} = h_{re} = 0$, $V_R = 5$ V, $h_{fe} = 19$, $h_{ie} = 1$ kΩ, $R_L = 10\ \Omega$, $R_m = 1\ \Omega$, and $R_S = 13\ \Omega$.

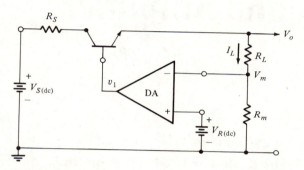

Figure P6.13

(a) Find an expression for, and evaluate numerically, the Thevenin output resistance R_L sees.

(b) Find an expression for, and evaluate numerically, the regulator's transconductance, $G_m = I_L/V_R$.

6.14 Find an expression for the sensitivity, $S_{K_v}^{G_m}$, of the VCCS of Problem 6.4.

6.15 Using the capacitance-neutralized amplifier schematic of Fig. 6.14, derive Eq. 6.103. Note that $\tau_s = R_s(C_1 + C_N)$.

Chapter 7

Feedback, Frequency Response, and Amplifier Stability

In this chapter, we will examine how feedback alters amplifier frequency response and see how negative voltage feedback, negative current feedback, or positive voltage feedback can lead to instability if applied incorrectly. However, practical use can be made of instability; tunable oscillators that generate sine waves can be designed using the same analytical techniques used in the design of linear feedback amplifiers.

   This chapter makes extensive use of the root-locus technique in the analysis and design of linear feedback amplifiers and oscillators. The root-locus technique allows an engineer to plot all possible locations of the closed-loop system's poles as a function of the dc loop gain, or mid-frequency gain if the system is reactively coupled (RC). Knowledge of the closed-loop system's poles and zeros allows the designer to predict the system's transient and frequency responses, as well as give information on stability. In the next section, we present a brief review of the rules for constructing root-locus plots.

7.1 Review of the Root-Locus Technique

Consider the general negative voltage feedback system signal flow graph shown in Fig. 7.1. The closed-loop gain of this NVF system is easily seen to be

$$\frac{V_o}{V_1} = A_v(s) = -\frac{\alpha K_v(s)}{1 + \beta(s)K_v(s)} = -\frac{\alpha K_v(s)}{1 + A_L(s)} = -\frac{\alpha K_v(s)}{RD} \tag{7.1}$$

A root-locus diagram is plotted in the s-plane using the poles and zeros of the system's loop gain, $A_L(s)$. It shows all complex s values for which

$$\mathbf{RD(s)} = 1 - \mathbf{A_L(s)} = 0 \tag{7.2}$$

as a function of the scalar portion of $\mathbf{A}_L(s)$. The graph of the points in the s-plane that satisfy Eq. 7.2 is also a plot of the locations of the poles of the closed-loop transfer function, $A_v(s)$, as the scalar part of $\mathbf{A}_L(s)$ is varied. The zeros of the closed-loop system are not predicted by the root-locus technique; they must be found by reducing the system's block diagram or signal flow graph to find the numerator of $A_v(s)$.

   To make a system's root-locus plot, we find all possible complex s values that satisfy the equations (note the minus signs)

$$-(\mathbf{A}_L(\mathbf{s})) = -1 = 1\underline{/-180°} \tag{7.3}$$

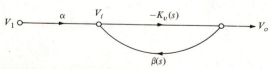

Figure 7.1 System with NVF.

For a NVF system, we have

$$-(-\beta(\mathbf{s})\mathbf{K}_v(\mathbf{s})) = 1\underline{/-180°} \tag{7.4}$$

In constructing and working with root-locus diagrams, it is a practical necessity to write the negative loop gain in vector form. For example, consider the loop gain

$$A_L(s) = \frac{-\beta_o K_{vo}(\tau_1 s + 1)}{(\tau_2 s + 1)(\tau_3 s + 1)} \tag{7.5}$$

Equation 7.5 is for a negative feedback system; it is in time-constant format. This loop gain can also be written in Laplace format:

$$-A_L(s) = \frac{\beta_o K_{vo}\tau_1}{\tau_2\tau_3} \frac{s + 1/\tau_1}{(s + 1/\tau_2)(s + 1/\tau_3)} \tag{7.6}$$

And finally it is put into vector format:

$$-\mathbf{A}_L(\mathbf{s}) = \frac{\beta_o K_{vo}\tau_1}{\tau_2\tau_3} \frac{\mathbf{s} - \mathbf{s}_1}{(\mathbf{s} - \mathbf{s}_2)(\mathbf{s} - \mathbf{s}_3)} \tag{7.7}$$

where

$$\mathbf{s}_1 = \frac{-1}{\tau_1} = -\omega_1 = \omega_1\underline{/-180°}$$

$$\mathbf{s}_2 = \frac{-1}{\tau_2} = -\omega_2 = \omega_2\underline{/-180°} \tag{7.8}$$

$$\mathbf{s}_3 = \frac{-1}{\tau_3} = -\omega_3 = \omega_3\underline{/-180°}$$

Figure 7.2 illustrates the graphical significance of Eq. 7.7. Systematic solution of Eq. 7.4 to plot the root loci requires satisfying the angle condition and the magnitude condition.

For a more general example,

$$-\mathbf{A}_L(\mathbf{s}) = -\left[\frac{(-\beta_o K_v')(\mathbf{s} - \mathbf{s}_1)(\mathbf{s} - \mathbf{s}_3)\cdots}{(\mathbf{s} - \mathbf{s}_2)(\mathbf{s} - \mathbf{s}_4)(\mathbf{s} - \mathbf{s}_6)\cdots}\right] = 1\underline{/-180°} \tag{7.9}$$

Thus,

$$\theta_1 = ARG(\mathbf{s} - \mathbf{s}_1) \tag{7.10}$$

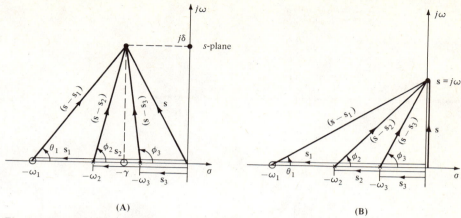

Figure 7.2   Vector interpretation of the negative loop gain, $-A_L(s)$.
(A) General case where $s = \gamma + j\delta$. **(B)** Frequency response case
where $s = j\omega$.

and

$$(\theta_1 + \theta_3 + \cdots) - (\phi_2 + \phi_4 + \phi_6 + \cdots) = -180° \qquad 7.11$$

$$\frac{|s - s_1||s - s_3| \cdots}{|s - s_2||s - s_4||s - s_6| \cdots} = \frac{1}{K_v'\beta_o} \qquad 7.12$$

All the root-locus plotting rules are derived from the angle criterion, Eq. 7.11, and the magnitude criterion, Eq. 7.12. Nine basic rules apply to constructing root-locus plots by hand:

1. *Number of Branches:* There is one branch for each pole of $A_L(s)$.
2. *Starting Points:* The locus branches start at the poles of $A_L(s)$ when $\beta_o K_v' = 0$.
3. *End Points:* The loci end at the zeros of $A_L(s)$ for $\beta_o K_v' = \infty$. Some of the zeros can be at $|s| = \infty$.
4. *Behavior of Loci on the Real Axis* (From the angle criterion, Eq. 7.11): Loci exist on the real axis of the s-plane for a negative feedback system $(0 \le \beta_o K_v' \le \infty)$ to the left of a total odd number of real, finite poles and zeros of $A_L(s)$. Conversely, if positive feedback is used $(0 \ge \beta_o K_v' \ge -\infty)$, the loci exist on the real axis to the left of a total even number of real, finite poles and zeros of $A_L(s)$. (Note that zero is an even number. See the examples in Fig. 7.3.)
5. *Symmetry of Points:* Loci are symmetrical around the real axis in the s-plane.

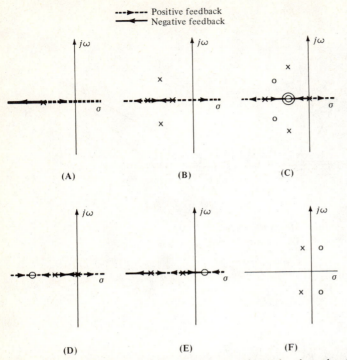

Figure 7.3   Examples of locus location on the s-plane's real axis for various pole and zero locations, for either negative feedback $(0 \leq \beta K_v' \leq \infty)$ or positive feedback $(0 \geq \beta K_v' \geq -\infty)$.

6. *Magnitude of Points:* A point on any valid locus branch must satisfy the magnitude condition. Hence, from Eq. 7.12,

$$\beta_o K_v' = \frac{|s_* - s_2||s_* - s_4| \overset{\text{poles of } A_L(s_*)}{\cdots}}{|s_* - s_1||s_* - s_3| \underset{\text{zeros of } A_L(s_*)}{\cdots}} \qquad 7.13$$

A graphical solution is frequently convenient in finding the scalar gain, $\beta_o K_v'$, at the point s_*. From Fig. 7.4, $\beta_o K_v'$ at $s = s_*$ is given by

$$\beta_o K_v' = \frac{\overset{\text{B}}{|s_* - s_2|} \overset{\text{C}}{|s_* - s_4|}}{\underset{\text{A}}{|s_* - s_1|}} \qquad 7.14$$

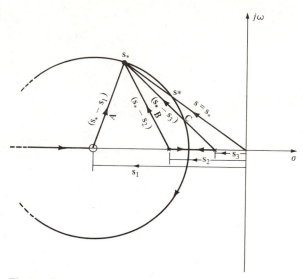

Figure 7.4 Graphical construction required to calculate the scalar gain, $\beta K'_v$, at $s = s_*$ for a NVF system with $A_L(s)$ poles and zeros as shown.

7. *Points Where Locus Branches Leave or Join the Real Axis:* A breakaway point is the point where two real-axis loci meet with increasing K'_v and leave the real axis at right angles to it. The breakaway point is also the point of maximum K'_v on that real-axis segment. It may be found by trial and error using Eq. 7.13, or by solving the equation

$$\frac{d(A_L(s_P))^{-1}}{ds} = 0 \qquad\qquad 7.15$$

or by a graphical-algebraic approach using the angle criterion, Eq. 7.11. An example of the latter approach is shown in Fig. 7.5 for a negative feedback system with a loop gain having three real poles and no zeros. The geometry of the vectors in Fig. 7.5 and Eq. 7.11 allows us to write

$$-(\theta_1 + \theta_2 + \theta_3) = -180° \qquad\qquad 7.16$$

or, equivalently,

$$\tan^{-1}\left(\frac{\varepsilon}{\omega_1 - P}\right) + \tan^{-1}\left(\frac{\varepsilon}{\omega_2 - P}\right) + \pi - \tan^{-1}\left(\frac{\varepsilon}{P - \omega_3}\right) = \pi$$

$$7.17$$

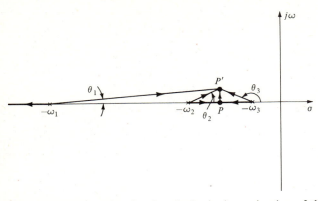

Figure 7.5 Construction for algebraic determination of the point P where the two locus branches meet and leave the real axis. PP' is a very small distance; P' is on the locus.

For small angles, the arctan function equals its argument (in radians), so we have

$$\frac{\varepsilon}{\omega_1 - P} + \frac{\varepsilon}{\omega_2 - P} - \frac{\varepsilon}{P - \omega_3} = 0 \qquad \text{7.18}$$

which can be solved as a quadratic for P, where P will be the negative real root lying between $-\omega_2$ and $-\omega_3$:

$$0 = 3P^2 - 2(\omega_1 + \omega_2 + \omega_3)P + (\omega_2\omega_3 + \omega_1\omega_2 + \omega_1\omega_3) \qquad \text{7.19}$$

8. *Breakaway or Reentry Angles of Loci with the Real Axis:* The loci are separated by angles of $180°/n$, where n is the number of loci intersecting the real axis. In most cases, $n = 2$, so the loci approach the real axis perpendicular to it.

9. *Asymptotic Behavior of the Loci for Large $\beta_o K'_v$:*

   (a) The number of asymptotes is $N_A = (N - M)$, where $M = $ the number of finite zeros of $A_L(s)$ and $N = $ the number of finite poles of $A_L(s)$.

   (b) The angles of the asymptotes with the real axis are ψ_k (from the angle criterion). See Table 7.1.

   (c) The intersection of the asymptotes with the real axis occurs at a point I_A.

$$I_A = \frac{\sum (\text{real parts of finite poles}) - \sum (\text{real parts of finite zeros})}{N - M}$$

$$\text{7.20}$$

Table 7.1 Asymptote angles in the s-plane

Negative Feedback		Positive Feedback	
$N - M$	ψ_k	$N - M$	ψ_k
1	180°	1	0°
2	90°, 270°	2	0°, 180°
3	60°, 180°, 300°	3	0°, 110°, 120°
4	45°, 135°, 225°, 315°	4	0°, 90°, 180°, 270°

See the example of finding I_A in Fig. 7.6. A negative feedback system with a quartic denominator and one zero is shown. Thus there are $N_A = (4 - 1) = 3$ asymptotes. They depart their intersection point, I_A, with the real axis at angles of 60°, 180°, and 300°, I_A is found to be at

$$I_A = \frac{(-1.5 - 1.5 - 3 - 9) - (-6)}{4 - 1} = -3 \qquad\qquad 7.21$$

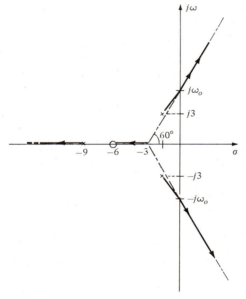

Figure 7.6 Illustration of root-locus asymptotes for an $N - M = 3$ loop gain with negative feedback.

Note that at some critical $\beta_o K_v'$, two locus branches cross the $j\omega$ axis, denoting an unstable system. The points ($\pm j\omega_o$) where the loci cross the $j\omega$ axis are not precisely known in this case; they are not necessarily where the asymptotes cross the $j\omega$ axis. Means of finding the $\beta_o K_v'(\text{crit})$ and ω_o are illustrated in the following sections of the chapter.

We now examine some examples of root-locus plots for typical negative and positive feedback electronic systems. The arrows indicate the direction of increasing $\beta_o K_v'$ magnitude.

An important special case is illustrated in Fig. 7.7. It may be shown, with some algebraic complexity, that the off-axis locus branches in Fig. 7.7(A) and (B) are indeed

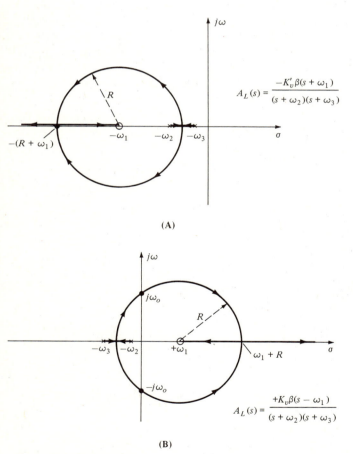

$$A_L(s) = \frac{-K_v'\beta(s + \omega_1)}{(s + \omega_2)(s + \omega_3)}$$

(A)

$$A_L(s) = \frac{+K_v\beta(s - \omega_1)}{(s + \omega_2)(s + \omega_3)}$$

(B)

Figure 7.7   Circle root-locus diagrams. **(A)** Negative feedback system. **(B)** Positive feedback system.

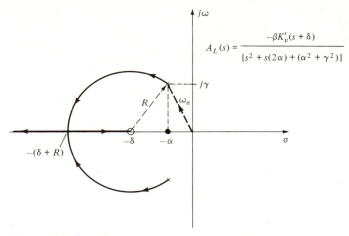

$$A_L(s) = \frac{-\beta K_v'(s + \delta)}{[s^2 + s(2\alpha) + (\alpha^2 + \gamma^2)]}$$

Figure 7.8   Circle root-locus where the poles of $A_L(s)$ are complex-conjugate.

circles centered on the zero (Angelo, 1969). The circle's radius (R) is the geometrical mean distance from the zero to the real poles. That is, in Fig. 7.7(A) and (B),

$$R = \sqrt{|(\omega_1 - \omega_2)|\,|(\omega_1 - \omega_3)|}$$ 7.22

If the poles of $A_L(s)$ are complex-conjugate and lie to the right of a real zero, we obtain an interrupted circle for off-axis loci, as shown in Fig. 7.8 for a negative feedback system. R is given by the Pythagorean theorem in this case:

$$R = \sqrt{(\delta - \alpha)^2 + \gamma^2}$$ 7.23

Figure 7.9 illustrates an open-loop unstable system made stable by negative feedback. Note that there is a critical, minimum $\beta_o K_v'$, below which the closed-loop system is unstable (has one real pole in the right-half s-plane). In this example, the two asymptotes intersect the real axis at

$$I_A = \frac{(+1 - 3 - 3) - (-3)}{3 - 1} = -1$$ 7.24

The critical minimum gain for stability is found by assuming that the locus of the critical pole lies at the origin (point P). From application of the Pythagorean theorem and Eq. 7.13,

$$\beta_o K_v'(\text{crit}) = \frac{1\sqrt{3^2 + 2^2}\sqrt{3^2 + 2^2}}{3} = \frac{13}{3}$$ 7.25

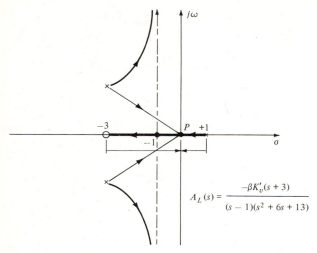

Figure 7.9 Root-locus diagram for a negative feedback system that is open-loop unstable.

Thus the system of Fig. 7.9 must have $\beta_o K'_v > 13/3$ in order to be closed-loop stable with negative feedback.

Many control systems texts have illustrations of root-locus plots and instruct in their use. See, for example, Chapter 4 in Franklin *et al.* (1986), Chapter 4 in Truxal (1955), or Chapter 8 in Ogata (1970). Reviewing these figures can give valuable insight into some of the less obvious forms of root-locus plots. Often locus behavior is not clear, even given rules for plotting, and it is prudent to rely on a computer software root-locus plotting routine to obtain quantitative details. For example, an excellent root-locus plotting subroutine can be found in the "Control Systems Toolbox" package available with PC-Mathlab software.* However, most of the negative feedback systems encountered in electronic circuit design lend themselves to easy graphical and pencil-and-paper root-locus solutions for stability or gains required for optimum closed-loop pole placement.

7.2 Use of Root-Locus in the Design of Feedback Amplifiers

Root-locus can be an effective tool in predicting what scalar gain, $\beta_o K'_v$, will be required to give a desired closed-loop frequency or transient response (pole placement). Of great importance is the behavior of the complex-conjugate pole pair as the scalar feedback gain is varied. A complex-conjugate pole pair can be represented as in Fig.

* Math Works, Inc., 20 N. Main St. Sherborn, MA 01770.

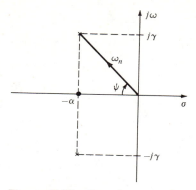

Figure 7.10   Complex-conjugate pole pair.

7.10. The pole pair can be written

$$H(s) = \frac{K_v}{s^2 + s(2\xi\omega_n) + \omega_n^2} = \frac{K_v}{s^2 + s(2\alpha) + (\alpha^2 + \gamma^2)} \qquad 7.26$$

in which the damping factor, ξ, for complex-conjugate poles always ranges between zero and one. The damping factor can also be interpreted geometrically in the s-plane:

$$\xi = \cos \psi \qquad 7.27$$

Often we must pick a $\beta_o K_v'$ value to place the closed-loop system's complex-conjugate poles at a desired ξ value to make the feedback system's transient or frequency response meet some design criterion.

EXAMPLE 7.1

Use of Root-Locus in the Design of a Feedback Amplifier

As a first example of this sort of design task, we will connect two IC differential amplifiers with NVF and adjust β so that $\xi = 0.707$ for the closed-loop system. The amplifiers are shown in Fig. 7.11. The system's loop gain is written in vector form:

$$A_L(s) = \frac{-\beta_o(1.1 \times 10^{15})}{(s - s_1)(s - s_2)} \qquad 7.28$$

where $s_1 = -10^6$ rad/s and $s_2 = -1.1 \times 10^7$ rad/s.

Next, the root-locus diagram is drawn to scale for the system in Fig. 7.12. Since we want the closed-loop system's complex-conjugate poles to have a damping factor of 0.707, we use Eq. 7.27 and construct a line of constant ξ from the origin at an angle $\psi = 45°$ with the negative real axis. This line of constant damping intersects

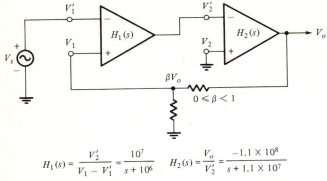

$$H_1(s) = \frac{V_2'}{V_1 - V_1'} = \frac{10^7}{s + 10^6} \qquad H_2(s) = \frac{V_o}{V_2'} = \frac{-1.1 \times 10^8}{s + 1.1 \times 10^7}$$

Figure 7.11 NVF amplifier made from two IC differential amplifiers.

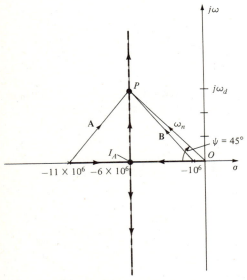

Figure 7.12 Root-locus diagram for the NVF system of Fig. 7.11.

the off-axis locus at point P. The length OP of this line is the closed-loop, undamped natural frequency of the system, ω_n. It is easy to see from the geometry that $\omega_n = (6 \times 10^6)\sqrt{2} = 8.49 \times 10^6$ rad/s. The feedback attenuation, β_o, required to place the closed-loop poles at P and P' is found from Eq. 7.13:

$$|-A_L(s = P)| = 1 = \frac{\beta_o(1.1 \times 10^{15})}{|A||B|} \tag{7.29}$$

$$|A| = |B| = \sqrt{(5 \times 10^6)^2 + (6 \times 10^6)^2} = 7.810 \times 10^6 \text{ rad/s} \tag{7.30}$$

Therefore, the value

$$\beta_o = \frac{(7.810 \times 10^6)(7.810 \times 10^6)}{1.1 \times 10^{15}} = 5.545 \times 10^{-2} \qquad 7.31$$

will give the desired closed-loop amplifier behavior.

Finally, we examine the closed-loop gain of the system:

$$A_v(s) = \frac{\dfrac{1.1 \times 10^{15}}{(s + 10^6)(s + 11 \times 10^6)}}{1 + \dfrac{\beta_o(1.1 \times 10^{15})}{(s + 10^6)(s + 11 \times 10^6)}} \qquad 7.32$$

Letting $s = 0$ and $\beta_o = 5.545 \times 10^{-2}$ in Eq. 7.32, we obtain the closed-loop dc gain:

$$A_v(0) = \frac{11 \times 10^{14}}{11 \times 10^{12} + 5.545 \times 10^{-2} \times 11 \times 10^{14}} = 15.28 \qquad 7.33$$

●

EXAMPLE 7.2

Use of Root-Locus to Obtain a Desired Closed-Loop System Damping Factor

For a second example, we consider the NVF system shown in Fig. 7.13. Figure 7.14(A) illustrates the root-locus diagram for this system. Its loop gain is just

$$A_L(s) = \frac{-\beta_o(3.5 \times 10^{15})}{(s - s_1)(s - s_2)} \qquad 7.34$$

where $s_1 = -10^6$ rad/s and $s_2 = -7 \times 10^6$ rad/s. We will find β_o to give the closed-loop, complex-conjugate poles a $\xi = 0.6$. A line of constant ξ has an angle of 53.13° with the real axis in the s-plane and intersects the locus branch at P. To find β_o at

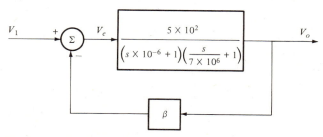

Figure 7.13   Another NVF system.

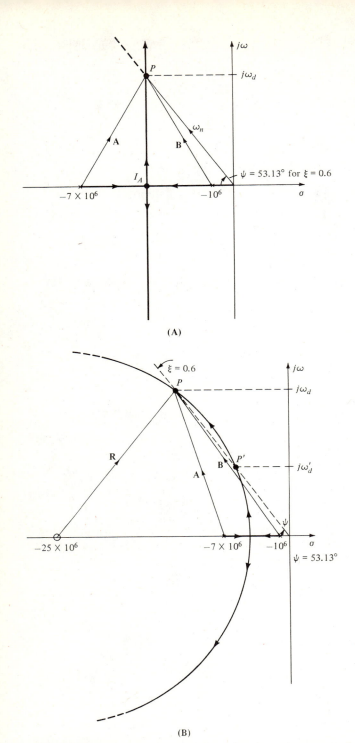

Figure 7.14 **(A)** Root-locus diagram for the uncompensated system of Fig. 7.13. **(B)** Root-locus diagram for the system of Fig. 7.13 compensated with a real zero in its feedback path (not the same scale as [A]).

P, we measure the distance $I_A P$ to scale, then use the Pythagorean theorem to find $|\mathbf{A}|$ and $|\mathbf{B}|$:

$$I_{AP} = \omega_d = 5.28 \times 10^6 \text{ rad/s} \tag{7.35}$$

$$|\mathbf{A}| = |\mathbf{B}| = \sqrt{(5.28 \times 10^6)^2 + (3 \times 10^6)^2} = 6.037 \times 10^6 \text{ rad/s} \tag{7.36}$$

Now, from the magnitude criterion,

$$|A_L(\mathbf{s} = \mathbf{P})| = 1 = \frac{\beta_o(3.5 \times 10^{15})}{|\mathbf{A}|\,|\mathbf{B}|} = \frac{\beta_o(3.5 \times 10^{15})}{3.688 \times 10^{13}} \tag{7.37}$$

Therefore,

$$\beta_o = 1.054 \times 10^{-2} \tag{7.38}$$

to obtain a damping factor of 0.6 in the closed-loop system. This system's dc gain is just

$$A_v(0) = \frac{5 \times 10^2}{1 + \beta_o(5 \times 10^2)} = 79.76 \tag{7.39}$$

Now we consider the same system as in Fig. 7.13 except we attempt to extend the closed-loop system's bandwidth through the addition of a zero in the feedback path. Instead of a scalar, β is now a function of frequency:

$$\beta(s) = \beta_o\left(\frac{s}{2.5 \times 10^7} + 1\right) \tag{7.40}$$

The new loop gain expression now includes Eq. 7.40 in the numerator. The root-locus plot for the compensated amplifier is shown in Fig. 7.14(B). The off-real axis locus is a circle of radius $R = \sqrt{(25 - 1)(25 - 7)} \times 10^6 = 20.78 \times 10^6$ rad/s. A line of constant $\xi = 0.6$ intersects this circle at two points, P and P'. We choose point P because it gives the closed-loop system the greatest bandwidth and hence the greatest speed of response. From scaled measurement of the root-locus diagram, we find $\omega_n = 21.14 \times 10^6$ rad/s and $\omega_d = 16.84 \times 10^6$ rad/s. The gain β_o is calculated using the magnitude criterion:

$$|A_L(\mathbf{s} = \mathbf{P})| = 1 = \frac{\beta_o(1.4 \times 10^8)(20.78 \times 10^6)}{(17.8 \times 10^6)(18.27 \times 10^6)} \tag{7.41}$$

From Eq. 7.41, we find that $\beta_o = 0.112$ to give the closed-loop, compensated system a damping factor of 0.6. The effects of using the zero at -25×10^6 rad/s to compensate the amplifier are summarized in Table 7.2.

Table 7.2 Comparison of compensated and uncompensated NVF amplifiers

Parameter	Compensated	Uncompensated
ξ	0.6	0.6
ω_n	6.6×10^6 rad/s	21.14×10^6 rad/s
ω_d	5.28×10^6 rad/s	16.84×10^6 rad/s
β_o	0.0105	0.112
$A_v(0)$	79.8	8.79

EXAMPLE 7.3

Use of Root-Locus in the Design of a Negative Current Feedback Amplifier

As a third and final example, consider the NCF power amplifier described by the circuit of Fig. 6.11 in Chapter 6 and the signal flow graph of Fig. 6.12. It is easy to write this amplifier's loop gain in time-constant form using Mason's rule:

$$A_L(s) = \frac{-\left(\dfrac{R_1}{R_1 + R_f}\right)K_{v_{(1)}}R_c(sRC + 1)}{(R_L + R_c)(s\tau_1 + 1)(s\tau_L + 1)} \qquad 7.42$$

The loop gain must be put into vector form to use root-locus as a design tool:

$$\mathbf{A_L(s)} = \frac{-\beta_o K_{v_{(1)}}R_c(RC)(\mathbf{s} - \mathbf{s_1})}{\tau_1 L(\mathbf{s} - \mathbf{s_2})(\mathbf{s} - \mathbf{s_3})} \qquad 7.43$$

where

$$\beta_o = \frac{R_1}{R_1 + R_f} \qquad 7.44$$

$$\mathbf{s_1} = -\frac{1}{RC} \text{ rad/s} \qquad 7.45$$

$$\mathbf{s_2} = -\frac{1}{\tau_1} \text{ rad/s} \qquad 7.46$$

$$\mathbf{s_3} = -\frac{1}{\tau_L} = -\frac{R_L + R_c}{L} \text{ rad/s} \qquad 7.47$$

The root-locus diagram of the NCF power amplifier is shown in Fig. 7.15. It, too, is seen to be the classical circle format. Given system parameter values, Eq. 7.43

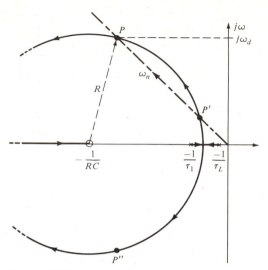

Figure 7.15 Root-locus diagram for the NCF power amplifier.

can be solved using the magnitude criterion to find the β required to fix the closed-loop system's poles at points P and P'' on the locus. The vector magnitude values required to solve for β_o at point P are best found graphically (have faith in a scaled root-locus diagram as a means of eliminating complex algebraic calculations.) ●

7.3 Stability

There have been many approaches to the study of stability in feedback systems, all of them designed to tell if one or more closed-loop system poles lie on or to the right of the $j\omega$ axis in the s-plane. If one or more poles lie in the right-half s-plane, a small disturbance will cause an exponentially growing output which is limited practically by amplifier saturation. If a single pole lies to the right of the positive real axis, then the amplifier's output grows smoothly exponentially; if a complex-conjugate pole pair lies in the right half of the s-plane, sinusoidal oscillations occur with an exponentially growing envelope. In the special case where a complex-conjugate pole pair lies directly on the $j\omega$ axis, we have an *ideal* oscillator; once oscillations are started, they neither grow nor shrink.

The root-locus technique is particularly useful in the study of feedback amplifier stability because it shows the loci of the closed-loop poles as the scalar gain is varied from zero to infinity. It also provides an easy means to find the scalar gain required for oscillation, and, by inspection of the locus slope where it crosses the $j\omega$ axis, gives an indication of frequency stability with gain. In the next section, we examine oscillator designs and analyze several *linear* oscillators using the root-locus technique.

7.4 Oscillators

As will be seen, some oscillators use negative voltage feedback, others positive voltage feedback. The purpose of a linear oscillator is to generate a pure sine wave of fixed frequency and amplitude. Unfortunately, all linear oscillators must contain some soft saturation nonlinearity whether they use negative or positive voltage feedback. Their scalar gain magnitude is made slightly larger than the value that puts their pole pair exactly on the $j\omega$ axis. This scalar gain locates their (closed-loop) poles slightly in the right-hand s-plane, causing their oscillations to grow slowly in amplitude until the soft saturation nonlinearity limits their output voltage. This design protocol ensures the buildup of oscillations at turn-on.

7.4.1 Single-Op-Amp Wien Bridge Oscillator

The Wien bridge oscillator, which uses PVF, is widely used in audio-frequency and ultrasonic applications. The circuit is shown in Fig. 7.16. There are two ways of introducing the soft saturation in this oscillator. A pair of *nose-to-nose* zener diodes (ZDs) in series with a resistor can be placed across the gain-setting feedback resistor, R_f. When the voltage difference ($v_o - v_i'$) exceeds the conduction value for the zener circuit ($V_z + 0.6$), the shunt path lowers the loop gain and moves the closed-loop poles toward the left-half s-plane, causing the oscillation amplitude to drop. The alternate means of introducing the soft saturation is to replace resistor R_1 with a positive temperature coefficient thermistor or a tungsten-filament light bulb. Now, when v_i' gets too large, the light bulb or thermistor heats up, raising the effective R_1 value, lowering the oscillator's loop gain, and so on.

To examine the conditions for oscillation, we must write the loop gain expression for the Wien bridge oscillator. A convenient place to break the loop is between the V_2 node and the V_i input. We first examine the gain between v_i and v_o. If we assume

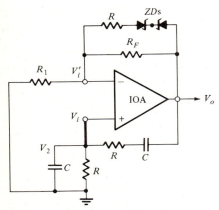

Figure 7.16 Wien bridge oscillator using one op-amp.

.t the zeners are not conducting, and that the op-amp is ideal, this is

$$\frac{V_o}{V_i} = 1 + \frac{R_f}{R_1} \qquad\qquad 7.48$$

The feedback ratio, $\beta(s)$, is found from the RC network considered as a voltage divider:

$$\frac{V_2}{V_o} = \frac{\dfrac{1}{G + sC}}{\dfrac{1}{G + sC} + R + \dfrac{1}{sC}} \qquad\qquad 7.49$$

Equation 7.49 may be reduced to Laplace format:

$$\frac{V_2}{V_o} = \frac{\dfrac{s}{RC}}{s^2 + s\dfrac{3}{RC} + \dfrac{1}{(RC)^2}} \qquad\qquad 7.50$$

The oscillator's loop gain is easily written in vector form by combining Eqs. 7.48 and 7.50:

$$A_L(s) = \frac{\left(1 + \dfrac{R_f}{R_1}\right)\left(\dfrac{1}{RC}\right)s}{(s - s_1)(s - s_2)} \qquad\qquad 7.51$$

where s_1 and s_2 are found by factoring the denominator of Eq. 7.50:

$$s_1 = \frac{-(3 + 5)}{RC} = \frac{-2.618}{RC} \qquad\qquad 7.52$$

$$s_2 = \frac{-(3 - 5)}{RC} = \frac{-0.382}{RC} \qquad\qquad 7.53$$

The root-locus diagram for the Wien bridge oscillator is shown in Fig. 7.17. Note that the frequency of oscillation is $\omega_o = 1/RC$ rad/s. The minimum gain required for oscillation is found from the magnitude criterion:

$$|A_L(s = j1/RC)| = 1 = \frac{\left(1 + \dfrac{R_f}{R_1}\right)\left(\dfrac{1}{RC}\right)\left(\dfrac{1}{RC}\right)}{\sqrt{\left(\dfrac{1}{RC}\right)^2 + \left(\dfrac{0.382}{RC}\right)^2}\sqrt{\left(\dfrac{1}{RC}\right)^2 + \left(\dfrac{2.618}{RC}\right)^2}} \qquad\qquad 7.54$$

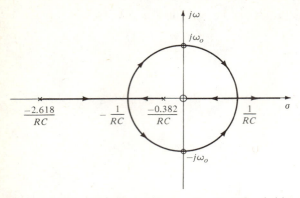

Figure 7.17 Root-locus diagram for the Wien bridge oscillator, which uses PVF. The circle radius is $R = \sqrt{(0.382/RC)(2.618/RC)} = 1/RC$.

Solution of Eq. 7.54 for $(1 + R_f/R_1)$ yields the well-known number

$$1 + \frac{R_f}{R_1} = 3.000 \qquad\qquad 7.55$$

Thus a gain of about 4 should be used to insure the growth of oscillations when the Wien bridge oscillator circuit is turned on.

7.4.2 Phase Shift Oscillator

The phase shift oscillator, like the Wien bridge oscillator, is primarily an audio-frequency system. This oscillator, which uses NVF, is illustrated in Figure 7.18. Op-amp 1 is used to obtain the noninverting gain required for oscillation; op-amp 2 acts as a saturating inverter to limit oscillation amplitude. The loop gain calculation can be broken into two parts. The gain of the two op-amps is

$$\frac{V_o}{V_i} = -K_v = -\left(1 + \frac{R_f}{R_1}\right) \qquad\qquad 7.56$$

The gain of the triple, RC, high-pass filter is found by writing three node equations and solving for $V_o/V_i = \beta(s)$:

$$V_1(s2C + G) \quad - V_2sC \qquad\quad + 0 \qquad\qquad = V_osC \qquad\qquad 7.57$$

$$- V_1sC \qquad\quad + V_2(s2C + G) \quad - V_isC \qquad\quad = 0 \qquad\qquad 7.58$$

$$0 \qquad\qquad\quad - V_2sC \qquad\quad + V_i(sC + G) \quad = 0 \qquad\qquad 7.59$$

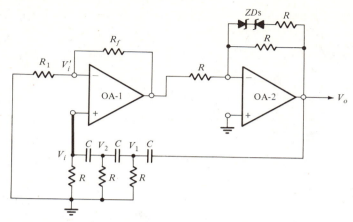

Figure 7.18   Phase shift RC oscillator. This circuit uses NVF.

Solving for $\beta(s)$, we find, after some algebra,

$$\beta(s) = \frac{V_o}{V_i} = \frac{s^3}{s^3 + s^2 \dfrac{6}{RC} + s \dfrac{5}{(RC)^2} + \dfrac{1}{(RC)^3}} \qquad \text{7.60}$$

It is not practical, or necessary, in this case to factor the denominator of Eq. 7.60 to find the exact loop gain pole positions. The loop gain will have negative real poles and be of the form

$$\mathbf{A}_L(\mathbf{s}) = \frac{-K_v(\mathbf{s})^3}{(\mathbf{s} - \mathbf{s}_1)(\mathbf{s} - \mathbf{s}_2)(\mathbf{s} - \mathbf{s}_3)} \qquad \text{7.61}$$

The approximate root-locus diagram for this oscillator is shown in Fig. 7.19. Even though we do not know the exact pole positions for $A_L(s)$, we can still use the root-locus criterion to find the minimum K_v required for oscillation, and the frequency of oscillation, ω_o. Note that $\mathbf{s} = j\omega_o$ is a point on the root-locus; therefore,

$$\mathbf{A}_L(j\omega_o) = \frac{-K_v(-j\omega_o^3)}{-j\omega_o^3 - \omega_o^2 \dfrac{6}{RC} + j\omega_o \dfrac{5}{(RC)^2} + \dfrac{1}{(RC)^3}} = 1\underline{/0^\circ} \qquad \text{7.62}$$

Because the numerator has a j factor, and $\mathbf{A}_L(j\omega_o)$ is real, the real terms in the denominator of Eq. 7.62 must sum to zero; that is,

$$-\omega_o^2 \frac{6}{RC} + \frac{1}{(RC)^3} = 0 \qquad \text{7.63}$$

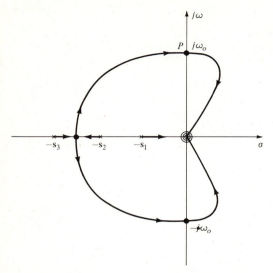

Figure 7.19   Approximate root-locus diagram for the phase shift oscillator.

from which we find the frequency of oscillation:

$$\omega_o = \frac{1}{\sqrt{6RC}} \qquad\qquad 7.64$$

Consideration of Eq. 7.62 with the real terms set to zero allows us to find $K_v = 29$. Hence, good design would have

$$K_v = \left(1 + \frac{R_f}{R_1}\right) > 29 \qquad\qquad 7.65$$

to insure oscillation buildup.

7.4.3 Colpitts Oscillator

The Colpitts oscillator is generally used to generate sine waves at radio frequencies and is often found in communications systems. A BJT or an FET may be used as the active device in this NVF circuit. Figure 7.20(A) shows a JFET Colpitts oscillator circuit, and Fig. 7.20(B) gives the MFSSM. (A detailed analysis would include the high-frequency capacitors C_{gs}, C_{gd}, and C_{ds} in the JFET model. However, because we wish to keep the analysis algebraically reasonable, we use the MFSSM.)

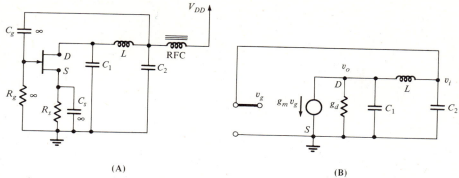

Figure 7.20 JFET Colpitts oscillator. The radio-frequency choke (RFC) is assumed to have an infinite impedance at ω_o. **(A)** Circuit. **(B)** MFSSM.

The loop gain is found by first calculating V_o/V_g, then V_i/V_o. Thus we write

$$\frac{V_o}{V_g} = \frac{-g_m}{g_d + sC_1 + \dfrac{1}{sL + 1/sC_2}} \qquad 7.66$$

which reduces to

$$\frac{V_o}{V_g} = \frac{\left(\dfrac{-g_m}{LC_1C_2}\right)(s^2LC_2 + 1)}{s^3 + s^2\dfrac{1}{C_1r_d} + s\left(\dfrac{C_1 + C_2}{LC_1C_2}\right) + \dfrac{1}{Lr_dC_1C_2}} \qquad 7.67$$

However, V_i/V_o is given by the voltage divider

$$\frac{V_i}{V_o} = \frac{1/sC_2}{1/sC_2 + sL} = \frac{1}{s^2LC_2 + 1} \qquad 7.68$$

Combining Eqs. 7.67 and 7.68, we obtain the loop gain:

$$A_L(s) = \frac{V_i}{V_g} = \frac{\dfrac{-g_m}{LC_1C_2}}{s^3 + s^2\dfrac{1}{C_1r_d} + s\left(\dfrac{C_1 + C_2}{LC_1C_2}\right) + \dfrac{1}{Lr_dC_1C_2}} \qquad 7.69$$

$A_L(s)$ has three poles; two are complex-conjugate in the left-half s-plane, and one is real in the left-half s-plane. This configuration means that the oscillator's root-locus diagram has three asymptotes; two locus branches cross the $j\omega$ axis at $\pm j\omega_o$

into the right-half s-plane as g_m is increased. As in the previous example, we do not need to factor the cubic denominator of $A_L(s)$ to solve for the conditions for oscillation. We merely set

$$A_L(j\omega_o) = \frac{\dfrac{-g_m}{LC_1C_2}}{-j\omega_o^3 - \omega_o^2 \dfrac{1}{C_1 r_d} + j\omega_o \left(\dfrac{C_1 + C_2}{LC_1C_2}\right) + \dfrac{1}{Lr_dC_1C_2}} = 1\underline{/0°} \qquad 7.70$$

Of course, the imaginary terms in the denominator must sum to zero:

$$-j\omega_o^3 + \frac{j\omega_o(C_1 + C_2)}{LC_1C_2} = 0 \qquad 7.71$$

from which we find that

$$\omega_o = \sqrt{\frac{C_1 + C_2}{LC_1C_2}} \text{ rad/s} \qquad 7.72$$

If Eq. 7.72 for ω_o is substituted into the real terms of Eq. 7.70 (the imaginary terms are zero), we can solve for the minimum g_m required for oscillation:

$$g_m = \frac{C_2}{r_dC_1} \qquad 7.73$$

7.4.4 Hartley Oscillator

The Hartley oscillator is another NVF radio-frequency oscillator that can use either a BJT or an FET as the active element. Hartley oscillators are also often found in communications electronic systems. A Hartley oscillator schematic is shown in Fig. 7.21(A); its MFSSM is given in Fig. 7.21(B). Analysis procedes exactly as for the Colpitts oscillator: First V_o/V_g is found, then by a voltage divider, V_g/V_o. It is left as an exercise to find the frequency of oscillation and the minimum g_m required for the Hartley oscillator to oscillate.

7.4.5 Piezoelectric Crystal Oscillators

When a fixed-frequency oscillator is to be designed to have very good frequency stability (low frequency drift with temperature, power supply voltage, etc.), a quartz piezoelectric crystal is used in the oscillator circuit as a resonant element. Piezoelectric crystals are reciprocal transducers; a dc voltage applied to electrodes on the crystal's surface causes a certain mechanical deformation of the crystal (twisting, bending, etc.), depending on the orientation of the crystal element when it was cut from the naturally occurring "mother" quartz crystal. Conversely, if the crystal is mechanically deformed (twisted, bent, etc.), a physical separation of charges will occur within the crystal,

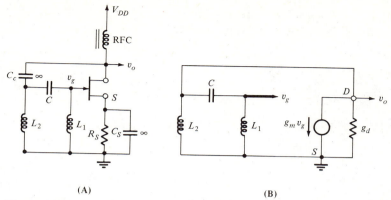

(A) (B)

Figure 7.21 **(A)** JFET Hartley oscillator. The inductances L_1 and L_2 are assumed not to be mutually coupled. RFC is a radio-frequency choke assumed to have an impedance much greater than r_d at the frequency of oscillation. **(B)** MFSSM for the Hartley oscillator. C_c, C_S, and the RFC do not appear in the model.

causing a voltage to appear across the electrodes on the crystal's surface. A crystal is also a mechanical resonator; much as a thin wine glass will ring when tapped, a quartz crystal given a mechanical displacement and released will vibrate with very little mechanical damping. The periodic mechanical vibration of the crystal induces an ac voltage at the electrodes. This voltage can be amplified and fed back to the crystal in such a manner as to keep it vibrating mechanically. This behavior is the basis for electronic crystal oscillators used in measurements, communications, and computer applications. A quartz crystal allowed to vibrate at or near its mechanical resonant frequency has an equivalent circuit given by Fig. 7.22.

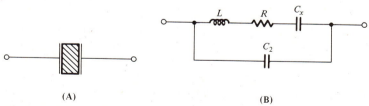

(A) (B)

Figure 7.22 Symbol **(A)** and electrical equivalent circuit **(B)** for piezo-electric quartz crystal resonators. The series RLC circuit is the electrical analog of the crystal's mechanical properties seen near resonance. The equivalent inductance, L, is due to the crystal's mass. R is due to mechanical losses from the mount, internal friction, and viscosity. C_X is proportional to the crystal's mechanical compliance. C_2 is the passive shunt capacitance seen between the crystal's electrodes.

The impedance seen looking into the terminals of the crystal near its resonant frequencies can be found from the equivalent circuit in Fig. 7.22. It is

$$Z_x(s) = \frac{s^2 LC_x + sC_1 R + 1}{s\left[s^2\left(\frac{LC_xC_2}{C_x + C_2}\right) + s\left(\frac{RC_xC_2}{C_x + C_2}\right) + 1\right](C_x + C_2)} \qquad 7.74$$

Both the numerator and the denominator of Eq. 7.74 have complex-conjugate roots. R is relatively small in most quartz crystals, giving the device very low losses and poles and zeros very near the $j\omega$ axis. The damping factor, ξ, of roots near the $j\omega$ axis is very low (see Eq. 7.27), corresponding to a high-Q tuned circuit. (Q can be shown to be given by $1/(2\xi)$ in a simple quadratic bandpass circuit.) Because R is small and Q is high, we can simplify the expression for Z_x by setting the R terms equal to zero:

$$\hat{Z}_x(s) = \frac{s^2 LC_x + 1}{s\left[s^2\left(\frac{LC_xC_2}{C_x + C_2}\right) + 1\right](C_x + C_2)} \qquad 7.75$$

If $s = j\omega$ (sinusoidal excitation), $\hat{\mathbf{Z}}_x$ can be written

$$\hat{\mathbf{Z}}_x(j\omega) = \frac{\omega_s^2 - \omega^2}{j\omega C_2(\omega_p^2 - \omega^2)} \qquad 7.76$$

The zeros of $Z_x(s)$ are at

$$\pm j\omega_s = \frac{\pm j}{\sqrt{LC_x}} \text{ rad/s} \qquad 7.77$$

and the poles are at

$$\pm j\omega_p = \frac{\pm j}{\sqrt{\dfrac{LC_xC_2}{C_x + C_2}}} \text{ rad/s} \qquad 7.78$$

Figure 7.23(A) illustrates the s-plane for $\hat{Z}_x(s)$ given by Eq. 7.75, and Fig. 7.23(B) shows a plot of $|\hat{\mathbf{Z}}_x(j\omega)|$.

Table 7.3 gives typical parameters for several crystals that oscillate in the ranges shown.

A variety of electronic oscillator circuits use quartz crystals; some use op-amps, others logic gates, and of course, FETs and BJTs are often used as the active elements. The Pierce oscillator (a variation on the Colpitts design) is frequently used with crystals. A JFET Pierce oscillator is shown in Fig. 7.24(A); its MFSSM is illustrated

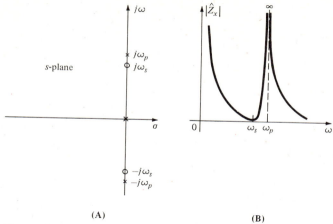

Figure 7.23 **(A)** s-plane for $\hat{Z}_x(s)$. **(B)** $|\hat{Z}_x(j\omega)|$ vs. ω.

Table 7.3 Selected quartz crystal parameters

| Parameter | Crystal | | | |
	A	B	C	D
f_s	427.4 kHz	88.70 kHz	1.998 MHz	32.78 kHz
f_p	430.1 kHz	89.00 kHz	2.001 MHz	32.81 kHz
f_p/f_s	1.00632	1.003	1.002	1.001
L	3.3 H	137 H	0.52 H	5800 H
R	——	15 kΩ	82 Ω	40 kΩ
C_x	0.042 pF	0.0235 pF	0.0122 pF	0.00491 pF
C_2	5.8 pF	3.5 pF	4.27 pF	2.85 pF
Q	23,000	5500	80,000	25,000
Cut	——	——	AT	XY

in Fig. 7.24(B). If the loop is broken at the gate, $A_L(s)$ can be found to be

$$
A_L(s) = \frac{V_g'}{V_g}
$$

$$
= [-g_m C_2(s^2 + \omega_p^2)]/[s^3(C_1 C_3 + C_2 C_3 + C_1 C_2) + s^2(C_1 + C_2)G_L
$$
$$
+ s(\omega_p^2 C_2 C_3 + \omega_s^2 C_1 C_3 + \omega_p^2 C_1 C_2) + G_L(\omega_p^2 C_2 + \omega_s^2 C_1)] \qquad 7.79
$$

Factoring the cubic denominator of Eq. 7.79 is out of the question. However, we do know that $A_L(s)$ has a pair of zeros on the $j\omega$ axis at $\pm j\omega_p$ rad/s and a real pole

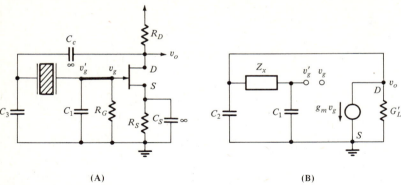

(A)   (B)

Figure 7.24 **(A)** Pierce crystal oscillator circuit (sometimes called a Clapp crystal oscillator). In some designs, R_D is replaced with a parallel RLC tuned circuit. The crystal behaves as a high-Q series inductance at the oscillation frequency, ω_o. **(B)** MFSSM of the Pierce oscillator. $G_L' = g_d + G_D$. $\hat{Z}_x$ is given by Eq. 7.76. C_1 includes the FET's C_{gs}; C_2 includes C_3 and the FET's output capacitance, C_{ds}.

plus a complex-conjugate pole pair in the left-half s-plane. The complex-conjugate pole pair is near the zeros. We also know that the system oscillates at some ω_o. It is left as an exercise to find ω_o and the g_m required for oscillation. [*Hint:* Solve $\mathbf{A}_L(j\omega_o) = 1\underline{/0°}$.]

Some other crystal oscillator circuits are shown in Fig. 7.25. In cases where exceptional oscillator stability is required, the crystal (and often the electronics) is housed in a thermostatted oven, run at a constant temperature well above ambient. Crystal ovens are found in many electronic instruments and communications equipment. Note that a crystal oscillator's output must be isolated by a buffer amplifier to prevent circuit loading from changing ω_o.

SUMMARY

We have seen how the use of voltage or current negative feedback can alter an amplifier's frequency response. The root-locus technique was shown to be a useful tool for the design of electronic feedback systems. It illustrates how the poles of the closed-loop transfer function of a system with a single feedback loop change with varying scalar loop gain. In some cases, the scalar loop gain can be set to a value that will make the system unstable. Oscillators are a special case of unstable feedback systems in which a pair of complex-conjugate poles are forced to lie on the $j\omega$ axis in the s-plane. Although the root-locus method must be applied to *linear* electronic feedback systems, oscillators were seen to contain an amplitude nonlinearity in the loop gain, which causes the oscillation amplitude to stabilize and forces the poles to remain on the $j\omega$ axis.

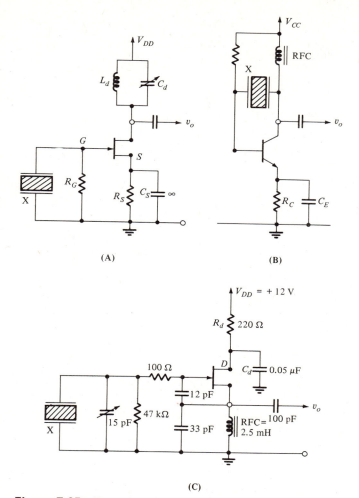

Figure 7.25 Examples of crystal oscillator (XO) circuits. **(A)** Miller XO. The crystal is used in its parallel resonant mode. Feedback is through the FET's C_{gd}. **(B)** Pierce XO using a BJT. The crystal is used in its series resonant mode. The RFC is assumed to have infinite impedance at ω_o. **(C)** Variation on the Miller XO. R_D and C_d are for decoupling.

Once the system's loop gain is known, the root-locus diagram is constructed, and the position of the closed-loop system's poles is specified, it is easy to use the root-locus method to find the required scalar loop gain to achieve the desired pole position. This situation is true whether amplifiers or oscillators are being considered.

Compensation of closed-loop amplifiers to permit broader bandwidth with no decrease in complex-conjugate pole damping was illustrated.

PROBLEMS

7.1 An ideal op-amp is connected as shown. $R = 1\ \mathrm{M\Omega}$ and $C = 1\ \mu\mathrm{F}$.

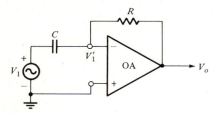

Figure P7.1

(a) Find an expression for $(V_o/V_1)(s)$. What time-domain operation does this circuit perform on $v_1(t)$?

(b) Now assume that the op-amp is nonideal so that

$$V_o = \frac{-1 \times 10^5}{1 \times 10^{-2}\,s + 1}\,V_1'$$

Find an expression for $A_v(s) = (V_o/V_1)(s)$ in time-constant form. Make a dimensioned Bode magnitude plot for $\mathbf{A}_v(j\omega)$. Note that this may be most easily done using an ECAP.

7.2 Negative current feedback is used to improve the frequency response of a magnetic deflection yoke driver circuit (Fig. P7.2). Assume that the amplifier is ideal except for finite gain, $K_v = -V_o/V_1'$. Also, $R_1 = R_2 = 10\ \mathrm{k\Omega} \gg R_c$. The coil's time constant is $T_c = L/R_c = 0.01$ s.

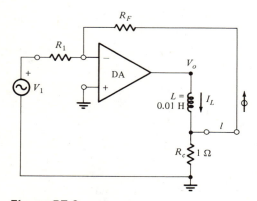

Figure P7.2

(a) Derive an algebraic expression for $(I_L/V_1)(s)$ in time-constant form.

(b) Use root-locus to show how the closed-loop system's frequency response changes as K_v is varied between zero and infinity. [*Hint:* Break the loop at l; find $A_L(s)$. Also, $R_1 = R_F = 10 \text{ k}\Omega \gg R_c$.]

(c) Using root-locus, what must K_v be in order to give a 1000-fold decrease in the coil's time constant? (That is, make $T'_c = 10^{-5}$ s.)

(d) Find an algebraic expression for the output resistance the coil "sees" in this system.

7.3 Two differential amplifiers are used to make the feedback amplifier in Fig. P7.3. Assume the following:

$$H_1(s) = \frac{V_2}{V_1 - V_o} = \frac{A_{v(1)}}{10^{-4}s + 1}$$

$$H_2(s) = \frac{V_o}{V_2} = \frac{A_{v(2)}}{s/(4 \times 10^4) + 1}$$

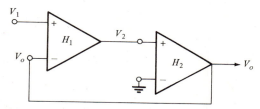

Figure P7.3

(a) Find the system's $A_L(s)$ in vector form.

(b) Draw the system's root-locus diagram to scale.

(c) Find $A_{v(1)}A_{v(2)}$ to give the closed-loop system's poles a damping factor of $\xi = 0.707$.

(d) Using the $A_{v(1)}A_{v(2)}$ value found in part (c), find the system's dc gain, $(V_o/V_1)(j_0)$, and f_T.

7.4 Negative voltage feedback is used to extend the bandwidth of a differential amplifier. The differential amplifier's gain is given by

$$V_o = \frac{10^3(V_i - V'_i)}{(10^{-3}s + 1)(2 \times 10^{-4}s + 1)}$$

(a) Draw the system's root-locus to scale.

(b) Find β to give the closed-loop system's poles a ξ of 0.707.

(c) What is the system's dc gain with the β value found in part (b)? Find the system's f_T.

(d) Find f_T for the differential amplifier without feedback ($\beta = 0$).

(e) The same amplifier is compensated with a zero in the feedback path so that

$$\beta(s) = \frac{V'_i}{V_o} = \beta_o(3.333 \times 10^{-5}s + 1)$$

Repeat parts (a) through (c).

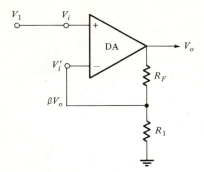

Figure P7.4

7.5 **(a)** Use root-locus to find the range of β (+ and − values) over which the cubic system shown is stable.

(b) At what ω_o will the system oscillate? What β value is required for oscillation?

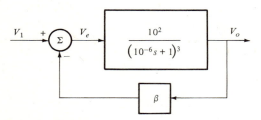

Figure P7.5

7.6 A nonideal op-amp amplifier circuit is shown in Fig. P7.6. Assume $R_2 \gg R_o$, $R_2 = 90$ kΩ, and $R_1 = 10$ kΩ. Use $R_o = 100$ Ω and an open-circuit voltage gain of

$$\frac{10^4}{10^{-3} s + 1} = H_{oc}(s)$$

(a) Find an expression for the system's Thevenin output impedance, $Z_o(s)$, in time-constant form.

(b) Sketch and dimension the Bode plot asymptotes for $|Z_o(j\omega)|$.

7.7 A negative feedback amplifier has a loop gain given by

$$A_L(s) = \frac{-K_v \beta (s + 6 \times 10^5)}{(s + 3 \times 10^5)(s + 2 \times 10^5)}$$

(a) Using a root-locus diagram drawn to scale, find the minimum possible ζ of the closed-loop system's quadratic poles.

(b) Find the $K_v \beta$ required to give this minimum ζ.

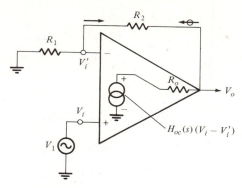

Figure P7.6

7.8 Two ideal op-amps are used to build an oscillator, as shown in Fig. P7.8.

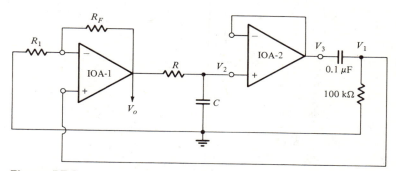

Figure P7.8

(a) Design the oscillator to oscillate at 1000 Hz. Use a root-locus graphical approach; note that the oscillator uses PVF. Specify numerical values of R_F, R, and C.

(b) Illustrate a nonlinear means to limit the peak-to-peak amplitude of V_o.

7.9 The system shown in Fig. P7.9 is a PVF oscillator. Let $\omega_n = 10^6$ rad/s, $\xi = 0.4$, and $K > 0$.

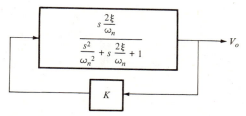

Figure P7.9

(a) Draw the oscillator's root-locus diagram to scale.
(b) Find an expression and value for the frequency of oscillation, ω_o.
(c) Find the minimum $K > 0$ required for oscillation.

7.10 Four ideal op-amps are used to make a phase shift oscillator. Assume $R = 22 \text{ k}\Omega$ and $C = 0.0062 \text{ μF}$.

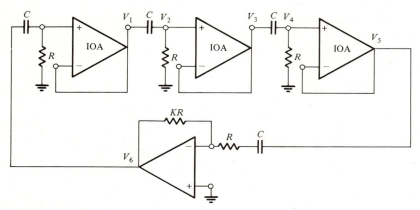

Figure P7.10

(a) Find the system's $A_L(s)$ in vector form.
(b) Draw the system's root-locus diagram to scale.
(c) Use $\mathbf{A}_L(j\omega_o) = 1\underline{/0°}$ to find ω_o algebraically.
(d) Find the minimum K required for oscillation.

7.11 A BJT transistor Colpitts oscillator is shown in Fig. P7.11. The BJT is characterized by the h-parameter MFSSM with $h_{re} = h_{oe} = 0$. Capacitors C' are considered to be short circuits at the oscillation frequency, ω_o. Assume $R_1 \parallel R_2 \gg h_{ie}$.

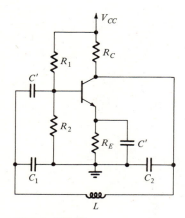

Figure P7.11

(a) Find $A_L(s)$ in Laplace format.

(b) Sketch an approximate root-locus diagram.

(c) Use the Barkhausen criterion, $\mathbf{A}_L(j\omega_o) = 1\underline{/0°}$, to find ω_o and the transistor h_{fe} required for oscillation.

7.12 An oscillator design is shown in Fig. P7.12.

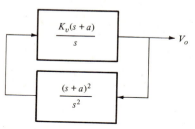

Figure P7.12

(a) Sketch the oscillator's root-locus diagram to scale. Should K_v be a positive or a negative number for oscillatory behavior?

(b) Use the Barkhausen criterion, $\mathbf{A}_L(j\omega_o) = 1\underline{/0°}$, to find ω_o of oscillation.

(c) Find the range of K_v over which the amplifier will oscillate. Why is this a poor design for an oscillator?

7.13 A JFET is used in a *tuned-input* oscillator shown in Fig. P7.13(B). A capacitor, C, is resonated with the transformer's secondary inductance, L_2. The transformer is described by the following equations and the model shown in Fig. P7.13(A).

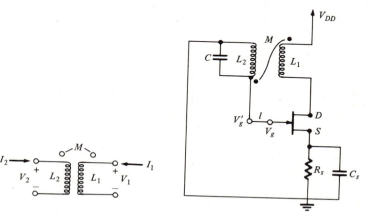

FET: Mid-frequency (μ, r_d) small-signal model

(A)   (B)

Figure P7.13

$$V_1 = I_1 j\omega L_1 + I_2 j\omega M$$

$$V_2 = I_1 j\omega M + I_2 j\omega L_2$$

$$M = k\sqrt{L_1 L_2}, \ 0 < k \leq 1$$

(a) Derive an expression for $A_L(s) = (V_g'/V_g)(s)$ in time-constant format. (Break link l.) Note that this is a positive feedback system.

(b) Use the Barkhausen criterion, $A_L(j\omega_o) = 1\underline{/0°}$, to find ω_o of oscillation and the minimum μ required for oscillation.

7.14 Negative current feedback is used to make a power op-amp a constant-current source. Assume that all the load current, i_L, flows through R_m to ground. Assume, $R_s = R_m = 1\ \Omega$, $L = 2 \times 10^{-4}$ H, and the op-amp's gain is given by

$$\frac{V_o}{V_i'} = \frac{-10^5}{10^{-2} s + 1}$$

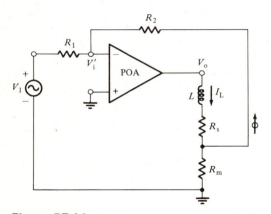

Figure P7.14

(a) Draw a signal flow graph or block diagram describing the system.

(b) Use the root-locus technique to find the $\beta = R_1/(R_1 + R_2)$ value to give the closed-loop system a damping factor of 0.707. Find ω_n for the closed-loop system.

7.15 A certain negative feedback system has the loop gain

$$A_L(s) = -\frac{K_v \beta}{(s + \omega_o)^4}$$

(a) Draw the system's root-locus diagram to scale. Find expressions for the minimum K_v value to make the system oscillate, and the frequency of oscillation.

(b) Describe the system's behavior when it is given PVF. Does it oscillate?

7.16 An op-amp is characterized by the gain

$$\frac{V_o}{V_i'}(s) = \frac{-10^5}{10^{-2}s + 1}$$

What is the maximum dc gain the circuit can be run at to get a 50 kHz -3 dB frequency? That is, find R_F.

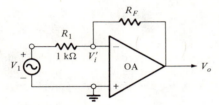

Figure P7.16

7.17 Positive voltage feedback is used around an op-amp circuit (Fig. P7.17). The op-amp's gain is given by

$$\frac{V_o}{V_i}(s) = \frac{+K_v}{\tau_a s + 1}$$

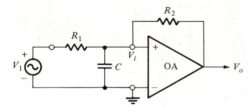

Figure P7.17

(a) Sketch the system's root-locus diagram to scale. Can this system oscillate?
(b) Derive an expression for $(V_o/V_1)(s)$ in time-constant form for the system. What conditions on R_1 and R_2 determine stability?

7.18 An amplifier with positive voltage feedback is characterized by the block diagram shown in Fig. P7.18.

(a) Sketch the amplifier's root-locus diagram to scale. Assume $K_v\beta > 0$.
(b) Find the value of K_v that will give the closed-loop system a ξ of 0.5.

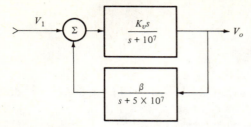

Figure P7.18

7.19 Positive voltage feedback is used to "neutralize" a JFET grounded-gate/source-follower amplifier to extend its high-frequency response. Assume $Q_1 = Q_2$, $g_d = 0$, $g_m > 0$, $C_{gd(1)} > 0$, and $C_{gd(2)} = C_{gs(2)} = C_{gs(1)} = 0$. Also, use the unilateral approximation on the v_o node; that is, $V_o = g_m V_s R_D$.

(a) Find an algebraic expression for $(V_o/V_1)(s)$ in time-constant form.

(b) Find an expression for the C_N value that will "neutralize" the amplifier. Let (V_o/V_1) $(0) = A_o$.

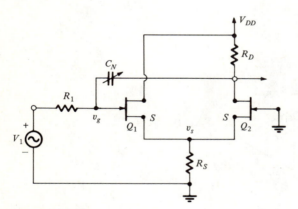

Figure P7.19

7.20 Two LM741A op-amps are to be connected in an *active negative feedback* circuit designed to extend amplifier bandwidth. Use an ECAP such as MicroCap II to compare the frequency response of a single noninverting stage (A) with the active feedback system (B). Compare the circuits' dc gains, -3 dB frequencies, and f_T's.

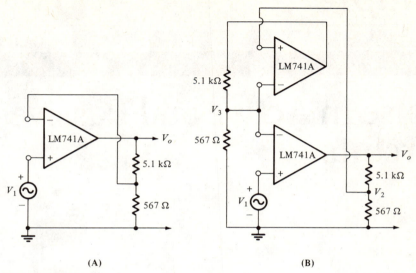

(A)

(B)

Figure P7.20

Chapter 8

Noise in Active and Passive Circuits

Noise is an important consideration in the application of electronic circuits and systems because it provides a fundamental limitation to the minimum useful signal input that can be processed (measured, detected, encoded, etc.) by an electronic system. Noise also determines the statistical frequency of errors in transmitting and receiving information. It is important to understand the sources of noise in electronic circuits, how noise is described, and how low-noise circuits may be designed and used.

In this chapter, we will consider only electrical noise caused by random phenomena in the circuits in question, and not "noise" caused by coherent or quasi-coherent sources external to the circuits, such as that caused by 60 Hz or 120 Hz power line phenomena, or radio-frequency interference (RFI), ignition (spark) noise, or noise arising in motor brushes (also spark-induced noise). The latter sources of noise may indeed contribute to reduced system output signal-to-noise ratios, and their reduction often uses as much art as science, generally involving electrostatic or magnetic shielding, special cables, grounding schemes, and RFI filters for power supplies and the like.

The random noise that we will examine in electric and electronic circuits in this chapter is treated as stationary noise. That is, the physical characteristics of the process or processes generating the noise are assumed not to change with time, so the ensemble-averaged or expected properties of the noise (mean, standard deviation, etc.) do not change with time. When stationarity is assumed for a random process, averages over time are equivalent to ensemble averages.

8.1 Descriptions of Noise in Circuits

Several statistical methods of describing random noise waveforms are available. The first, the probability density function (PDF), considers only the amplitude statistics of a noise voltage waveform, $n(t)$, and not how $n(t)$ varies with time. The probability density function, $p(x)$, is defined as follows; note that $n(t)$ is the amplitude of the noise sampled at some time t:

$$p(x) = \frac{\text{Probability } x < n \leq (x + dx)}{dx} \qquad 8.1$$

The next two equations illustrate well-known properties of the PDF:

$$\int_{-\infty}^{v} p(x)\, dx = \text{Prob}(n \leq v) \qquad 8.2$$

$$\int_{-\infty}^{\infty} p(x)\, dx = 1 = \text{Prob}(n \leq \infty) \qquad 8.3$$

Also,

$$\int_{v_1}^{v_2} p(x)\, dx = \text{Prob}(v_1 < n \leq v_2) \qquad 8.4$$

Several widely used PDFs are useful for describing the amplitude characteristics of electrical noise. These include the Gaussian or normal density,

$$p(x) = \left(\frac{1}{\sigma_x \sqrt{2\pi}}\right) \exp\left[-\frac{1}{2}\left(\frac{x - \bar{x}}{\sigma_x}\right)^2\right] \qquad 8.5$$

the rectangular density,

$$p(x) = \frac{1}{2a} \quad \text{for } -a < x < a$$

$$p(x) = 0 \quad \text{for } |x| > a \qquad 8.6$$

and the Rayleigh density,

$$p(x) = \left(\frac{x}{\alpha^2}\right) \exp\left[-\frac{1}{2}\left(\frac{x}{\alpha}\right)^2\right] \qquad 8.7$$

The assumption that circuit noise can be characterized by the Gaussian PDF is widely made because this PDF is often a valid approximation to the actual case, and many mathematical benefits follow this assumption. For example, if Gaussian noise is the input to a linear transfer function, Gaussian noise will be seen at the output, although with a different variance, σ_x^2, than for the input noise. This is generally not the case for noises described by other PDFs.

Another important descriptor of noise is the power density spectrum (PDS). The PDS characterizes the noise in the frequency domain, regardless of the PDF. We present an operational definition of the power density spectrum that describes the distribution of "power" in a noise source versus frequency.

Consider the system of Fig. 8.1. A noise voltage source is connected to the input of an ideal low-pass filter (ILPF), the output of which is measured by a broadband, true rms voltmeter. The operation of the true rms voltmeter can be represented by the block diagram of Fig. 8.2. $v_n(t)$, the filter input, is squared, then the dc component in $v_n^2(t)$ is extracted by a low-pass filter with a long time constant. The output of this filter, $\overline{v_n^2}$, is then square-rooted, yielding, in inverse order of operation, the rms value of $v_n(t)$. Note that a true rms meter must always be used in noise measure-

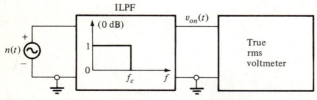

Figure 8.1   System for measuring the cumulative mean-squared noise voltage of a noise source.

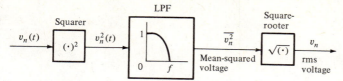

Figure 8.2 rms operation on $v_n(t)$.

ments because it is the only type of ac meter calibrated for random signals, as well as for all forms of periodic, deterministic waveforms (sine waves, triangle waves, etc.). One type of true rms voltmeter performs the operations indicated in Fig. 8.2 with active analog circuit elements. Another, more common type of true rms meter uses a vacuum thermocouple. In this case, the input, $v_{on}(t)$, causes a temperature rise in a small heater element with resistance R_h. This ΔT is proportional to Ph, the average power dissipated in R_h. A thermocouple is bonded to the heater, and it generates an emf, E_o, proportional to ΔT. Hence $E_o = kv_{on}^2$. A dc meter movement with a square-root scale responds to E_o, giving an indicated rms value of $v_{on}(t)$. There are also sophisticated electronic feedback-type true rms meters that balance the dc power in a reference thermocouple with the input signal's average ac power in an input thermocouple.

Let us now return to the circuit of Fig. 8.1 and perform the following operations. We start with the variable low-pass filter's cutoff frequency, f_c, set to zero. We then increase f_c in small increments and record the square of the rms meter's reading, that is, the mean-squared output voltage, along with f_c. A typical plot of these measurements is shown in Fig. 8.3. Note that $\overline{v_{on}^2}$ increases rapidly and then reaches a limit, $\overline{v_{on(max)}^2}$, because the source contains no further noise power at high frequencies (the source in this case is considered to be bandwidth-limited noise).

The one-sided PDS, which is often used to characterize noise, is simply found by taking the derivative of the $\overline{v_{on(f)}^2}$ plot in Fig. 8.3. That is,

$$S_n(f) = \frac{\overline{d v_{on}^2(f)}}{df} \text{ mean squared volts/Hz} \qquad (\text{MSV/Hz}) \qquad 8.8$$

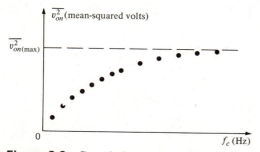

Figure 8.3 Cumulative mean-squared noise voltage plot for the source $n(t)$ in Fig. 8.1.

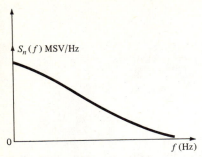

Figure 8.4   One-sided power density spectrum.

A plot of a one-sided PDS based on the curve of MSV versus frequency of Fig. 8.3 is shown in Fig. 8.4. Note that the $S_n(f)$ falls off to zero at high frequencies.

  We note that the total mean-squared nosie voltage in a given frequency band is found from

$$\overline{v_{on}^2(f_1, f_2)} = \int_{f_1}^{f_2} S_n(f) \, df \tag{8.9}$$

and also

$$\overline{v_{on(max)}^2} = \int_0^\infty S_n(f) \, df \tag{8.10}$$

Often noise is specified or described using *root* PDSs that are plots of $\sqrt{S_n(f)}$ versus f, and that have the units of rms units per root Hertz. Root spectrums are commonly used to characterize the equivalent input noise voltage and current of transistors and amplifiers.

  It is important to comment on certain special PDSs frequently used in noise analysis and calculations. They are shown in Figs. 8.5 through 8.7.

  White noise is characterized by a PDS spectrum flat from 0 to ∞ Hz (Fig 8.5); thus,

$$\int_0^\infty S_{n\omega}(f) \, df = \infty \tag{8.11}$$

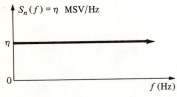

Figure 8.5   White noise power density spectrum.

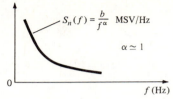

Figure 8.6 $1/f$ noise power density spectrum.

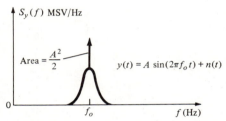

Figure 8.7 Power density spectrum of a coherent signal mixed with narrowband noise.

The assumption that system input noise is Gaussian and white is often made for mathematical convenience and is justifiable in many cases where the spectrum of the input noise is flat or constant over the effective bandwidth of the system for which it is an input.

One-over-f ($1/f$), or flicker noise (Fig. 8.6), is a reasonable approximation describing the PDS of low-frequency noise from electrodes, surface imperfections affecting emission and diffusion phenomena in semiconductor devices, and carbon composition resistors carrying direct current (metallic resistors are substantially free of $1/f$ noise). It can present a real problem in electronic systems used in low-level, low-frequency (and dc) measurements.

A coherent signal has a line spectrum, as illustrated in Fig. 8.7. The smooth line shows the PDS of incoherent, narrowband noise centered on f_o. Often we can add the various types of PDSs together to describe a noise waveform whose origins lie in several simultaneous physical processes. It is important to note that spectrums can be added in the mean-squared sense only when the sources described by the component spectrums are statistically independent.

8.2 Noise in Resistors

Statistical mechanics tells us that any pure resistor R at temperature T K will have a zero mean noise voltage associated with it. This noise voltage is seen in series with R and has a Gaussian PDF. This noise is called thermal or Johnson noise, and it has a white PDS given by

$$S_n(f) = 4kTR \quad \text{MSV/Hz} \qquad\qquad 8.12$$

where $k =$ Boltzmann's constant $= 1.38 \times 10^{-23}$. From Eqs. 8.9 and 8.12, we see that the mean-squared voltage in a bandwidth $B = (f_2 - f_1)$ is

$$\overline{v_{on}^2} = 4kTR(f_2 - f_1) = 4kTRB \quad \text{MSV} \qquad\qquad 8.13$$

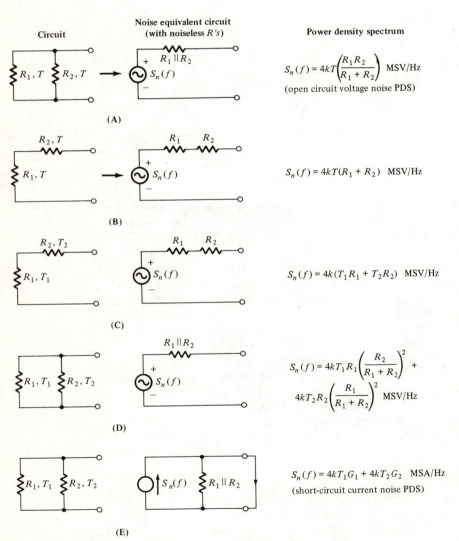

Figure 8.8 Ways of combining resistor Johnson noise spectrums.

It is important to realize that there is a practical limit to the white PDS given by Eq. 8.12. This limit is set by stray capacitances shunting the resistor to ground, and, at ultra-high frequencies, by the inductances of the resistor's leads and body. We will not consider these distributed parameter effects in our treatment of resistor thermal noise.

It is important to know how to combine noises arising in simple passive resistive circuits to arrive at a grand noise source seen at the circuit port in question. Figures 8.8(A) through (E) show how the individual noises are combined. Note that we assume the noise arising in physically separate resistors is statistically independent, regardless of the resistors' temperatures. In every case, we use addition of mean-squared quantities to find the net output white PDS in mean-squared volts per hertz, or in MSA/Hz if the Norton noise model is being used.

It is also known that when direct current passes through a resistor, the resistor's voltage noise spectrum will include a $1/f$ component whose magnitude is proportional to the power dissipated in the resistor. This can be described as

$$S_n(f) = 4kTR + \frac{AI^2}{f} \quad \text{MSV/Hz} \qquad 8.14$$

where A is a constant that depends on the material and the construction of the resistor. The PDS given by Eq. 8.14 is shown in Fig. 8.9. Note that a crossover frequency can be defined for the resistor where the $1/f$ power density equals the white noise density. This occurs at

$$f_c = \frac{AI^2}{4kTR} \qquad 8.15$$

It is possible to show that an increase in the wattage rating of a resistor of a given type and value results in a corresponding reduction in the $1/f$ noise spectrum component. To illustrate this, we note that the noise voltage PDS of a single resistor R of a given wattage is given by Eq. 8.14. We now construct a resistor of nine times this wattage having the same resistance R by using the series-parallel configuration

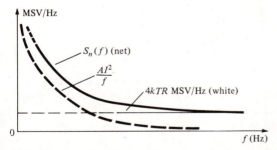

Figure 8.9 Composite noise power density spectrum from a resistor conducting direct current

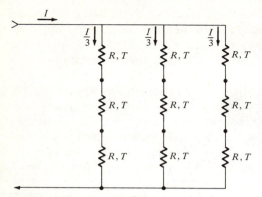

Figure 8.10   9 W resistor of net resistance R made from nine 1 W resistors.

of Fig. 8.10. The noise voltage PDS in any of the nine resistors is found from Eq. 8.12:

$$S'_n(f) = 4kTR + \frac{A(I/3)^2}{f} \quad \text{MSV/Hz} \qquad 8.16$$

Each of the nine spectrums of Eq. 8.16 contributes an MSV/Hz component to the net noise PDS of the nine-resistor network. Each resistor's noise source "sees" a voltage divider formed by the other eight resistors in the network. It is easy to see that these dividers each have an attenuation given by

$$\frac{3R/2}{3R/2 + 3R} = \frac{1}{3} \qquad 8.17$$

Thus the net noise voltage PDS at the terminals of the 9 W resistor is

$$S_{n(9)}(f) = \sum_{j=1}^{9} \left[4kTR + A\left(\frac{I}{3}\right)\frac{1}{f} \right]\frac{1}{9} \qquad 8.18$$

which easily reduces to

$$S_{n(9)}(f) = 4kTR + \frac{AI^2}{9f} \qquad 8.19$$

Note that the Johnson noise term for the 9 W resistor is unchanged; it depends only on the net resistance, R. The $1/f$ term, however, enjoys a ninefold reduction due to the reduced dc current in the individual elements. It is safe to generalize that the use of high-wattage resistors of a given type and resistance will result in reduced $1/f$ noise generation when the resistor carries dc current. The dc current density in the high-wattage resistor will be lower, contributing to a lower $1/f$ spectral component.

Figure 8.11   Noise conditioned by a linear filter.

8.3 Propagation of Noise Through Linear Filters

In a rigorous treatment of noise in linear systems, it is possible to show how the PDS of a Gaussian random input is modified by conditioning it with a linear filter. In general, for the system of Fig. 8.11, it can be shown that

$$S_o(f) = S_i(f)|\mathbf{H}(2\pi f j)|^2 \qquad\qquad 8.20$$

That is, the PDS of the filter's output noise is given by the product of the input noise spectrum times the magnitude squared of the linear system's transfer function. This concept can be extended to include two or more cascaded systems as shown in Fig. 8.12. Thus,

$$S_o(f) = S_i(f)|\mathbf{H}(2\pi f j)|^2|\mathbf{G}(2\pi f j)|^2 \qquad\qquad 8.21$$

or

$$S_o(f) = S_i(f)|\mathbf{H}(2\pi f j)\mathbf{G}(2\pi f j)|^2 \qquad\qquad 8.22$$

Note that all PDS functions are always positive and real.

   If white noise with a PDS $S_n(f) = D$ MSV/Hz is a linear filter input, then, by relation 8.21, we have

$$S_o(f) = D|\mathbf{H}(2\pi f j)|^2 \qquad\qquad 8.23$$

The total mean-squared output noise of the filter is found by substituting Eq. 8.23 into Eq. 8.10:

$$\overline{v_{on}^2} = \int_0^\infty S_o(f)\,df = D\int_0^\infty |\mathbf{H}(2\pi f j)|^2\,df \qquad\qquad 8.24$$

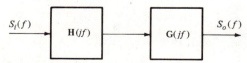

Figure 8.12   Noise conditioned by two cascaded linear filters.

Table 8.1   Gain squared-bandwidth products for common transfer functions

Transfer function $H(j\omega)$	Bode plot	Gain squared-bandwidth (Hz)
1. $\dfrac{K_v}{j\omega\tau + 1}$	20 log K_v; 0 dB; $1/\tau$; −6 dB/octave; ω	$K_v^2\left(\dfrac{1}{4\tau}\right)$
2. $\dfrac{K_v}{(j\omega\tau_1 + 1)(j\omega\tau_2 + 1)}$	−6 dB/octave; 0 dB; $1/\tau_1$; $1/\tau_2$; ω; −12 dB/octave	$K_v^2\left[\dfrac{1}{4(\tau_1 + \tau_2)}\right]$
3. $\dfrac{K_v}{\left(1 - \dfrac{\omega^2}{\omega_n^2}\right) + j\omega\,\dfrac{2\xi}{\omega_n}}$ $0 < \xi < 1$	20 log K_v; 0 dB; ω_n; ω; −12 dB/octave	$K_v^2\left(\dfrac{\omega_n}{8\xi}\right)$
4. $\dfrac{j\omega\left(\dfrac{2\xi}{\omega_n}\right)K_v}{\left(1 - \dfrac{\omega^2}{\omega_n^2}\right) + j\omega\,\dfrac{2\xi}{\omega_n}}$ $0 < \xi < 1$	+6 dB/octave; −6 dB/octave; ω_n; ω	$K_v^2(\omega_n \xi)$
5. $\dfrac{j\omega\tau_1 K_v}{(j\omega\tau_1 + 1)(j\omega\tau_2 + 1)}$	−20 log K_v; +6 dB/octave; 0; $1/\tau_1$; $1/\tau_2$; −6 dB/octave; ω	$K_v^2\left[\dfrac{1}{4\tau_2(1 + \tau_2/\tau_1)}\right]$

The right-hand integral of Eq. 8.24 may be shown to be, for any transfer function with more poles than zeros, equal to the product of the system's mid-band (or peak) gain squared times the transfer function's equivalent noise bandwidth in hertz. In other words,

$$G^2BW = \int_0^\infty |H(2\pi fj)|^2 \, df$$ 8.25

Gain squared-bandwidth products have been calculated using complex variable theory for a number of common transfer functions; they are valid for $S_i(f)$ defined for $0 < f < \infty$ Hz, that is, for one-sided PDS functions. Some of the more common gain squared-bandwidth functions are illustrated in Table 8.1. (Gain squared-bandwidth products for other, more complex transfer functions have been evaluated for two-sided power density spectrums in the reference.*) Note that the units of the noise bandwidth are hertz, despite the absence of 2π factors.

The equivalent noise Hz bandwidth for a signal transmission system can be found from Eq. 8.25. It is simply equal to the right-hand integral in the equation divided by the gain magnitude squared (for a dc system) or the mid-frequency gain magnitude squared (for a bandpass system).

8.4 General Treatment of Noise in Amplifiers

All electronic amplifiers and passive circuits containing resistances contribute noise to the signals they are conditioning. Hence the signal-to-noise ratio (SNR) at an amplifier's output depends on the internal sources of noise contributed by the amplifier as well as on the action of the amplifier's transfer function on the input signal and on input noise. In all future references to SNRs, we will use the ratio of mean-squared quantities.

It is conventional to refer all noise generated within an amplifier to the amplifier's input, as shown in Fig. 8.13. Here the amplifier is assumed to have an input resistance $R_1 \gg R_s$, which is the Thevenin source resistance, and to have a transfer function $H(j\omega)$. From Fig. 8.13, we see that it is necessary to define two independent

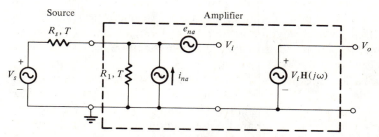

Figure 8.13   Two-noise-source model for a noisy amplifier.

* James, Nichols, and Phillips, 1947.

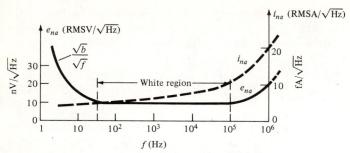

Figure 8.14   Typical $e_{na}(f)$ and $i_{na}(f)$ root power density spectrums for an amplifier with a JFET headstage.

noise sources at the amplifier's input; e_{na} and i_{na}. e_{na} is given as a root PDS and has units of rms volts per $\sqrt{\text{Hertz}}$ (RMSV/$\sqrt{\text{Hz}}$). e_{na} represents the short-circuited input equivalent noise voltage. i_{na} is the root PDS of the input equivalent noise current; it has units of rms amperes per $\sqrt{\text{Hertz}}$ (RMSA/$\sqrt{\text{Hz}}$). Plots of typical $e_{na}(f)$ and $i_{na}(f)$ are shown in Fig. 8.14. Note that both e_{na} and i_{na} have flat regions in their spectrums where they may be considered to be white noises. At low frequencies, e_{na} behaves with $1/\sqrt{f}$ characteristics. Individual transistors (BJTs and FETs) as well as amplifiers can have their noise behavior characterized by the use of e_{na} and i_{na}.

To characterise the virtues of low-noise amplifiers and transistors, manufacturers often specify a parameter called the noise figure (NF), given in decibels. Figure 8.13 shows how the noise figure is defined. First we note the definitions

$$\text{NF} = 10 \log(F) \qquad\qquad 8.26$$

and

$$F = \frac{\text{SNR}_{\text{in}}}{\text{SNR}_{\text{out}}} \qquad\qquad 8.27$$

The noise factor, F, is defined as the ratio of the input SNR (a ratio of mean-squared quantities) to the output SNR (also a ratio of mean-squared quantities). Because of the internal noise, SNR_{in} will always be greater than SNR_{out}. Hence F is always greater than one, and NF > 0 dB. Ideally, we would like the noise factor F to equal one (noiseless amplifier), or NF = 0 dB.

We may find a basic expression for F using the model of Fig. 8.13. We find the mean-squared quantities N_i, S_i, N_o, and S_o and then use Eq. 8.27 to get F. N_i is due to the mean-squared noise voltage arising in the Thevenin equivalent source resistor, and also includes any additive mean-squared noise contamination of V_s. The bandwidth used in evaluating N_i is the amplifier's equivalent noise bandwidth in hertz, here designated B. Hence, for a noise-free signal source V_s, we can write

$$N_i = 4kTR_sB \quad \text{MSV} \qquad\qquad 8.28$$

S_i is simply the mean-squared input voltage. No bandwidth factor is necessary if V_s is assumed to be deterministic. In general, we can write

$$S_i = \overline{V_s^2} \quad \text{MSV} \qquad\qquad 8.29$$

The mean-squared output signal is simply S_i times the amplifier's mid-band gain squared:

$$S_o \cong S_i K_v^2 \quad \text{MSV} \qquad\qquad 8.30$$

The mean-squared output noise is found by assuming that e_{na}, i_{na}, and the thermal noise in R_s have white spectrums, and applying Eqs. 8.24 and 8.25:

$$N_o = (4kTR_s + e_{na}^2 + i_{na}^2 R_s^2) K_v^2 B \quad \text{MSV} \qquad\qquad 8.31$$

If Eqs. 8.28 through 8.31 are substituted into definition 8.27, we can write the noise factor:

$$F = 1 + \frac{e_{na}^2 + i_{na}^2 R_s^2}{4kTR_s} \qquad\qquad 8.32$$

Note that this general expression for the noise factor, F, contains no terms dependent on the amplifier's bandwidth; these terms cancel out. In specifying the noise figure, manufacturers should give values for R_s, the range of frequencies over which NF(F) was measured, and the system temperature, T. In most cases, T is taken to be 290 K. It is observed that NF(F) is a function of frequency and R_s, tending to rise, at constant R_s, at very low and very high frequencies as a result of the frequency dependence of e_{na} and i_{na}. On the other hand, at a fixed frequency, NF(F) generally has a minimum in some range of R_s.

Some manufacturers of low-noise preamplifiers give spot noise figure contours (NF in decibels versus R_s and f) to characterize their products' noise behavior over a wide range of conditions. Spot noise figure contours for a commercial, state-of-the-art low-noise preamplifier are shown in Fig. 8.15.

To measure the spot noise figure, a narrow bandpass (or high-Q) filter is used to select a specific portion of $S_o(f)$ at the amplifier's output. A true rms voltmeter is then used to measure the portion of $S_o(f)$'s power in the bandpass of the selective filter.

One system for measuring the spot noise figure is shown in Fig. 8.16. This system is used as follows. First, the filter is set to the desired center frequency, f_o, then the desired R_s at temperature T K is connected to ground across the input (this corresponds to $e_N = 0$). The mean-squared output noise is found from the rms meter reading. If we assume that white noise is generated in R_s, e_{na}, and i_{na}, then the observed mean-squared noise voltage at the meter can be written as

$$N_{o(1)} = (4kTR_s + e_{na}^2 + i_{na}^2 R_s^2) K_v^2 B \quad \text{MSV} \qquad\qquad 8.33$$

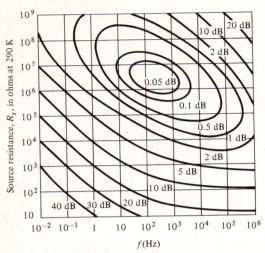

Figure 8.15   Spot noise figure contours for a typical commercial low-noise preamplifier. Note the ranges of operating frequency and source resistance over which NF$_{spot}$ is a minimum.

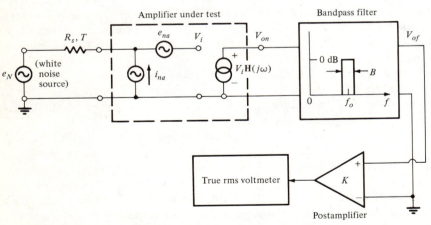

Figure 8.16   System for measuring an amplifier's spot noise figure. A narrowband filter, $\mathbf{F}(j2\pi f)$, with center frequency f_o and equivalent noise Hz bandwidth B, selects the spectral power in V_{on}. A flat bandpass postamplifier is generally needed to amplify V_{of} to a level the meter can read effectively. The noises of the filter and the postamplifier are generally negligible compared to that of the amplifier being measured.

where

K_v = the combined gain of the amplifier, the bandpass filter, and the postamplifier at $f = f_o$

B = the Hz noise bandwidth set by the narrow bandpass filter

Next, an external white noise source, e_N, is connected to the amplifier's input. The combined output resistance of the noise source plus a series external resistor must equal the R_s value used in the short-circuit noise measurement described previously. e_N is adjusted so that the rms voltmeter reads $\sqrt{2}$ times higher than the value obtained in the first, short-circuited input noise measurement. That is, the mean-squared output voltage, $N_{o(2)}$, will be $2N_{o(1)}$. Thus we can write

$$N_{o(2)} = 2N_{o(1)} = 2(4kTR_s + e_{na}^2 + i_{na}^2 R_s^2)K_v^2 B$$
$$= (4kTR_s + e_{na}^2 + i_{na}^2 R_s^2 + e_N^2)K_v B \quad \text{MSV} \qquad 8.34$$

Equation 8.34 is solved for the total mean-squared input noise spectrum:

$$e_N^2 = 4kTR_s + e_{na}^2 + i_{na}^2 R_s^2 \quad \text{MSV/Hz} \qquad 8.35$$

If Eq. 8.35 is substituted into Eq. 8.27 for the noise factor, F, we have

$$F_{\text{spot}} = \frac{e_N^2}{4kTR_s} \qquad 8.36$$

Note that in this expression for the spot noise factor, there are no terms for the bandwidth or center frequency, yet these parameters must be specified with F_{spot} for a complete description of this quantity. F_{spot} is actually found by determining the e_N to double the system's mean-squared output voltage, then dividing this value of e_N MSV/Hz by the calculated thermal noise spectrum for R_s, $4kTR_s$. Note that this method does not require a calibrated true rms voltmeter but does require an accurate knowledge of the noise generator's output spectrum, e_N RMSV/$\sqrt{\text{Hz}}$.

8.5 Transformer Optimization of Amplifier Output SNR and NF

Inspection of the spot noise figure contours illustrated in Fig. 8.15 shows that there is a region in (R_s, f) space where F_{spot} is a minimum. However, system requirements do not always give us the flexibility to choose R_s or the range of f in which to obtain the lowest noise figure values. In fact, a low F may be misleading; it may be the result of the ratio of two poor SNRs. It is better, therefore, to maximize the amplifier's output SNR than to minimize its F. This may be done given a fixed R_s value and amplifier e_{na} and i_{na}, by using an input transformer between the source and the amplifier. It should be noted that a practical transformer is a bandpass device; it loses efficiency at high and low frequencies, limiting the range of frequencies over

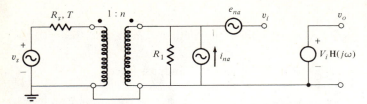

Figure 8.17 Improvement of output SNR using a transformer to couple an ac source, v_s, with low source resistance, R_s, to a noisy ac amplifier. We assume $R_1 \gg nR_s$.

which output SNR can be maximized. Figure 8.17 shows a Thevenin source (v_s, R_s) coupled to a noisy amplifier through an ideal (noiseless and lossless) transformer with a turns ratio of $1:n$.

The output signal for the tranformer-coupled amplifier is

$$S_o = \overline{V_s^2} n^2 K_v^2 \quad \text{MSV} \tag{8.37}$$

The mean-squared output noise is given by

$$N_o \cong [4kTR_s + e_{na}^2 + i_{na}^2(n^2 R_s)^2]K_v^2 B \quad \text{MSV} \tag{8.38}$$

Note that i_{na} sees a resistor of $n^2 R_s$ Ω looking toward the source. The output SNR is then found by dividing Eq. 8.37 by Eq. 8.38:

$$\text{SNR}_{\text{out}} = \frac{\overline{V_s^2}/B}{4kTR_s + e_{na}^2/n^2 + (i_{na}R_s n)^2} \tag{8.39}$$

The output SNR clearly has a single maximum for a nonzero value of the turns ratio, n. The value of the turns ratio that will give the maximum output SNR may be found by differentiating the denominator of Eq. 8.42 to find its minimum. When this is done, we find

$$n_o = \sqrt{\frac{e_{na}}{i_{na}R_s}} \quad \text{RMSV} \tag{8.40}$$

and the maximum output SNR possible is seen to be

$$\text{SNR}_{\text{out}} = \frac{\overline{V_s^2}/B}{4kTR_s + 2e_{na}i_{na}R_s} \tag{8.41}$$

In practice, we can never attain this value because transformers are not ideal; their windings have finite resistance and thus generate thermal noise, and their magnetic cores contribute noise from the Barkhausen effect. Use of a transformer to improve

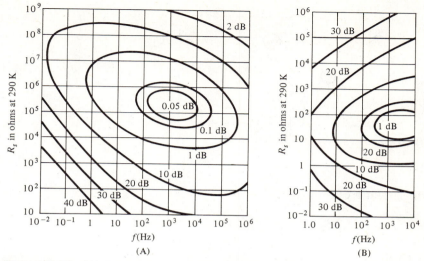

Figure 8.18 **(A)** Spot noise figure contours for a low-noise pre-amplifier with direct (no transformer) input. **(B)** Noise figure contours for the same preamplifier using a 1:100 turns ratio, low-noise transformer at its input. Note the shift of the minimum contour to a lower range of source resistance.

output SNR is generally worthwhile if

$$(e_{na}^2 + i_{na}^2 R_s^2) > 20 e_{na} i_{na} R_s \qquad 8.42$$

in the range of frequencies of interest.

It should be noted that transformer optimization of the output SNR also minimizes NF(F). n_o found in Eq. 8.40 gives a minimum noise factor,

$$F_{\min} = 1 + \frac{2 i_{na} e_{na}}{4kT} \qquad 8.43$$

Evidence of this minimization may be seen in Fig. 8.18 which shows the measured spot noise figure curves for a commercial low-noise preamplifier operating with its internal matching transformer having a 1:100 turns ratio. From Fig. 8.18, we see that this transformer shifts the locus of the minimum spot noise figure contour from $10^5\ \Omega < R_s < 10^6\ \Omega$ to a minimum of 1 dB from $20\ \Omega < R_s < 200\ \Omega$. The frequency range of the minimum spot noise figure contours is not changed significantly by the transformer.

8.6 Cascaded Noisy Amplifiers

In this section, we will illustrate a basic principle used in the design of low-noise amplifiers. Figure 8.19 shows three cascaded, noisy voltage amplifier stages. The i_{na}'s are not considered for simplicity (i_{na} noise terms are important only when $i_{na} R_s > e_{na}$).

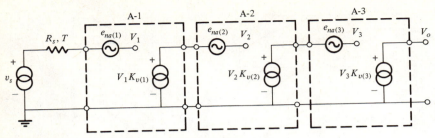

Figure 8.19   Three cascaded noisy amplifiers.

We write the output mean-squared signal as

$$S_{\text{out}} = \overline{V_s^2} K_{v(1)}^2 K_{v(2)}^2 K_{v(3)}^2 \quad \text{MSV}$$

8.44

The output mean-squared noise is

$$N_{\text{out}} = (4kTR_s + e_{na(1)}^2)(K_{v(1)}^2 K_{v(2)}^2 K_{v(3)}^2)B + (e_{na(2)}^2 K_{v(2)}^2 K_{v(3)}^2)B + (e_{na(3)}^2 K_{v(3)}^2)B$$

8.45

The input mean-squared noise is

$$N_{\text{in}} = 4kTR_s B$$

8.46

The noise factor for the three cascaded amplifiers is thus

$$F_{123} = 1 + \frac{e_{na(1)}^2}{4kTR_s} + \frac{e_{na(2)}^2}{K_{v(1)}^2 4kTR_s} + \frac{e_{na(3)}^2}{K_{v(1)}^2 K_{v(2)}^2 4kTR_s}$$

8.47

Inspection of Eq. 8.47 shows that if the first amplifier has a mid-band gain $K_{v(1)} > 5$, then the $e_{na(2)}$ and $e_{na(3)}$ terms will generally be negligible compared to the $e_{na(1)}$ term. That is, the cascaded system's noise factor will be set by the first stage, which should use low-noise components so that $e_{na(1)}$ and $i_{na(1)}$ will be a minimum. The other stages need not be of special low-noise design, as long as $K_{v(1)} > 5$. This is a fundamental principle in the design of low-noise voltage amplifiers.

8.7 Noise in Differential Amplifiers

A noisy, symmetrical differential amplifier is shown in Fig. 8.20. Because the amplifier has two active input devices, we will consider the noise sources arising in each input. From the Middlebrook equations (Chapter 5, Eqs. 5.7 and 5.8), we can write

$$v_o = \left(\frac{A_{dd} + A_{cc}}{2}\right) v_i + \left(\frac{A_{cc} - A_{dd}}{2}\right) v_i'$$

8.48

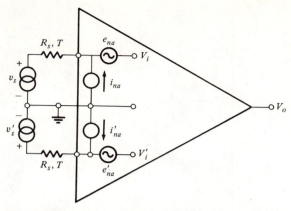

Figure 8.20 Noisy, symmetrical differential amplifier. Note that the root power spectrums are equal: $i_{na} = i'_{na}$ and $e_{na} = e'_{na}$.

Now the total mean-squared noise voltage PDS at the noninverting input is just

$$S_i(f) = e_{na}^2 + i_{na}^2 R_s^2 + 4kTR_s \quad \text{MSV/Hz} \qquad 8.49$$

and the noise voltage PDS at the inverting input is

$$S'_i(f) = e_{na}'^2 + i_{na}'^2 R_s^2 + 4kTR_s \quad \text{MSV/Hz} \qquad 8.50$$

Using Eqs. 8.49 and 8.50 in Eq. 8.48, and assuming that the root power spectra are white and equal, we find that the total mean-squared output noise is

$$N_{\text{out}} = (e_{na}^2 + i_{na}^2 R_s^2 + 4kTR_s)\left(\frac{A_{dd}^2 + A_{cc}^2}{2}\right)B \quad \text{MSV} \qquad 8.51$$

If we assume a purely difference-mode input signal, $v_1 = v_s$, $v'_s = -v_s$, and $v_{1(d)} = v_s$. Using Eq. 8.48, it is easy to show that

$$S_{\text{out}} = \overline{V_s^2} A_{dd}^2 \quad \text{MSV} \qquad 8.52$$

Hence the mean-squared SNR at the differential amplifier's output may be written

$$\text{SNR}_{\text{out}} = \frac{\dfrac{\overline{V_s^2}}{B}}{\left(\dfrac{e_{na}^2 + i_{na}^2 R_s^2 + 4kTR_s}{2}\right)(1 + \text{CMRR}^{-2})} \qquad 8.53$$

Note that the $1/\text{CMRR}^2$ term is much less than one, and may be neglected.

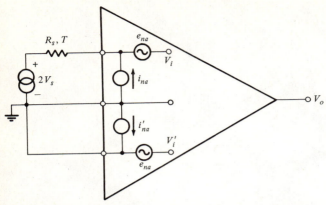

Figure 8.21 Differential amplifier with a single-ended input. Note that the source is two times the v_s used in Fig. 8.20 in order to get the same output voltage. Also, $e_{na} = e'_{na}$ RMSV/$\sqrt{\text{Hz}}$ and $i_{na} = i'_{na}$ RMSA/$\sqrt{\text{Hz}}$.

Often, differential amplifiers such as instrumentation amplifiers are used with one input grounded to condition single-ended signals. In Fig. 8.21, an input of $2v_s$ is used to give the same signal output as in the difference-mode input case just considered. The mean-squared output signal is found from Eq. 8.51 in this case to be

$$S_{\text{out}} = \overline{V_s^2}(A_{dd} + A_{cc})^2 \quad \text{MSV} \tag{8.54}$$

The mean-squared noise output voltage is again found from the superposition of mean-squared voltages:

$$N_{\text{out}} = (e_{na}^2 + i_{na}^2 R_s^2 + 4kTR_s)\left[\frac{(A_{dd} + A_{cc})^2}{4}\right]B + e_{na}'^2\left[\frac{(A_{cc} - A_{dd})^2}{4}\right]B \quad \text{MSV} \tag{8.55}$$

We finally can write the output SNR from the preceding results:

$$\text{SNR}_{\text{out}} = \frac{\dfrac{V_s^2}{B}}{\dfrac{i_{na}^2 R_s^2 + 4kTR_s}{4} + \dfrac{e_{na}^2(1 + \text{CMRR}^{-2})}{2(1 + 2\text{CMRR}^{-1} + \text{CMRR}^{-2})}} \tag{8.56}$$

Since the differential amplifier's CMRR $\gg 1$, the term multiplying $e_{na}^2/2$ is essentially unity, so we can finally write

$$\text{SNR}_{\text{out}} \cong \frac{\dfrac{\overline{V_s^2}}{B}}{\dfrac{i_{na}^2 R_s^2 + 4kTR_s}{4} + \dfrac{e_{na}^2}{2}}$$

8.57

That is, when the differential amplifier is used in a single-ended mode, there is an improved output SNR because the noises from R_s' and i_{na}' are not felt at the system output.

8.8 Noise in Feedback Amplifiers

There is a common misconception that negative voltage feedback, which is used to improve several factors in amplifier performance, also acts to improve an amplifier's noise factor and output SNR. The fact that negative feedback generally has little effect on a feedback amplifier's output SNR, and in some cases, makes it significantly lower than without feedback may be illustrated using the system of Fig. 8.22.

The differential amplifier in Fig. 8.22 is described by the Middlebrook equation

$$v_o = \left(\frac{A_{dd} + A_{cc}}{2}\right)v_i + \left(\frac{A_{cc} - A_{dd}}{2}\right)v_i'$$

8.58

The mean-squared output signal without feedback is

$$S_{\text{out}} = \frac{\overline{V_s^2}(A_{dd} + A_{cc})^2}{4} \quad \text{MSV}$$

8.59

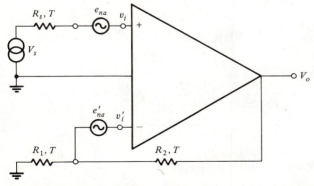

Figure 8.22   Differential amplifier with negative voltage feedback.

The mean-squared noise voltage at the output is given by

$$N_{\text{out}} = (e_{na}^2 + 4kTR)\left(\frac{A_{dd} + A_{cc}}{2}\right)^2 B + e_{na}'^2\left(\frac{A_{cc} - A_{dd}}{2}\right)^2 B \qquad 8.60$$

Hence the output SNR is

$$\text{SNR}_{\text{out}} = \frac{\dfrac{\overline{V_s^2}}{B}}{4kTR_s + \dfrac{2e_{na}^2(1 + \text{CMRR}^{-2})}{1 + 2\text{CMRR}^{-1} + \text{CMRR}^{-2}}} \qquad 8.61$$

which, because $\text{CMRR} \gg 1$, reduces to

$$\text{SNR}_{\text{out}} = \frac{\overline{V_s^2}/B}{4kTR_s + 2e_{na}^2} \qquad 8.62$$

Now if we assume that negative feedback is applied through the resistor network R_1 and R_2, a portion of v_o is subtracted from v_i, giving a gain

$$v_o = v_i \frac{\dfrac{A_{dd} + A_{cc}}{2}}{1 + \left(\dfrac{A_{dd} - A_{cc}}{2}\right)b} \qquad 8.63$$

where $b = R_1/(R_1 + R_2)$. Thus the mean-squared output signal voltage with negative feedback is

$$S_{oF} = \overline{V_s^2}\left[\frac{\dfrac{A_{dd} + A_{cc}}{2}}{1 + \dfrac{(A_{dd} - A_{cc})b}{2}}\right]^2 \qquad 8.64$$

The mean-squared noise output voltage with feedback can be shown to be

$$N_{oF} = (e_{na}^2 + 4kTR_s)B\left[\frac{\dfrac{A_{dd} + A_{cc}}{2}}{1 + \dfrac{A_{dd} - A_{cc}}{2}}\right]^2$$

$$+ (a^2 4kTR_1 + b^2 4kTR_2 + e_{na}^2)B\left[\frac{\dfrac{A_{cc} - A_{dd}}{2}}{1 + \dfrac{(A_{dd} - A_{cc})b}{2}}\right]^2 \qquad 8.65$$

where $a = R_2/(R_1 + R_2)$. Noting that CMRR $\gg 1$, and using a little algebra, we obtain

$$\text{SNR}_{\text{out}} = \frac{\dfrac{\overline{V_s^2}}{B}}{4kTR_s + 2e_{na}^2 + 4kT\left(\dfrac{R_1 R_2}{R_1 + R_2}\right)} \qquad 8.66$$

The output SNR for the amplifier with feedback is seen to be reduced by a noise term arising in the feedback resistors, R_1 and R_2. If these resistors were noiseless, the feedback amplifier would have the same SNR as one without feedback. This analysis assumes that the mean-squared input signal remains the same in both cases.

A further demonstration of this principle may be obtained by calculating the noise figure of the differential amplifier of Fig. 8.22 with and without feedback. F without feedback is found using Eq. 8.27:

$$F = 1 + \frac{2e_{na}^2}{4kTR_s} \qquad 8.67$$

assuming, as before, that CMRR $\gg 1$ and $e_{na} = e_{na}'$.

The noise factor with feedback is found to reduce to

$$F_{\text{with feedback}} = 1 + \frac{2e_{na}^2}{4kTR_s} + \frac{a^2 R_1 + b^2 R_2}{R_s} \qquad 8.68$$

where the third term on the right-hand side represents the increase in F caused by the noise in the feedback resistors. Here again, system noise performance is the same or worse when feedback is used, as measured by the system noise factor.

8.9 Examples of Calculating Threshold Input Signals in Noise-Limited Measurements

It is often necessary to be able to predict what minimum input signal level will produce a given SNR at the output of a signal-conditioning system. In this section, and in some of the end-of-chapter problems, we examine some typical situations in which a noise-limited, minimum input signal level is found.

EXAMPLE 8.1

Calculating the Minimum Resolvable dc Current in White and $1/f$ Noise

In the first example, an electrometer op-amp transresistor is used to detect a dc current, I_s, in the presence of thermal noise from resistors and the equivalent input noise sources of the op-amp. The system is shown in Fig. 8.23. The dc voltage output is

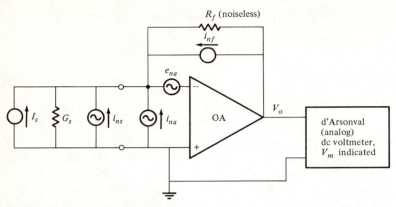

Figure 8.23 Noisy op-amp transresistor used to measure a low direct current, I_s. i_{nf} and i_{ns} are white root power spectrums arising in R_f and G_s, respectively. All of the noise sources are assumed to be white.

assumed to be indicated by a noiseless analog dc voltmeter whose mechanical deflection acts like an underdamped second-order low-pass filter. The indicated voltage, V_m, is related to the actual instantaneous output voltage, V_o, of the op-amp by the low-pass transfer function

$$\frac{V_m}{V_o}(s) = \frac{1}{1 + s^2 \xi/\omega_n + s^2/\omega_n^2} \qquad\qquad 8.69$$

where

ω_n = the undamped natural frequency in radians per second

ξ = the damping constant, usually around 0.4 for a typical analog voltmeter

To find the minimum resolvable I_s in this system, we must first set an output SNR criterion. This criterion is often arbitrary and is generally based on the application of the system. In this case, we will require $\text{SNR}_{\text{out}} = 9$. This means that the dc output voltage, V_m, at the threshold will be three times larger than the rms noise voltage perturbing the voltmeter.

To find the output mean-squared SNR, we note first that for dc,

$$\frac{V_o}{I_s} = \frac{V_m}{I_s} = -R_f \qquad\qquad 8.70$$

Hence the mean-squared output voltage is

$$\overline{V_m^2} = R_f^2 I_s^2 \qquad\qquad 8.71$$

The input noise arising in R_f and G_s is assumed to be white. This is also a good assumption for i_{na} of the amplifier. At low frequencies, e_{na}, however, is likely to have a strong $1/f$ component. That is, e_{na} at low frequencies can be approximated by

$$e_{na}^2(f) = \frac{b}{f} + e_{nao}^2 \qquad\qquad 8.72$$

where

e_{nao}^2 = the white noise component of $e_{na}^2(f)$

b = the $1/f$ coefficient of $e_{na}(f)$ in mean-squared volts

The amplifier output noise voltage PDS may be shown to be

$$S_n(f) = \left(\frac{b}{f} + e_{nao}^2\right)\left(1 + \frac{R_f}{R_s}\right)^2 + [4kT(G_f + G_s) + i_{na}^2]R_f^2 \qquad\qquad 8.73$$

The meter transfer function (Eq. 8.69) imposes an equivalent Hz noise bandwidth on the white noise terms in $S_n(f)$, and also on the $1/f$ term. Evaluation of the total output noise due to the white noise terms has been shown to be a simple process where the gain squared-bandwidth of the system is multiplied by the white noise terms (cf. Eq. 8.27). However, evaluation of the mean-squared output noise from the $1/f$ term involves use of Eq. 8.10, which is algebraically messy. To avoid this problem, we integrate the $1/f$ term in Eq. 8.66 between a lower frequency limit, f_1, which is taken arbitrarily as the inverse of the time it takes to read the meter, say 4 s. Thus, $f_1 = 0.25$ Hz. The upper frequency bound for the integration of the $1/f$ term is taken to be the actual Hz noise bandwidth of the meter's electromechanical filter. From example 3 in Table 8.1, we see that this bandwidth is $\omega_n/8$ Hz. Thus the mean-squared output SNR at V_m can be written

$$\text{SNR}_{\text{out}} = \frac{I_s^2}{[i_{na}^2 + 4kT(G_f + G_s) + e_{nao}^2(G_f + G_s)^2]\left(\dfrac{\omega_n}{8\xi}\right) + b\ln\left(\dfrac{\omega_n}{8\xi}4\right)(G_f + G_s)^2}$$

$$8.74$$

Now let us assume $\text{SNR}_{\text{out}} = 9$, and the following typical system parameters, and find $I_{s(\min)}$: $\omega_n/8 = 8/8(0.4) = 2.5$ Hz, $f_1 = 0.25$ Hz, $R_s = 10^7$ Ω, $R_f = 10^9$ Ω, $4kT = 1.66 \times 10^{-20}$ at 300 K, $e_{nao} = 10^{-7}$ RMSV/$\sqrt{\text{Hz}}$, $i_{na} = 2 \times 10^{-14}$ RMSA/$\sqrt{\text{Hz}}$, and $b = 4 \times 10^{-12}$ MSV. From these values, we find $I_{s(\min)}$ to be 9.548×10^{-13} A dc for the conditions given. $V_{o(\min)}$ would then be 945.8 μV, so the electronic dc voltmeter would have to have sufficient sensitivity to work in this range. ●

EXAMPLE 8.2

Calculating the Minimum Resolvable ac Input Voltage to a Noisy Inverting Op-Amp Amplifier

We will find the minimum coherent voltage input to an inverting op-amp that will give a mean-squared output SNR of 100. The system is shown in Fig. 8.24. i_{na} is not

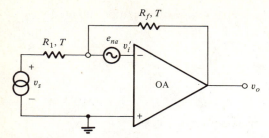

Figure 8.24 Inverting op-amp amplifier used to condition a sinusoidal signal, v_s.

considered, since we note that the voltages $i_{na}R_1$ and $i_{na}R_f$ are much less than e_{na}. The mean-squared output signal is

$$S_o = \left(\frac{V_s^2}{2}\right)\left(\frac{R_f}{R_1}\right)^2 \tag{8.75}$$

and the total mean-squared output noise is

$$N_o = 4kT(G_1 + G_f)R_f^2 B + e_{na}^2\left(1 + \frac{R_f}{R_1}\right)^2 B \tag{8.76}$$

The equivalent Hz noise bandwidth, B, may be found from the op-amp's specified unity-gain frequency, f_T, and the assumption that the gain-bandwidth product for the op-amp is a constant (this assumption is not strictly true for an inverting op-amp at low gains but is valid when $R_f/R_1 \gg 10$). If the op-amp has $f_T = 5$ MHz, and the gain $R_f/R_1 = 100$, then the closed-loop amplifier's corner frequency, f_c, is $(5 \times 10^6)/100 = 50$ kHz. The radian corner frequency, ω_c, is then $2\pi(5 \times 10^4)$ rad/s, and from example 1 in Table 8.1, we find the equivalent Hz noise bandwidth for a simple single-real-pole, low-pass filter to be $B = \omega_c/4 = 7.854 \times 10^4$ Hz.

Thus the output mean-squared SNR is given by

$$\text{SNR}_{\text{out}} = \frac{\left(\frac{V_s^2}{2}\right)\left(\frac{R_F}{R_1}\right)^2}{\left[4kT(G_1 + G_F)R_F^2 + e_{na}^2\left(1 + \frac{R_F}{R_1}\right)^2\right]B} \tag{8.77}$$

In this expression, we assume that e_{na} is white over the bandwidth, B, and that $\omega_s \ll \omega_c$.

Let us substitute typical numerical values for the parameters: $R_f = 10^5 \ \Omega$, $R_1 = 10^3 \ \Omega$, $B = 7.854 \times 10^4$ Hz, $4kT = 1.66 \times 10^{-20}$, $e_{na}^2 = 10^{-16}$ MSV/Hz, and $\text{SNR}_{\text{out}} = 100$. $V_{s(\text{min})}$ is found to be 43.18 μV peak. This value is relatively large because of the large noise bandwidth, B, and the requirement that the output mean-squared SNR be large (100). ●

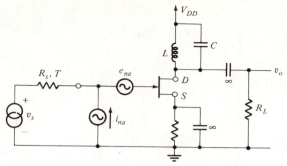

Figure 8.25   Tuned amplifier using a noisy JFET.

EXAMPLE 8.3

Calculating the Minimum Resolvable ac Input Voltage to a Noisy, Tuned JFET Amplifier

As a third example, we will calculate the minimum signal to give a specified output SNR for a simple tuned FET amplifier. The circuit is shown in Fig 8.25. The small-signal model for this FET tuned amplifier is given in Fig. 8.26. The input signal, v_s, is assumed to be a sine wave at the tuned center frequency, ω_n, of the RLC circuit. The transfer function for the amplifier is easily written as

$$\frac{V_o}{V_g}(j\omega) = \frac{-g_m L(j\omega)}{(j\omega)^2 LC + (j\omega)L(g_d + G_L) + 1} \qquad 8.78$$

which can be put in the general form

$$\frac{V_o}{V_g}(j\omega) = \frac{-\left(\dfrac{g_m}{g_d + G_L}\right)L(g_d + G_L)(j\omega)}{\dfrac{(j\omega)^2}{\omega_n^2} + j\omega\left(\dfrac{2\xi}{\omega_n}\right) + 1} \qquad 8.79$$

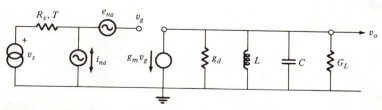

Figure 8.26   Small-signal model for the tuned amplifier of Fig. 8.25.

where

ω_n = the resonant frequency of the tuned amplifier

ξ = the damping factor

Note that it is possible to show that the Q of the tuned filter is given by

$$Q = \frac{\omega_n}{\Delta\omega} = \frac{1}{2\xi} \qquad 8.80$$

in which $\Delta\omega$ is the bandwidth between the two frequencies where $|(V_o/V_g)(j\omega)|$ is down by a factor of 0.707 from its peak value at $\omega = \omega_n$.

Because the frequency of the input signal is at ω_n, the mean-squared output signal is easily found to be

$$S_o = \left(\frac{V_s^2}{2}\right)\left(\frac{g_m}{g_d + G_L}\right)^2 \qquad 8.81$$

The total mean-squared noise voltage at the amplifier's output is assumed to come from the white noise PDSs at the input. Note that no thermal noise is assumed to arise in g_d; noise from g_d is accounted for in e_{na}. Johnson noise does enter the circuit from G_1, but we will show that its contribution is negligible.

To find the mean-squared output noise from these white sources, we make use of example 4 in Table 8.1 for transfer function gain-bandwidth products. In this case, the transfer function is in the form of the bandpass filter given in Eq. 8.79. It is used to find N_o:

$$N_o = (4kTR_s + e_{na}^2 + i_{na}^2R_s^2)\left(\frac{g_m}{g_d + G_L}\right)^2(\omega_n\xi) + 4kTG_L\left(\frac{1}{g_d + G_L}\right)^2(\omega_n\xi) \quad \text{MSV}$$
$$8.82$$

$\omega_n\xi$ is the equivalent Hz noise bandwidth for the tuned circuit. The resonant frequency of the tuned circuit is given by the well-known relation

$$\omega_n = \frac{1}{\sqrt{LC}} \text{ rad/s} \qquad 8.83$$

and the damping factor is found to be

$$\xi = \sqrt{\frac{L}{C}}\left(\frac{g_d + G_L}{2}\right) \qquad 8.84$$

Using the preceding relations, we may write a compact expression for the system's mean-squared output SNR:

$$\text{SNR}_{\text{out}} = \frac{\left(\dfrac{V_s^2}{2}\right)}{(4kTR_s + e_{na}^2 + i_{na}^2 R_s^2 + 4kTG_L g_m^{-2})\left(\dfrac{g_d + G_L}{2}\right)\left(\dfrac{1}{C}\right)} \qquad 8.85$$

The $4kTG_L$ is the white noise current PDS arising in the load resistor, R_L.

Now let us find $V_{s(\min)}$ required to give an output mean-squared SNR of unity. The following parameters are assumed: $R_s = 1\ \text{k}\Omega$, $R_1 = 100\ \text{k}\Omega$, $T = 300\ \text{K}$, $4kT = 1.66 \times 10^{-20}$, $g_m = 4 \times 10^{-3}$, $g_d = 13.33 \times 10^{-6}$, $f_n = 25\ \text{kHz}$, $\omega_n = 1.571 \times 10^5\ \text{rad/s}$, $L = 27.28\ \text{mH}$, $C = 1.485\ \text{nF}$, $\xi = 0.05$, $Q = 10$, and $e_{na} = 12\ \text{nV RMS}/\sqrt{\text{Hz}}$. Substituting these figures into Eq. 8.79, we find that $V_{s(\min)} = 1.588\ \mu\text{V}$. Use of a higher-$Q$ resonant circuit would, of course, further restrict the output mean-squared noise, and allow a lower $V_{s(\min)}$ to be found. ●

Additional examples of noise and bandwidth calculations in low-noise headstage design are given in the problems section for this chapter.

8.10 Noise Consideration in FETs, BJTs, and Low-Noise IC Amplifiers to Be Used in Headstage Design

In this section, we will examine some of the available low-noise transistors, op-amps, and instrumentation amplifiers used in the design of signal-conditioning systems used to amplify threshold signals. Threshold signals arise in many biomedical (electrophysiological) measurements, in sonar hydrophones, in ultrasound transducers, in geophones used in mineral prospecting, and, of course, in communications.

Certain JFETs are widely used in the design of discrete low-noise headstages where signals containing energy in the subsonic and audio regions of the spectrum are to be conditioned. Other special radio-frequency JFETs are used in video-frequency amplifiers, oscillators, and mixers where low noise is required.

Van der Ziel (1974) has shown that the theoretical thermal noise generated in the conducting channel of a JFET has a white spectrum which can be referred to the input of the JFET, that is, put in series with the gate lead. The PDS of this noise voltage can be written

$$\begin{aligned}
e_{na}^2 &= \frac{4kT}{g_m} = \frac{4kT}{g_{mo}} \sqrt{\frac{I_{DSS}}{I_{DQ}}} \\
&= \frac{(4kT/g_{mo})V_P}{V_P - V_{GSQ}} \quad \text{MSV/Hz}
\end{aligned} \qquad 8.86$$

where

g_{mo} = the FET's transconductance measured for $I_D = I_{DSS}$ at $V_{GS} = 0$, and for $|V_{DS}| > |V_P|$

I_{DSS} = the dc drain current measured at $V_{GS} = 0$ and $|V_{DS}| > |V_P|$

I_{DQ} = the quiescent drain current at the JFET's operating point, where $V_{GS} = V_{GSQ}$

In practice, the observed or measured device e_{na} in the flat or white region may differ significantly from that predicted from Eq. 8.86. Also, all measured e_{na} spectrums show a rising trend with decreasing frequency. This low-frequency behavior can be modeled by a $1/f$ term, as shown in

$$e_{na}^2(f) = \frac{4kT}{g_m}\left(1 + \frac{f_l}{f^n}\right) \tag{8.87}$$

where the exponent n has the range $1 < n < 1.5$ and is device- and lot-oriented. We usually take $n = 1$ for algebraic simplicity and speak of the low-frequency portion of the e_{na} spectrum as $1/f$ noise. The origins of the $1/f$ effect are poorly understood.

From Eq. 8.86, we can predict that a minimum e_{na} will occur for $V_{GS} = 0$ or $I_D = I_{DSS}$. This turns out not to be the case, however, because as $V_{GS} \to 0$, the power dissipation, and hence the temperature of the device, increases, causing the low-frequency e_{na}^2 spectrum to increase. If the JFET is cooled, e_{na} will decrease, however.

The parameter f_l in Eq. 8.87 is the corner frequency of the $1/f$ noise spectrum. It typically ranges from 10 Hz to 1 kHz, depending on the device design. f_l should be as low as possible for FETs used in low-frequency, low-noise amplifiers. The f_l value is generally quite high in JFETs used for video amplifiers, because in this type of device, $e_{na}(f)$ dips to values around $2\ \text{nV}/\sqrt{\text{Hz}}$ in the region of interest (10^5 Hz to 10^7 Hz).

JFET gate current noise is largely *shot noise,* which results from random fluctuations in the dc (average) component of gate leakage current, I_{GL}. The fluctuations in I_{GL} are due to the random occurrence of charge carriers having enough energy to cross the reverse-biased gate-channel diode junction. The PDF of shot noise can be shown to have a Gaussian form. In the JFET, the gate current noise has a white PDS and is given by

$$i_{na}^2 = 2qI_{GL} \quad \text{MSA/Hz} \tag{8.88}$$

where

$q = 1.602 \times 10^{-19}$ C (electron charge)

I_{GL} = the dc gate leakage current in amperes

Obviously, I_{GL} is small because the JFET gate-source junction is normally operated as a reverse-biased pn junction. I_{GL} is typically around 2 pA to 10 pA. An $I_{GL} = 10$ pA gives an $i_{na}^2 = 3.204 \times 10^{-30}$ MSA/Hz, or $i_{na} = 1.79$ fARMS$/\sqrt{\text{Hz}}$. $i_{na}(f)$ is seen to

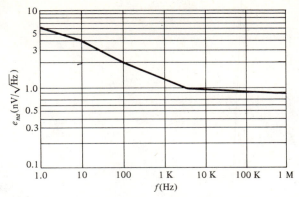

Figure 8.27 Short-circuit voltage noise root PDS for a 2N6550 JFET.

remain flat from below 10 Hz up to a break frequency of about 10 kHz in most quality audio-range, low-noise JFETs. Thus it is safe to assume that $i_{na}(f)$ has a white roots PDS in the audio range of frequencies. The low-frequency variation in i_{na}^2 can be described by Eq. 8.89:

$$i_{na}^2 = 2qI_{GL}\left(\frac{f_l + f}{f_l}\right) \quad \text{MSA/Hz} \tag{8.89}$$

where $f_l \cong 10$ kHz.

One of the best low-frequency, discrete, low-noise JFETs is the 2N6550. Its e_{na} spectrum is shown in Fig. 8.27. The i_{na} is given as 10 fA/$\sqrt{\text{Hz}}$ at 1 kHz for this FET. Other low-frequency, low-noise JFETs often used in discrete-component headstage designs include devices in the 2N4867A–2N4869A series, as well as the 2N5521, 2N4393, 2N5434, AD840, and 2N3821. In selecting such devices, the designer must consider trade-offs involving cost (low-noise JFETs are expensive), noise performance, frequency response (C_{iss}), and gain (g_m).

Low-noise FETs are also used for radio-frequency oscillators, mixers, and radio-frequency tuned amplifiers. Devices such as the 2N4416, 2N4417, E-300, 2N4223, and 2N4224 are used in the 100 MHz range and higher.

The treatment of noise in BJTs is more complex than in JFETs because the equivalent shot noise sources associated with the base and collector nodes vary with transistor operating point. It is possible to show that bias conditions can be found for a maximum output SNR in a conventional grounded-emitter BJT amplifier operating at mid-frequencies. In general, there will be some Q-point for a BJT amplifier of any configuration that will yield a maximum output SNR.

Figure 8.28(A) illustrates a conventional discrete, grounded-emitter, BJT amplifier. The hybrid-pi MFSSM for this amplifier is shown in Fig. 8.28(B). The base spreading resistance, r_x, is included in the model. Note that $R_s' = R_s + r_x$. We also assume that $R_b = [R_1 R_2/(R_1 + R_2)] \gg r_\pi$ and may be neglected in terms of its thermal noise and effect on the circuit's gain. Resistors R_L and R_s' are assumed to make

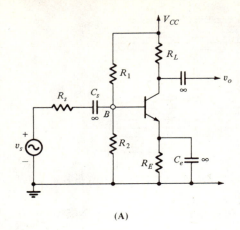

(A)

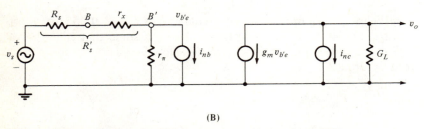

(B)

Figure 8.28 **(A)** Conventional discrete, grounded-emitter, BJT amplifier operated at mid-frequencies. **(B)** MFSSM for the amplifier showing the white shot noise current sources (root spectrums) associated with the base and collector nodes.

thermal noise. Current noise PDSs are associated with the base and collector nodes in the small-signal model. They are assumed to be white spectrums of shot noise origin and can be written

$$i_{nb}^2 = 2qI_{BQ} \quad \text{MSA/Hz} \tag{8.90}$$

$$i_{nc}^2 = 2qI_{CQ} = 2q(\beta I_{BQ}) \quad \text{MSA/Hz} \tag{8.91}$$

The mean-squared output signal voltage is given by

$$S_o = \overline{v_s^2} \left(\frac{r_\pi}{r_\pi + R_s'} \right)^2 (g_m R_L)^2$$

$$= \overline{v_s^2} \left(\frac{\beta R_L}{\dfrac{V_T}{I_{BQ}} + R_s'} \right)^2 \tag{8.92}$$

where it is evident that $g_m = \beta/r_\pi$. There are four terms in the mean-squared output noise voltage:

$$N_o = \left(4kTR_L + 2q(\beta I_{BQ})R_L^2 + \frac{4kTR_s'(\beta R_L)^2}{(V_T/I_{BQ} + R_s')^2} + \frac{2qI_{BQ}(g_m R_L)^2(r_\pi R_s')^2}{(r_\pi + R_s')^2} \right)B$$

8.93

Note that in the BJT small-signal model, $r_\pi = V_T/I_{BQ}$. The mean-squared output voltage SNR can be found from Eqs. 8.92 and 8.93:

$$\text{SNR}_{\text{out}} = \frac{\dfrac{\overline{v_s^2}}{B}}{4kTR_s' + 4kTR_L \dfrac{(V_T/I_{BQ} + R_s')^2}{\beta R_L'} + \dfrac{2qV_T^2}{\beta I_{BQ}} + 2qI_{BQ}R_s'^2}$$

8.94

Inspection of this relation shows that it has a maximum for some optimum I_{BQ}. The denominator of Eq. 8.94 can be differentiated and set to zero to find the $I_{BQ\,(\text{max})}$ for peak output SNR. (We neglect the denominator term for the noise from R_L because it is numerically small.)

$$I_{BQ\,(\text{max})} \cong \frac{V_T}{R_s'\sqrt{\beta + 1}}$$

8.95

If $V_T = 0.026$ V, $R_s' = 1000\ \Omega$, and $\beta = 50$, then $I_{BQ\,(\text{max})} = 3.64\ \mu\text{A}$. This development neglects high-frequency effects and the presence of low-frequency $1/f$ noise. It also assumes BJT beta is constant.

As we have seen in the case of JFETs, there are a number of excellent discrete low-noise BJTs. They are usually grouped into low-frequency and video-frequency types. Matched pairs of low-noise BJTs are offered by Analog Devices and include the AD818, AD820–AD822, and AD814–AD816 devices. The AD818 has an $e_{na}(f)$ of less than 2 nVRMS/$\sqrt{\text{Hz}}$ from 100 Hz to over 100 kHz.

Many excellent IC low-noise instrumentation amplifiers and op-amps are also available. Precision Monolithics puts out two op-amps with excellent low-frequency, low-noise characteristics: the OP-27 and the OP-37. Both of these IC devices have $e_{na}(f)$ spectrums with corner frequencies at 2.7 Hz and flat portions at 3 nVRMS/$\sqrt{\text{Hz}}$. The equivalent wideband input noise taken from 0.1 Hz to 100 kHz is only 1 μVRMS for these op-amps; this is state-of-the-art op-amp noise performance. The current noise root spectrum for these op-amps is flat beyond 1 kHz at 0.3 pARMS/$\sqrt{\text{Hz}}$; $i_{na}(f)$ has a low-frequency $1/f$ characteristic with a corner at 140 Hz. $i_{na}(f)$ rises to about 1.5 pARMS/$\sqrt{\text{Hz}}$ at 10 Hz because of its $1/f$ behavior.

Many manufacturers of IC instrumentation amplifiers and op-amps do not give $e_{na}(f)$ and $i_{na}(f)$ curves. Instead, spot values of these parameters may be provided, one in the flat e_{na} region and another in the low-frequency $1/f$ region. Other manufacturers specify equivalent input noise voltage, usually as volts peak-to-peak in a low-frequency range, and also as rms volts in a broad range of frequencies where $e_{na}(f)$ is presumably flat. For example, the Datel/Intersil AM-435 instrumentation amplifier is said to have $(1.3 + 670/G)\,\mu V$ peak-to-peak of input noise in a 0.1 Hz to 10 Hz bandwidth, where G is the overall amplifier gain. Its noise is also specified as $(8 + 450/G)\,\mu VRMS$ at its input in a 10 Hz to 10 kHz bandwidth. Another Datel instrumentation amplifier, the AM-201, is claimed to have an equivalent input noise of 1 $\mu VRMS$ measured over a 10 Hz to 10 kHz bandwidth. The equivalent white e_{na} for this amplifier is 10 $nVRMS/\sqrt{Hz}$. This is state-of-the art performance for e_{na}.

The Burr-Brown INA101 instrumentation amplifier has spot values of e_{na} and i_{na} specified. At $G = 1000$, $e_{na} = 18\,nVRMS/\sqrt{Hz}$ at $f = 10$ Hz, 15 $nVRMS/\sqrt{Hz}$ at 100 Hz, and 13 $nVRMS/\sqrt{Hz}$ at 1 kHz; $i_{na} = 0.8\,pARMS/\sqrt{Hz}$ at 10 Hz, 0.46 $pARMS/\sqrt{Hz}$ at 100 Hz, and 0.35 $pARMS/\sqrt{Hz}$ at 1 kHz.

One of the more phenomenal series of low-noise preamplifiers is made by Ferranti. The ZN459 and ZN460 series has flat e_{na} regions of 0.8 $nVRMS/\sqrt{Hz}$ specified from about 2 kHz to 50 kHz. The corner frequency for $1/f$ noise is about 1 kHz. e_{na} is given as 3 $nVRMS/\sqrt{Hz}$ at 25 Hz. Input noise current is specified as white at 1 $pARMS/Hz$. This amplifier clearly has superior noise performance, a voltage gain of 1000, and a 15 MHz bandwidth. It is a single-ended input device, however, and has a low input impedance of 7 kΩ.

The choice of which preamplifier to use to condition low-level signals is a complex one. As we have seen, output SNR depends on a number of parameters, including the signal's spectrum, $i_{na}(f)$, $e_{na}(f)$, R_s, T, R_{in}, and the amplifier's frequency response (gain-bandwidth). Normally, the signal's spectrum, R_s, T, and a desired amplification are given. The design process consists of finding a suitable IC amplifier at "reasonable cost" that will give the highest output SNR. This, in turn, requires a practical knowledge of what good, low-noise, IC amplifiers and discrete transistors are available. The reader is encouraged to build up-to-date files on the products of analog IC and discrete device-manufacturers. Most engineers will develop a bias toward products that in their experience have provided satisfactory solutions to past design problems; the author is no exception.

8.11 Quantization Noise

An important source of what may be considered to be random noise in the analog-to-digital and digital-to-analog conversion processes will now be treated. This noise, called quantization noise, arises when a signal with a broadband PDS containing no noise above one-half the sampling frequency (Nyquist criterion) is converted to digital form by an N-bit analog-to-digital converter (ADC). To illustrate the properties of this noise, we examine the quantization error associated with digital conversion. Because the input signal strictly obeys the Nyquist criterion, we show an error-generating model (Fig. 8.29) in which the analog input signal is not sampled, and in which the conversion processes are assumed to be instantaneous. Use of this model is valid because it is theoretically possible to reconstruct a Nyquist-limited signal back to analog form with an ideal low-pass filter. The ADC-DAC system has a step-wise or quantized transfer function, as illustrated in Fig. 8.30. Note that a quantizer

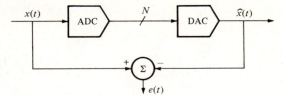

Figure 8.29 Model used to define quantization error, *e*, (quantization noise). The analog signal, *x(t)*, is assumed to obey the Nyquist criterion.

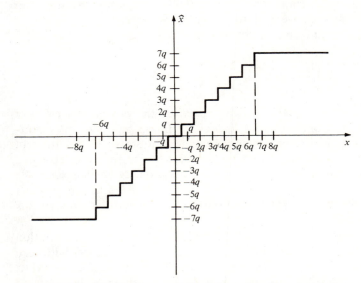

Figure 8.30 Nonlinear transfer function of an $N = 4$ quantizer.

has $(2^N - 1)$ levels, and in this case, $N = 4$. Note that compared to the direct path, the quantizer generates an error that can range over $\pm q/2$. q is the least-significant-bit (LSB) quantization level, and is given by

$$q = \frac{V_{max}}{2N - 1} \text{ V} \qquad 8.96$$

where

N = the number of binary bits or output lines from the ADC

V_{max} = the peak-to-peak voltage span of the ADC-DAC system

For example, if an 8-bit ADC is used to convert analog signals ranging from -5 V to $+5$ V, then, by Eq. 8.96, $q = 3.922 \times 10^{-2}$ V.

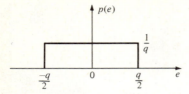

Figure 8.31   Probability density function assumed for quantization error (noise).

The signal $x(t)$ is assumed to have a PDF that is slowly varying over an interval in x, which is wide compared to q and which has a standard deviation much greater than q. Hence we can assume that the PDF for the error, $e(t)$, is rectangular, as shown in Fig. 8.31. The mean-squared error voltage is found by

$$E(e^2) = \overline{e^2} = \int_{-q/2}^{+q/2} e^2\left(\frac{1}{q}\right) de = \frac{(1/q)e^3}{3}\bigg|_{-q/2}^{+q/2} = \frac{q^2}{12} \quad \text{MSV} \qquad 8.97$$

That is, the error has zero mean and a variance given by the well-known result of Eq. 8.97. The quantization error, because of its low-amplitude, high-frequency characteristics, is usually treated as a broadband noise with a standard deviation of $q/\sqrt{12}$ RMSV, designated here as n_q. n_q may be considered to be added to the signal input, $x(t)$, which is undergoing analog-to-digital conversion. Note that once the signal is in digital form, additional equivalent noise may be added as the result of round-off in arithmetic operations. This round-off noise is beyond the scope of this text and will not be considered here.

It is possible to describe an inherent SNR for ADCs due to ADC quantization noise. We assume we have a bandwidth-limited Gaussian input signal, $x(t)$, with zero mean and an rms value of σ_x. We also assume that the dynamic range of the ADC is large enough to include ± 3 standard deviations of the signal. From Eq. 8.85, we can write

$$V_{\text{max}} = (2^N - 1)q = 6 \qquad 8.98$$

from which we see that

$$q = \frac{6\sigma_x}{2^N - 1} \qquad 8.99$$

Now this relation may be substituted into Eq. 8.98 to give an expression for the mean-squared quantization noise voltage contaminating the input signal to a noiseless ADC:

$$N_o = \frac{q^2}{12} = \frac{36\sigma_x^2}{12(2^N - 1)} \quad \text{MSV} \qquad 8.100$$

Table 8.2 SNR values
for an N-bit ADC
treated as a quantizer

N	dB SNR$_q$
4	18.8
6	31.2
8	43.4
10	55.4
12	67.5
14	79.5
16	91.6

Note: The total dynamic range
of the ADC is assumed to be
± 3 standard deviations of the
Gaussian input signal.

Hence the equivalent quantization mean-squared SNR can be written

$$SNR_q = (2^N - 1)0.333 \qquad\qquad 8.101$$

Note that the quantization SNR is independent of σ_x as long as σ_x is held constant under the dynamic range constraint given in Eq. 8.98. Table 8.2 presents SNR$_q$ values for some typical N values.

Since the outputs of many analog signal-conditioning systems ultimately interface with computers for one reason or another, it is important to realize that the conversion from analog to digital format is not without cost in terms of noise. Quantization noise is broadband and appears after the ADC's anti-aliasing filters; hence it will be subject to the problems of aliasing in digital processing. The advent of digital storage and playback of high-fidelity audio signals at maximum SNR requires the use of 14- and 16-bit ADCs to ensure that quantization effects are minimal.

SUMMARY

In this chapter, we have examined some sources of random noise arising in passive and active linear circuits. Our approach has been from the standpoint of the noise power density spectrum. We have shown how the spectrums arising from various noisy circuit components can be added in a mean-squared sense and then integrated over a certain noise bandwidth to find the total mean-squared noise output voltage. The concept of gain squared-bandwidth for white noise sources was introduced. The input stage, or headstage, of an amplifier was shown to be the critical portion of the circuit in the design of a low-noise signal-conditioning system; the system's overall signal-to-noise ratio is generally set by headstage performance.

Choosing low-noise components for use in an amplifier's headstage is often difficult; the designer must frequently make trade-offs in such designs. For example, the lowest-noise transistor may not have the f_T required to meet circuit gain-bandwidth requirements, and a noisier device may have to be used.

We have also shown that maximization of an ac amplifier's output signal-to-noise ratio may be accomplished, in certain circumstances, by using an input-coupling transformer with an optimized turns ratio. Adjustment of the amplifier's equivalent noise bandwidth or manipulation of headstage transistor quiescent (dc) bias conditions may also lead to improved output SNR.

Quantization noise, inherent in analog-to-digital conversion, was also shown to be an important source of broadband noise in digital signal-processing systems, such as compact disk recording; a table of quantization noise SNRs versus number of bits in the ADC and DAC was derived.

PROBLEMS

8.1 **(a)** Find an expression for the output noise voltage power density spectrum, $S_n(f)$ MSV/Hz. Let

$$i_s = I_s\sin(2\pi f_o t)$$

(b) Use the gain squared-bandwidth relations in Table 8.1 to find an expression for the mean-squared output noise voltage, N_o.

(c) Find an expression for the mean-squared output signal-to-noise ratio. Sketch and dimension SNR$_{out}$ versus C.

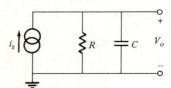

Figure P8.1

8.2 A parallel RLC circuit is at 300 K.

(a) Use Table 8.1 to find an algebraic expression for the total rms noise voltage at the circuit's terminals.

(b) Sketch and dimension a detailed Bode magnitude plot for the rms voltage per $\sqrt{\text{Hz}}$ at the circuit's terminals. Let $C = 10^{-8}$ F, $R = 1000\ \Omega$, and $L = 0.1$ mH. Assume 1 nVRMS/$\sqrt{\text{Hz}} = 0$ dB.

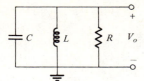

Figure P8.2

8.3 White noise with the power density spectrum $e_n^2 = \eta$ MSV/Hz is added to a sinusoidal signal (see Fig. P8.3). The noisy signal is low-pass filtered to improve the output SNR.

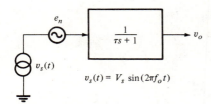

$$v_s(t) = V_s \sin(2\pi f_o t)$$

Figure P8.3

(a) Find an expression for the mean-squared output signal, S_o.
(b) Find an expression for the mean-squared output noise voltage, N_o.
(c) Find an expression for the optimum filter break frequency,

$$f_{b(opt)} = \frac{1}{2\pi \tau_{b(opt)}}$$

which will maximize SNR_{out}.
(d) Find an expression for the maximum output SNR.

8.4 A noisy op-amp is used to condition an ac current, $i_s(t)$. It is followed by a noiseless, ideal bandpass filter. Let

$$i_s(t) = I_s \sin(2\pi f_o t)$$

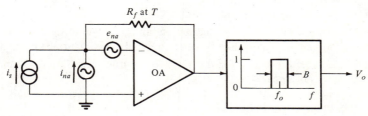

Figure P8.4

Assume $I_s = 10^{-11}$ A, $f_o = 10^4$ Hz, $R_f = 10^{10}$ Ω, $4kT = 1.66 \times 10^{-20}$, $e_{na} = 100$ nV/$\sqrt{\text{Hz}}$, and $i_{na} = 1$ pA/$\sqrt{\text{Hz}}$.

(a) Find an expression for the mean-squared output signal, S_o.

(b) Find an expression for the mean-squared output noise, N_o.

(c) Find B in hertz required to give a mean-squared $\text{SNR}_{\text{out}} = 10$.

8.5 A tuned FET amplifier (Fig. P8.5) is used to condition a sinusoidal signal, V_1. Assume the following:

$g_d = 0$ $V_1 = 10$ μV peak

$g_m = 0.004$ S $R_1 = 10^4$ Ω

R_L, R_s noiseless $4kT = 1.66 \times 10^{-20}$

$C_s \rightarrow$ short circuit at f_o $i_{na} = 1$ pA/$\sqrt{\text{Hz}}$ (white)

$v_1(t) = V_1 \sin(2\pi f_o t)$ $e_{na} = 10$ nV/$\sqrt{\text{Hz}}$ (white)

$R_L = 500$ Ω

$C = 10^{-8}$ F $f_o = \dfrac{1}{\sqrt{LC}2\pi}$

$L = 10^{-4}$ H

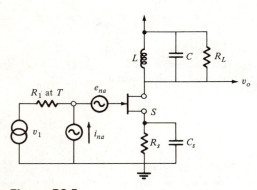

Figure P8.5

(a) Find $(V_o/V_1)(j\omega)$ in time-constant format.

(b) Find a numerical value for the output rms SNR.

8.6 A noisy but otherwise ideal op-amp is used to condition a low-level signal, $v_s(t)$. e_{na} is white noise, and R_1 and R_F make white thermal noise only. See Figure P8.6.

(a) Derive an algebraic expression for the amplifier's output rms SNR. (Assume the mean-squared input signal $= \overline{V_s^2}$.)

(b) Let $B = 10^3$ Hz, $e_{na} = 20$ nV/$\sqrt{\text{Hz}}$, $4kT = 1.66 \times 10^{-20}$, $\overline{V_s^2} = 1$ MSV, and $K_v = V_o/V_s = 100$. Plot the rms SNR_{out} versus R_F for 10^3 $\Omega \leq R_F \leq 10^7$ Ω.

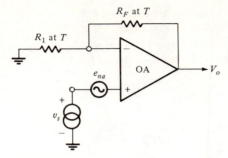

Figure P8.6

8.7 A low-noise, tuned video amplifier (VA) is used to amplify a sinusoidal 900 kHz ultrasound signal mixed with noise. The amplifier gain is 10^4 and its noise bandwidth is 10^4 Hz centered around 900 kHz. The system is shown in Fig. P8.7. Assume $i_{na} = 0$, $e_{na} = 10$ nV/$\sqrt{\text{Hz}}$ (white), $e_{ns} = 25$ nV/$\sqrt{\text{Hz}}$ (white), $R_s = 1200$ Ω, and $T = 300$ K.

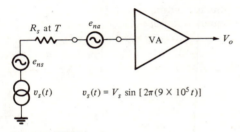

$$v_s(t) = V_s \sin [2\pi(9 \times 10^5 t)]$$

Figure P8.7

(a) Find the rms output noise.
(b) Find V_s required to give $\text{SNR}_{\text{out}} = 1$.

8.8 In the three-stage feedback amplifier shown in Fig. P8.8, K_1, K_2, and K_3 are scalar gains of cascaded stages, each with a different white noise equivalent input root spectrum, $e_{na(1)}$, $e_{na(2)}$, and $e_{na(3)}$, respectively.

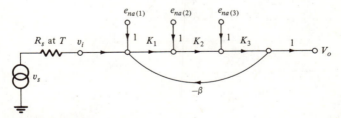

Figure P8.8

(a) Write an expression for the system's output mean-squared SNR.

(b) Which amplifier should be the headstage for maximum output SNR?

8.9 An Op-27 low-noise op-amp is to be connected to a Thevenin source, (V_1, R_1) through a noiseless, ideal transformer to maximize the mean-squared output SNR.

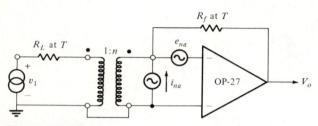

Figure P8.9

(a) Write an expression for the mean-squared output signal, S_o.

(b) Write an expression for the mean-squared output noise, N_o.

(c) Let $R_1 = 10\,\Omega$, $R_f = 10^4\,\Omega$, $e_{na} = 3 \times 10^{-9}$ RMSV/$\sqrt{\text{Hz}}$ (white), $i_{na} = 0.4$ pARMS/$\sqrt{\text{Hz}}$ (white), and $4kT = 1.66 \times 10^{-20}$. Find a numerical value for n_o to maximize the output SNR. (*Hint*: Minimize the denominator of SNR_{out} with respect to n.)

(d) What is the amplifier's gain, $K_v = V_o/V_1$, with this n_o?

8.10 K identical noisy JFETs are connected in parallel as shown in Fig. P8.10. Use the Norton MFSSM for the FETs with $g_d > 0$. Assume $e_{na(1)} = e_{na(2)} = \cdots = e_{nak} > 0$ RMSV/$\sqrt{\text{Hz}}$. Note that $e_{na(1)}(t) \neq e_{na(2)}(t) \neq \cdots \neq e_{nak}(t)$.

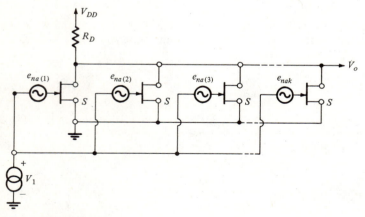

Figure P8.10

(a) Draw the MFSSM for the amplifier.
(b) Derive an expression for the mean-squared signal output, S_o.
(c) Assume R_d is noiseless. Write an expression for the mean-squared noise output, N_o.
(d) Find an expression for the mean-squared output signal-to-noise ratio.

8.11 Derive an algebraic expression for the spot noise factor of the system of Fig. 8.16 when the external white noise source, e_N, is replaced with a sinusoidal voltage,

$$v_s = V_s \sin(2\pi f_o t)$$

8.12 A noisy voltage amplifier drives a noisy power amplifier which is coupled to a load, R_L. The power amplifier has an output resistance, R_o.

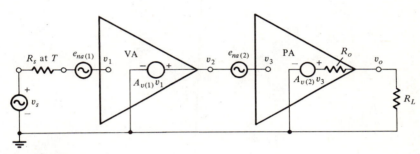

Figure P8.12

(a) Show that the output power in R_L (due to the source, V_s) is maximum when $R_o = R_L$. Find an expression for P_{max}.
(b) Show that the system's mean-squared voltage noise factor is independent of R_o and R_L. R_o and R_L are assumed to be noiseless. R_s, however, makes thermal white noise.

8.13 A noisy differential amplifier is used to condition a signal v_s. Assume the root spectrums are equal (i.e., $e_{na} = e'_{na}$ and $i_{na} = i'_{na}$), but certainly $e_{na}(t) \neq e'_{na}(t)$, and so on. Use

$$v_o = (A_{dd} + A_{cc})v_i + (A_{cc} - A_{dd})v'_i$$

Assume $R_s \ll R_{sg}$ at T K.

(a) Assuming differential connection, as shown in Fig. P8.13(a), develop an expression for the amplifier's output mean-squared signal-to-noise ratio. Assume a noise Hz bandwidth, B, and that all sources of noise are white.
(b) Now ground the inverting input at A (B is *not* connected to ground). Again, find the system's mean-squared output SNR. Compare it to the value found in part (a).

8.14 The system illustrated in Fig. P8.14 is used to measure a low-level dc signal in the presence of noise. A noiseless analog multiplier (AM) forms the product of the two noisy amplifier outputs. It is followed by a low-pass (averaging) filter with a noise bandwidth of B Hz. The filter output is then square-rooted. Assume $K_v = k'_v$, $e_{na} = e'_{na}$,

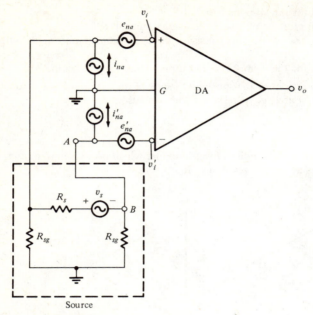

Figure P8.13(a)

and $i_{na} = i'_{na}$ (numerically equal root spectrums). Also it is obvious that $e_{na}(t) \neq e'_{na}(t)$ and $i_{na}(t) \neq i'_{na}(t)$; the input noise sources are statistically independent, uncorrelated, and have zero means.

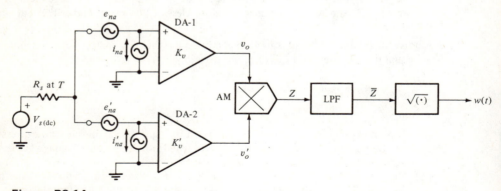

Figure P8.14

(a) Find an expression for the analog multiplier output, $w(t)$.
(b) Derive an expression for the mean-squared signal-to-noise ratio at w. Note that if $x(t)$ and $y(t)$ are statistically independent, then $\overline{x \cdot y} = \bar{x} \cdot \bar{y}$.

8.15 N identical noisy amplifiers, each with a short-circuit input white noise root spectrum, e_{naj} RMSV/$\sqrt{\text{Hz}}$, are connected as shown in Fig. P8.15. Write an expression for the system's noise factor, F. Assume all e_{naj} are equal.

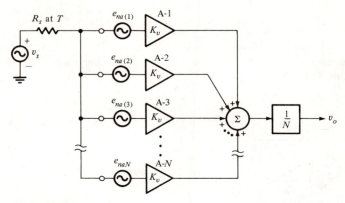

Figure P8.15

8.16 In the op-amp noninverting amplifier in Fig. P8.16, R_1 and R_2 at temperature T make thermal white noise, the op-amp's short-circuit input noise, e_{na}, is white, and the source, v_1, lies within the system's flat bandpass region of frequencies. The op-amp is characterized by the differential gain

$$V_o = \frac{K_v}{\tau s + 1}(V_i - V_i')$$

and $A_o = R_2/R_1$. Write an expression for the amplifier's mean-squared output SNR.

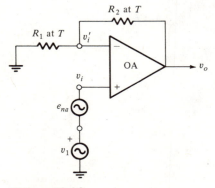

Figure P8.16

8.17 In the simple op-amp inverter circuit in Fig. P8.17, R_1 and R_2 at temperature T make thermal white noise, the op-amp's short-circuit input noise, e_{na}, is white, and the source, v_1, lies within the system's flat bandpass region of frequencies. The op-amp is characterized by the differential gain

$$V_o = \frac{K_v}{\tau s + 1}(V_i - V_i')$$

Define $\alpha = R_2/(R_1 + R_2)$, $\beta = R_1/(R_1 + R_2)$, and $A_o = \alpha/\beta = R_2/R_1$. Write an expression for the mean-squared output SNR. Assume $s_o = \overline{v_s^2} A_o^2$. Show what happens to the output SNR as A_o increases.

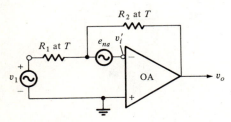

Figure P8.17

8.18 For each of the following, write an expression for the short-circuit current noise power density spectrum in mean-squared amperes per hertz.

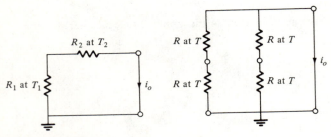

Figure P8.18

(a) The series resistor circuit
(b) The series-parallel resistor circuit

8.19 An RLC circuit is at 293 K.

(a) Use Table 8.1 to find an algebraic expression for the total rms noise voltage at the circuit's terminals.

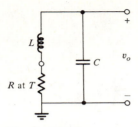

Figure P8.19

(b) Sketch and dimension a detailed Bode magnitude plot for the rms noise voltage per hertz at the terminals. Let $C = 100$ pF, $R = 750\ \Omega$, $L = 10^{-4}$ H, and $1\ \text{nV}/\sqrt{\text{Hz}} = 0$ dB.

8.20 A JFET is used as a low-noise headstage for an op-amp. Assume a FET MFSSM with g_m and $g_d > 0$. Also, R_1 and R_f make white thermal noise at 300 K. R_s is completely bypassed. The op-amp is ideal except for its short-circuit input white noise, $e_{na(2)}$ RMSV/$\sqrt{\text{Hz}}$. $i_{na(1)}$ and $e_{na(1)}$ of the JFET are also white. The FET's g_d is noiseless, and the system has an overall noise bandwidth of B Hz.

Find an expression for the system's mean-squared output SNR in terms of system parameters.

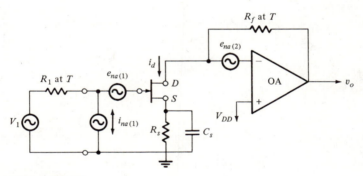

Figure P8.20

8.21 Two identical JFETs are used to make a grounded-gate/source-follower amplifier. Assume $g_{m(1)} = g_{m(2)}$, $g_{d(1)} = g_{d(2)} = 0$, $e_{na(1)} = e_{na(2)}$ (white), and $g_m \gg G_s$.

(a) Use MFSSM analysis to find expressions for v_o/v_g and v_o/v_g'.

(b) Find a general expression for the mean-squared output SNR of the amplifier; assume a noise bandwidth of B Hz.

(c) Now couple the Thevenin source (v_1, R_1) to the input node with an ideal, noiseless transformer (see Fig. 8.17) with turns ratio $1:n$. Find an expression for the turns ratio, n_o, that will maximize the output SNR.

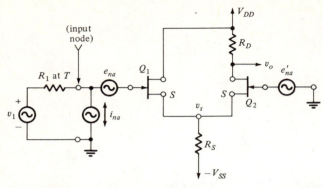

Figure P8.21

8.22 A transformer with a variable secondary winding is used to couple a Thevenin source to a noisy amplifier. The transformer primary has a dc resistance of r_p Ω. Its secondary has a dc resistance proportional to its turns ratio; that is, $r_s = n\rho$. Both r_p and r_s make thermal white noise at temperature T. The transformer is otherwise ideal.

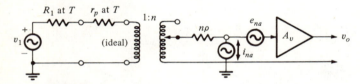

Figure P8.22

(a) Find an expression for the mean-squared output SNR. Assume a system noise bandwidth of B Hz.

(b) Does the system output SNR have a maximum with respect to n? Plot SNR_{out} versus n to illustrate this peak. Let $T = 300$ K, $R_1 = 100$ Ω, $r_p = 10$ Ω, $\rho = 10$ Ω/n, $e_{na} = 10$ nVRMS/$\sqrt{Hz}$, $i_{na} = 1$ pARMS/$\sqrt{Hz}$, $K_v = 1000$, $B = 1000$ Hz, and $v_1 = 1$ μV.

Chapter 9

Active Filters

Historically speaking, the first important use of operational amplifiers (op-amps) was in analog computers which were used to simulate the dynamics of various linear and nonlinear systems. These systems included mechanical, thermal, chemical, kinetic, and, of course, electromechanical dynamics. With the development of high-speed digital computers and compact, Fortran-based simulation languages such as CSMP, LISA, and PCAP, the use of analog computers declined sharply. At the same time, electronics engineers observed that many of the linear transfer function circuits using op-amps that were used in simulation studies could be adapted to applications in signal conditioning for communications and control. Thus active filters have evolved as a major application for op-amps.

A major advantage of op-amp active filters is that they generally do not require the use of inductances in their designs. It is far easier to build a nearly ideal resistor or capacitor than an "ideal" inductor. Also, inductors are bulky and heavy. Thus, modern active filter designs that use only resistors, capacitors, and op-amps can be made compact and inexpensive relative to passive *RLC* networks with the same pole-zero patterns. This is especially true when we consider filters designed to work in the subsonic spectral region. Also, design procedures for high-order, passive *RLC* filters are very complex, whereas the modular approach used in the synthesis of high-order active filters makes design more straightforward.

One major application of active filters involves an important class of sharp cutoff low-pass filter used to condition analog input signals that are being periodically sampled and converted to digital form as real-time inputs to digital computers. Such low-pass active filters, called anti-aliasing filters, are required to remove all significant signal spectral energy at and above one-half the sampling frequency.

Other uses of active filters include bandpass filtering of low-level signals to improve the signal-to-noise ratio at the filter output; also, sharply tuned bandpass filters used to select noisy coherent signals, and notch (band-reject) filters used to attenuate an unwanted coherent contamination (e.g., 60 Hz hum) of a broadband signal.

Active filters can also be made to be part of adaptive signal-processing systems. Here the filter's poles, zeros, or gain can be altered electronically in response to some criterion. A tracking active filter can be made to follow the frequency of a coherent input signal to preserve maximum signal-to-noise ratio at the filter's output.

The well-known Dolby B audio noise reduction system is an example of an adaptive active filter. In this system, the high-frequency bandpass characteristics are manipulated in response to the input signal's high-frequency spectral energy. Both tracking and adaptive active filters are treated in detail in Chapter 10.

Numerous types of active filter designs have evolved in the past twenty years. Many of them are unwieldy to design with, requiring odd, nonstandard, precision values of resistors and capacitors to achieve design goals. Also, in certain active filter designs, all components interact; that is, there can be no independent control of mid-band gain, damping factor, or break frequency. In the following sections, we will examine the designs of certain types of active filters, concentrating on types that have emerged as the easiest to use in design and discussing certain of their applications. We will focus on second-order (quadratic) filter realizations using one or more op-amps.

9.1 Types of Analog Active Filters

In the examples to follow, the op-amps are assumed to be ideal; that is, they have infinite differential gain, infinite input impedance, infinite bandwidth, no noise, and zero output resistance. The effects of departures from op-amp ideality will not be considered here. They are the subject of sensitivity analysis, which is beyond the scope of this text. It should be noted, however, that the open-loop frequency response of a practical op-amp does set a practical upper limit to an active filter's frequency response. The high-frequency asymptote of any active filter's frequency response cannot exceed the high-frequency asymptote of the op-amp's frequency response. In other words, the f_T of the active filter cannot exceed the f_T of the op-amp.

9.1.1 Infinite-Gain, Single-Feedback (IGSF) Active Filters

This active filter design uses two, two-port, passive R-C filters to derive a desired transfer function, which is generally quadratic (second-order) in form for simplicity and can be high-pass, bandpass, or low-pass. The basic IGSF active filter circuit is shown in Fig. 9.1. Note that the two two-port circuits shown in Fig. 9.1 can be described by the short-circuit admittance Y-parameters given in the equations

$$I_{1(a)} = Y_{11(a)}V_{1(a)} + Y_{12(a)}V_{2(a)} \qquad 9.1$$
$$I_{2(a)} = Y_{21(a)}V_{1(a)} + Y_{22(a)}V_{2(a)} \qquad 9.2$$

and

$$I_{1(b)} = Y_{11(b)}V_{1(b)} + Y_{12(b)}V_{2(b)} \qquad 9.3$$
$$I_{2(b)} = Y_{21(b)}V_{1(b)} + Y_{22(b)}V_{2(b)} \qquad 9.4$$

Now, because the op-amp is assumed to be ideal, its summing junction voltage, V_{sj}, is zero. Hence

$$V_{sj} = V_{2(a)} = V_{1(b)} = 0 \qquad 9.5$$

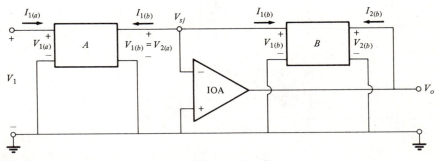

Figure 9.1   Infinite-gain, single-feedback active filter.

The filter's output voltage is $V_o = V_{2(b)}$, and its input is $V_1 = V_{1(a)}$. Also, it is obvious that $I_{2(a)} = -I_{1(b)}$. These observations, taken with the Y-equations for the two R-C filters, lead to the transfer function for the active filter:

$$\frac{V_o}{V_1} = -\frac{Y_{21(a)}}{Y_{12(b)}}$$

9.6

It is appreciated that the design of IGSF filters can proceed in a systematic but tedious manner. All components used in such designs interact, and no independent control of mid-band gain, break frequency, or Q is possible. IGSF filters offer no flexibility for adjustment of these parameters.

9.1.2 Infinite-Gain, Multiple-Feedback (IGMF) Active Filters

This class of active filter design suffers from the same drawbacks as do IGSF filters. Again, resistors and capacitors are used to synthesize quadratic high-pass, bandpass, and low-pass filters; all five components interact in the determination of mid-band gain, break frequency, and damping. Analysis of the IGMF filter proceeds using the parameters described in Fig. 9.2. If we assume that $V_{sj} = 0$ and note that $I_3 = I_5$, then we can write the node equation for V_2 as

$$V_2(Y_1 + Y_2 + Y_3 + Y_4) - V_o Y_4 - V_{sj} Y_3 = V_1 Y_1$$

9.7

Also,

$$I_3 = V_2 Y_3 = I_5 = -V_o Y_5$$

9.8

So,

$$V_2 = -\frac{V_o Y_5}{Y_3}$$

9.9

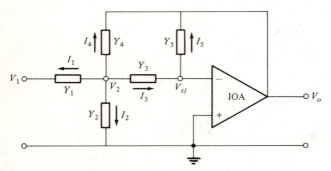

Figure 9.2   Infinite-gain, multiple-feedback active filter.

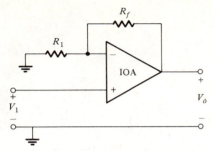

Figure 9.3 Noninverting VCVS. The voltage gain is $K_v = (1 + R_F/R_1)$. If $R_1 = \infty$ (open circuit) and $R_F = 0$ (short circuit), then obviously $K_v = +1$.

Substitution of Eq. 9.9 into Eq. 9.7 gives the transfer function

$$\frac{V_o}{V_1} = -\frac{Y_1 Y_3}{Y_5(Y_1 + Y_2 + Y_3 + Y_4) + Y_4 Y_3} \qquad 9.10$$

In addition to design complexity, it should be noted that it is difficult to get high-Q poles in bandpass designs without large spreads of element values, and designs with zeros other than at $s = 0$ or $|s| = \infty$ lead to excessive complexity.

9.1.3 Controlled-Source Active Filters

In this class of active filter, an op-amp is used to provide a VCVS with a low inverting or noninverting voltage gain, and along with various resistors and capacitors, to realize quadratic low-pass, bandpass, and high-pass filters. This class includes the well-known Sallen and Key designs, which are described in the following paragraphs.

A noninverting VCVS can easily be made with an op-amp and two resistors, as shown in Fig. 9.3. The Sallen and Key quadratic low-pass filter is illustrated in Fig. 9.4. Analysis of the behavior of this simple active filter design proceeds by noting

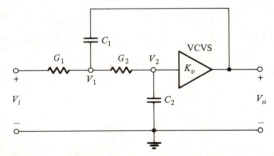

Figure 9.4 Sallen and Key, VCVS, low-pass active filter.

that $V_o = K_v V_2$ and

$$V_2 = \frac{V_1(1/sC_2)}{R_2 + 1/sC_2} = \frac{V_1}{1 + sR_2C_2} \qquad 9.11$$

and, by Kirchhoff's current law,

$$V_1\left(G_1 + \frac{1}{R_2 + 1/sC_2} + sC_1\right) - sC_1V_o = V_iG_1 \qquad 9.12$$

If we solve Eq. 9.11 for V_1 and substitute this expression into Eq. 9.12, we find

$$(sR_2C_2 + 1)\left(\frac{V_o}{K_v}\right)\left(G_1 + \frac{sC_2}{1 + sR_2C_2} + sC_1\right) - sC_1V_o = V_iG_1 \qquad 9.13$$

From which we can write the low-pass active filter's transfer function in time-constant format:

$$\frac{V_o}{V_i} = \frac{K_v}{s^2C_1C_2R_1R_2 + s[C_2(R_1 + R_2) + R_1C_1(1 - K_v)] + 1} \qquad 9.14$$

In the Sallen and Key format, $K_v = 1$, so the filter's transfer function reduces to

$$\frac{V_o}{V_i} = \frac{1}{s^2C_1C_2R_1R_2 + s[C_2(R_1 + R_2)] + 1} \qquad 9.15$$

This filter's break frequency is seen to be

$$\omega_n = \sqrt{\frac{1}{C_1C_2R_1R_2}} \qquad 9.16$$

and the damping factor is

$$\xi = \frac{C_2(R_1 + R_2)}{2}\sqrt{\frac{1}{C_1C_2R_1R_2}} \qquad 9.17$$

If R_1 is made to equal R_2, so that $R_1 = R_2 = R$, then the damping factor reduces to the simple expression

$$\xi = \sqrt{\frac{C_2}{C_1}} \qquad 9.18$$

Hence the ratio C_2/C_1 sets the Sallen and Key low-pass filter's damping factor, and simultaneous adjustment of R_1 and R_2 with a ganged, dual-variable resistor can set ω_n independently of ξ. The dc gain of this low-pass filter is, of course, unity.

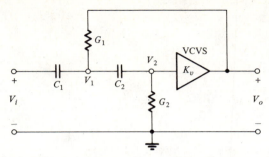

Figure 9.5   Sallen and Key, VCVS, high-pass active filter.

Figure 9.5 illustrates a Sallen and Key high-pass filter. Following a nodal analysis similar to that just used on the low-pass filter, we obtain the following transfer function, where we assume $K_v = 1$ and $C_1 = C_2 = C$:

$$\frac{V_o}{V_i} = \frac{s^2 C_1 C_2 R_1 R_2}{s^2 C_1 C_2 R_1 R_2 + s(C_1 R_1 + C_2 R_1) + 1} \qquad 9.19$$

Here again, the undamped natural frequency of the filter is

$$\omega_n = \sqrt{\frac{1}{C_1 C_2 R_1 R_2}} \qquad 9.20$$

and the damping factor is

$$\xi = \frac{R_1(2C)}{2} \sqrt{\frac{1}{C^2 R_1 R_2}} = \sqrt{\frac{R_1}{R_2}} \qquad 9.21$$

The high-frequency ($\omega \gg \omega_n$) gain for this filter is unity.

VCVS architecture can also be used to realize a quadratic bandpass filter, shown in Fig. 9.6. By solution of the node equations for V_1 and V_2, noting that $V_o = K_v V_2$,

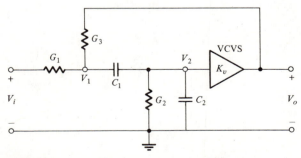

Figure 9.6   VCVS, band-pass active filter.

we find

$$\frac{V_o}{V_i} = \frac{s\left[\dfrac{K_vC_1G_1}{G_2(G_1 + G_3)}\right]}{s^2\left[\dfrac{C_1C_2}{G_2(G_1 + G_3)}\right] + s\left[\dfrac{C_1G_2 + (C_1 + C_2)(G_1 + G_3) - K_vC_1G_3}{G_2(G_1 + G_3)}\right] + 1}$$

$$\text{9.22}$$

If we let $K_v = 2$ and set $C_1 = C_2 = C$, we obtain the filter parameters

$$\omega_n = \frac{1}{C}\sqrt{G_2(G_1 + G_3)}$$

$$A_v = \frac{2G_1}{G_2 + 2G_1} \qquad \text{(at resonance)}$$

$$\text{9.23}$$

$$Q = \frac{G_2(G_1 + G_3)}{\sqrt{G_2(G_1 + G_3)(G_2 + 2G_1)}}$$

Note that the peak response (A_v) and Q are independent of C; $C_1 = C_2 = C$ can be used to set ω_n at constant Q and A_v.

9.1.4 State Variable Filter Synthesis

The state variable active filter design offers a compact and easy means for synthesizing filters of high order, providing low Q is used for complex-conjugate pole and zero pairs. If the pole and zero locations in the s-plane are known exactly, the numerator and denominator factors must be multiplied out to form a rational polynomial transfer function of the form

$$H(s) = \frac{V_o}{V_i} = \frac{s^mC_{m+1} + s^{m-1}C_m + \cdots + sC_2 + sC_1}{s^n + s^{n-1}a_n + \cdots + sa_2 + a_1}$$

$$\text{9.24}$$

A rational polynomial of this form can be written in the time domain as a set of state equations:

$$\dot{x} = \begin{bmatrix} 0 & 1 & 0 & 0 & \cdots & 0 \\ 0 & 0 & 1 & 0 & \cdots & 0 \\ 0 & 0 & 0 & 1 & \cdots & 0 \\ & & & \cdots & & \\ 0 & 0 & 0 & 0 & \cdots & 1 \\ -a_1 & -a_2 & -a_3 & -a_4 & & -a_n \end{bmatrix} x + \begin{bmatrix} 0 \\ 0 \\ 0 \\ \vdots \\ 1 \end{bmatrix} V_i$$

$$\text{9.25A}$$

$$V_o = \begin{bmatrix} c_1 & c_2 & c_3 & \cdots & c_m & c_{m+1} & 0 & 0 \end{bmatrix} x$$

$$\text{9.25B}$$

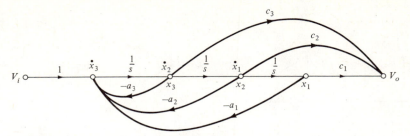

Figure 9.7 Signal flow graph for a third-order, state variable active filter.

From Fig. 9.7, we see that the polynomial coefficients (a_j and c_k) can be realized directly as gains. To better illustrate the state variable method of active filter synthesis, we use the following third-order system:

$$\dot{\mathbf{x}} = \mathbf{A}\mathbf{x} + \mathbf{b}V_i$$

$$V_o = \mathbf{c}_T\mathbf{x}$$

9.26A

$$\mathbf{A} = \begin{bmatrix} 0 & 1 & 0 \\ 0 & 0 & 1 \\ -a_1 & -a_2 & -a_3 \end{bmatrix} \quad \mathbf{b} = \begin{bmatrix} 0 \\ 0 \\ -1 \end{bmatrix} \quad \mathbf{c}_T = \begin{bmatrix} c_1 & c_2 & c_3 \end{bmatrix}$$

9.26B

From these equations we can construct a signal flow graph giving the transfer function Eq. 9.24 with $n = 3$ and $m = 2$. This signal flow graph is shown in Fig. 9.7. An op-amp realization of this filter is shown in Fig. 9.8.

Note that if we wish to realize a state variable low-pass filter of order n, the $\mathbf{c}_T$-matrix must be

$$\mathbf{c}_T = \begin{bmatrix} c_1 & 0 & 0 & 0 & \cdots & 0 \end{bmatrix}$$

9.27

To make a high-pass filter with an nth-order zero at the origin, we take the output from the x_n node. This cannot be represented with the nth-order $\mathbf{c}_T$ matrix notation just presented.

Using the state variable approach to active filter synthesis is most advantageous when the filter's denominator has an order of 4 or greater. Excellent quad op-amp ICs are available, such as the Texas Instruments TL-074, which allow compact realization of high-order transfer functions. One caveat in state variable active filter design is the possibility of using summing resistor values in the state summing amplifier that exceed the normal range for these components. Most op-amps will source or sink no more than 10 mA; hence input resistor values less than 1000 Ω are not advised. The upper bound on the summing op-amp's resistors is set by the dc offset voltage resulting from bias current flowing through the parallel combination of all the resistors connected to the summing junction.

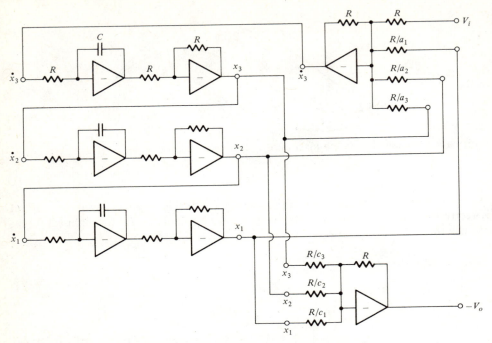

Figure 9.8 Op-amp realization of a third-order, state variable active filter.

One disadvantage of state variable filter design is seen when high-Q filters are synthesized. As an example, consider a fourth-order, high-Q, bandpass filter. It has two zeros at the origin, and poles at $s_1 = -10 + j990$, $s_2 = -10 - j990$, $s_3 = -10 + j1010$, and $s_4 = -10 - j1010$. This pole placement approximates a Q of 50. The denominator of the transfer function is found by multiplying out the two pairs of complex-conjugate roots, that is, forming the product $(s - s_1)(s - s_2)(s - s_3)(s - s_4)$. This is

$$[s^4 + s^3(40) + s^3(2 \times 10^6) + s(4 \times 10^7) + (10^{12})]$$
$$\quad\ a_4 \qquad\quad\ a_3 \qquad\quad\ a_2 \qquad\quad\ a_1$$

Because the gains to be realized vary over an enormous range (2.5×10^{10}:1), it is not practical to attempt synthesis of this transfer function with the state variable method.

9.1.5 The Biquad Active Filter

The biquad filter is widely used in modern active filter designs. The architecture of a biquad active filter using three op-amps is shown in Fig. 9.9. The transfer function,

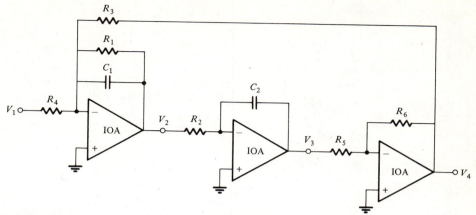

Figure 9.9 Basic biquad active filter. Note that when $R_5 = R_6$, then $V_4 = -V_3$.

V_3/V_1, can be written by inspection using Mason's rule:

$$\frac{V_3}{V_1} = \frac{\left[\dfrac{-1/(sC_1 + G_1)}{R_4}\right]\left(\dfrac{-1/sC_2}{R_2}\right)}{1 - \left[\left(\dfrac{-1/(sC_1 + G_1)}{R_4}\right)\left(\dfrac{-1/sC_2}{R_2}\right)\left(\dfrac{-R_6}{R_5}\right)\right]}$$

9.28

This transfer function can be put in time-constant form and is seen to be that for a quadratic low-pass filter:

$$\frac{V_3}{V_1} = \frac{\dfrac{R_5R_3}{R_4R_6}}{s^2\left(\dfrac{C_1C_2R_2R_3R_5}{R_6}\right) + s\left(\dfrac{C_2R_2R_3R_5}{R_1R_6}\right) + 1}$$

9.29

It is expedient to let $R_5 = R_6$. If this is done, the filter's parameters are found to be

$$\omega_n = \sqrt{\frac{1}{C_1C_2R_2R_3}}$$

$$K_v = \frac{R_3}{R_4}$$

$$\xi = \frac{1}{2R_1}\sqrt{\frac{C_2R_2R_3}{C_1}}$$

9.30

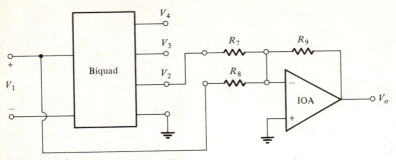

Figure 9.10   Biquad all-pass filter.

Note that R_1 can be used to set the filter's damping coefficient, that R_4 sets the dc gain, and that the break frequency and the damping factor are jointly interdependent on $C_1 C_2 R_2 R_3$. If the filter output is taken from the V_2 node, it is easy to show that the transfer function is bandpass, given by

$$\frac{V_2}{V_1} = \frac{-s\left(\dfrac{R_2 C_2 R_3}{R_4}\right)}{s^2 C_1 C_2 R_2 R_3 + s\left(\dfrac{C_2 R_2 R_3}{R_1}\right) + 1} \tag{9.31}$$

Here again, we let $R_5 = R_6$, and the relations for ω_n and ξ in Eqs. 9.30 are valid. The peak response of this filter for $\omega = \omega_n$ is easily seen to be R_1/R_4.

Other variations of the basic biquad active filter architecture can be used to synthesize all-pass, high-pass, and notch filters of the second order. A fourth op-amp summer is used to add system voltages in various ways to realize these filter characteristics.

A biquad *all-pass filter,* which has a flat magnitude response and a phase that varies with frequency, is shown in Fig. 9.10. All-pass filters are used to modify the phase response of electronic signal-conditioning systems without introducing frequency-dependent gain changes. The filter output, V_o, can be written as

$$V_o = -R_9\left(\frac{V_1}{R_8} + \frac{V_2}{R_7}\right) \tag{9.32}$$

Substituting Eq. 9.31 into Eq. 9.32 and rearranging terms gives

$$\frac{V_o}{V_1} = -\frac{\left(\dfrac{R_9}{R_8}\right)\left[s^2 C_1 C_2 R_2 R_3 + s\left(\dfrac{C_2 R_2 R_3}{R_1} - \dfrac{C_2 R_2 R_3 R_8}{R_4 R_7}\right) + 1\right]}{s^2 C_1 C_2 R_2 R_3 + s\left(\dfrac{C_2 R_3 R_2}{R_1}\right) + 1} \tag{9.33}$$

To make the numerator equal to the denominator but have roots (zeros) with positive real parts, we must set

$$\frac{C_2 R_2 R_3}{R_1} - \frac{C_2 R_2 R_3 R_8}{R_4 R_7} = -\frac{C_2 R_2 R_3}{R_1} \qquad 9.34$$

A simplification occurs if we let $R_7 = R_8 = R$. From Eq. 9.34, we find that R_4 must be

$$R_4 = \frac{R_1}{2} \qquad 9.35$$

Using the preceding assumptions, the transfer function for the all-pass filter is found to be

$$\frac{V_o}{V_1} = -\frac{\left(\dfrac{R_9}{R}\right)\left[s^2 C_1 C_2 R_2 R_3 - s\left(\dfrac{C_2 R_2 R_3}{R_1}\right) + 1\right]}{s^2 C_1 C_2 R_2 R_3 + s\left(\dfrac{C_2 R_2 R_3}{R_1}\right) + 1} \qquad 9.36$$

In this transfer function, the poles and zeros have the same natural frequency and are symmetrically placed around the imaginary ($j\omega$) axis in the s-plane. The natural frequency is seen to be

$$\omega_n = \sqrt{\frac{1}{C_1 C_2 R_2 R_3}} \ \text{rad/s} \qquad 9.37$$

and the damping factor is

$$\xi = \frac{1}{2R_1}\sqrt{\frac{C_2 R_2 R_3}{C_1}} \qquad 9.38$$

The filter's dc gain is just

$$K_v = -\frac{R_9}{R_8} = -\frac{R_9}{R} \qquad 9.39$$

When the damping factor equals 0.707 for this filter, the pole-zero plot appears as in Fig. 9.11.

The phase of this filter's frequency response can be found from the s-plane geometry in Fig. 9.11:

$$\psi = +180° - 2\tan^{-1}\left(\frac{\omega - \omega_d}{\omega_d}\right) - 2\tan^{-1}\left(\frac{\omega + \omega_d}{\omega_d}\right) \qquad 9.40$$

where the damped natural frequency, ω_d, in this case is $0.707\omega_n$.

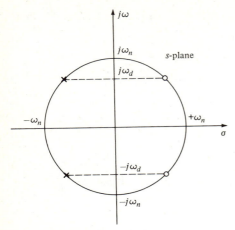

Figure 9.11   Pole-zero plot for the biquad all-pass active filter with $\xi = 0.707$.

The biquad architecture can also be used to realize a notch or bandstop filter. These filters are used to attenuate selectively a specific frequency, such as 60 Hz contamination of a signal. An example is shown in Fig. 9.12. Inspection of this figure shows that the output can be written

$$V_o = -R_9\left(\frac{V_1}{R_8} + \frac{V_2}{R_7} + \frac{V_4}{R_6}\right) \qquad 9.41$$

When we substitute the appropriate biquad transfer functions into Eq. 9.41 and rearrange terms, we find

$$\frac{V_o}{V_1} = -\frac{\left(\dfrac{R_9}{R_8}\right)\left[s^2C_1C_2R_2R_3 + s\left(\dfrac{C_2R_2R_3}{R_1} - \dfrac{C_2R_2R_3R_8}{R_4R_7}\right) + 1 - \dfrac{R_3R_8}{R_4R_6}\right]}{s^2C_1C_2R_2R_3 + s\left(\dfrac{C_2R_2R_3}{R_1}\right) + 1} \qquad 9.42$$

In this transfer function, we wish to place the zeros on the $j\omega$ axis at the stop frequency, ω_o; this configuration requires setting the numerator's s^1 coefficient to zero. Thus,

$$\frac{C_2R_2R_3}{R_1} - \frac{C_2R_2R_3R_8}{R_4R_7} = 0 \qquad 9.43$$

We see that Eq. 9.43 will be satisfied, giving purely imaginary zeros when $R_7 = R_8 = R$ and $R_1 = R_4$. Another algebraic requirement for this filter's design is that

$$\left(1 - \frac{R_3R_8}{R_4R_6}\right) > 0 \qquad 9.44$$

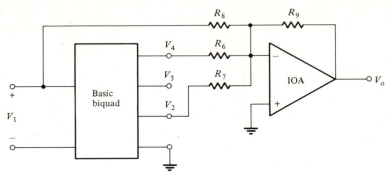

Figure 9.12   Biquad notch filter.

The bandstop zeros are located at $s = \pm j\omega_o$. ω_o is given by

$$\omega_o = \omega_n \sqrt{1 - \frac{R_3 R_8}{R_4 R_6}}$$ 9.45

ω_o is less than ω_n by an amount set by the square-root term in Eq. 9.45, giving the pole-zero plot shown in Fig. 9.13.

This filter's dc gain is

$$K_v = -\left(\frac{R_9}{R_8}\right)\left(\frac{R_4 R_6 - R_3 R_8}{R_4 R_6}\right)$$ 9.46

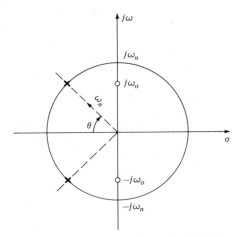

Figure 9.13   Pole-zero plot for the biquad notch filter.

The denominator's break frequency and damping factor are, as before,

$$\omega_n = \sqrt{\frac{1}{C_1 C_2 R_2 R_3}} \text{ rad/s}$$

$$\xi = \frac{1}{2R_1} \sqrt{\frac{C_2 R_2 R_3}{C_1}} \qquad\qquad 9.47$$

In this complex filter design, it is not possible to obtain independent control of the parameters ω_o, ω_n, and ξ. K_v can be altered with R_9, however.

9.1.6 Capacitor-Free Active Filters

In this unusual design of active filter, quadratic bandpass or low-pass forms are created using the natural dominant, open-loop poles of two frequency-compensated op-amps. The filters thus created use no external capacitors and work best as high-frequency bandpass filters. Analysis of the operation of a capacitor-free filter proceeds from inspection of Fig. 9.14. The output is taken as V_2. Because both op-amps are nonideal and compensated, their open-loop gains are given by

$$V_2 = -\frac{V_i K_{v(2)}}{T_1 s + 1} \qquad V_3 = \frac{\beta V_2 K_{v(2)}}{\tau_2 s + 1} \qquad\qquad 9.48$$

where

$\quad K_v = $ the op-amp's dc gain

$\quad \tau = $ the op-amp's time constant

$\quad \beta = $ an attenuation constant $(0 < \beta < 1)$

Analysis of this active filter requires solution of the node equation

$$V_i(G_1 + G_2 + G_3) - G_3 V_3 - G_2 V_2 = G_1 V_1 \qquad\qquad 9.49$$

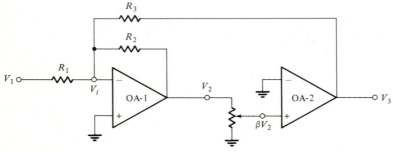

Figure 9.14   Quadratic bandpass active filter using no external capacitors.

Substituting from Eqs. 9.48, we can eliminate V_i and V_3 in Eq. 9.49. After some algebra, the transfer function, letting $G_1 + G_2 + G_3 = G_T$, is found to be

$$\frac{V_2}{V_1} = -\frac{\dfrac{K_{v(1)}G_1(\tau_2 s + 1)}{G_T + G_2 K_{v(1)} + G_3 \beta K_{v(1)} K_{v(2)}}}{s^2 \left(\dfrac{\tau_1 \tau_2 G_T}{G_T + G_2 K_{v(1)} + G_3 \beta K_{v(1)} K_{v(2)}}\right) + s \left(\dfrac{G_T(\tau_1 + \tau_2) + G_2 K_{v(1)} \tau_2}{G_T + G_2 K_{v(1)} + G_3 \beta K_{v(1)} K_{v(2)}}\right) + 1}$$

$$9.50$$

If we assume matched op-amps, where $\tau_1 = \tau_2 = \tau$ and $K_{v(1)} = K_{v(2)} = K_v$, and also $K_{v(1)} K_{v(2)} \beta \gg G_2 K_{v(1)} \gg G_T$, the transfer function reduces to

$$\frac{V_2}{V_1} \cong -\frac{\left(\dfrac{G_1}{G_3 \beta K_v}\right)(\tau s + 1)}{s^2 \left(\dfrac{\tau^2 G_T}{G_3 \beta K_v^2}\right) + s \left(\dfrac{G_2 \tau}{G_3 \beta K_v}\right) + 1} \qquad\qquad 9.51$$

where

$$\omega_n = \frac{K_v}{\tau} \sqrt{\frac{G_3 \beta}{G_\tau}} \quad \text{and} \quad Q = \frac{\sqrt{G_\tau G_3 \beta}}{G_2} \qquad\qquad 9.52$$

In this filter, ω_n and Q cannot be set independently, and there is a real zero at $1/\tau$ rad/s (a relatively low frequency). At denominator resonance, the transfer function peak gain is $-R_2/R_1$. If a true bandpass filter is to be realized, the transfer function, Eq. 9.51, should be multiplied by the high-pass function, $s\tau/(s\tau + 1)$. This multiplication may be accomplished by inserting a capacitor, C_1, in series with R_1, and making the $R_1 C_1$ time constant equal to τ_2.

9.1.7 Generalized Impedance Converter Active Filters

The generalized impedance converter (GIC) circuit, shown in Fig. 9.15, can be used as a module in the design of various types of biquad filters. The simplest application uses the GIC module to realize an equivalent high-Q inductance to ground. This equivalent inductance can be used in conjunction with a resistor and capacitor to make either a bandpass or a high-pass active filter. The GIC can also be used in more complex designs to obtain notch and low-pass filters.

Analysis of the general GIC circuit of Fig. 9.15 begins by noting that the input impedance of the GIC to ground is just

$$Z_{11} = \frac{V_1}{I_1} = \frac{V_1}{(V_1 - V_2)/Z_1} \qquad\qquad 9.53$$

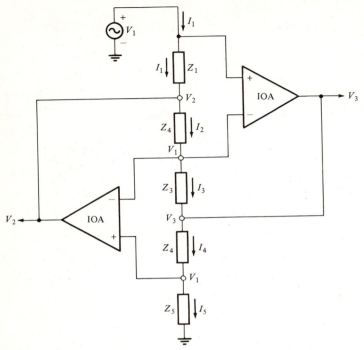

Figure 9.15   Generalized impedance converter circuit (GIC) using two ideal op-amps. It is easy to show that $Z_{11} = V_1/I_1 = Z_1Z_3Z_5/Z_2Z_4$.

Also, from the ideal op-amp assumption and Ohm's law, we have

$$I_2 = \frac{V_2 - V_1}{Z_2} = I_3 = \frac{V_1 - V_3}{Z_3} \qquad\qquad 9.54$$

$$I_4 = \frac{V_3 - V_1}{Z_4} = I_5 = \frac{V_1}{Z_5} \qquad\qquad 9.55$$

Equations 9.54 and 9.55 can be manipulated to find that

$$V_3 = V_1\left(1 + \frac{Z_4}{Z_5}\right) \qquad\qquad 9.56$$

$$V_2 = V_1\left(\frac{Z_3Z_5 - Z_2Z_4}{Z_3Z_5}\right) \qquad\qquad 9.57$$

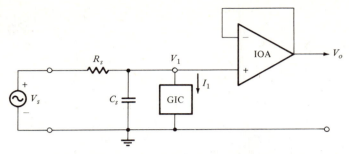

Figure 9.16 GIC-derived quadratic bandpass filter with a buffered output. The GIC emulates a pure inductance.

And, from the definition of Z_{11}, it is easy to show that

$$Z_{11} = \frac{V_1}{I_1} = \frac{Z_1 Z_3 Z_5}{Z_2 Z_4}$$

9.58

Now if Z_2 is made a capacitor, and the other impedances are made resistors, we can write

$$Z_{11}(s) = \frac{s(C_2 R_1 R_3 R_5)}{R_4}$$

9.59

Thus the GIC with $Z_2 = 1/sC_2$ appears like an inductance of $L = C_2 R_1 R_3 R_5 / R_4$ H from the V_1 node to ground. As long as the operating frequencies of the GIC are below the f_T's of the op-amps, this equivalent inductance appears to have negligible loss (series resistance) and therefore can be used to realize very high Q filters. Figure 9.16 illustrates a GIC bandpass filter. A third op-amp is used to buffer the filter output. However, reference to Eq. 9.56 shows that a buffered output is also available at the V_3 node, although with a gain greater than unity.

Other examples of GIC-derived active filter designs are given in the problems at the end of the chapter.

9.2 Switched-Capacitor Filters

Switched-capacitor filters (SCFs) are a unique class of active filter used primarily in voice band telecommunication applications. These filters are ideally suited for MOS VLSI implementation. They have low power consumption and occupy very little space, because nearly all their circuit elements are on the IC chip. SCFs can be used to realize high-Q bandpass, high-pass, low-pass, and bandstop filter designs using cascaded, basic biquad format active filters in which resistors are replaced with switched capacitors.

SCFs are inherently noisy, having only an 80 to 90 dB dynamic range. Their noise comes from aliasing in the sampling process, and from the MOS op-amps used which are noisier than BJT or JFET-BJT designs. In spite of their higher noise, SCFs are well suited to process telephone signals, and have found use in speech synthesis and recognition systems, PCM (CODEC) systems, echo cancellation on transmission lines, among other applications.

SCFs are, in reality, analog sampled data filters; many aspects of their exact analysis require the use of sampled data mathematics (difference equations and z-transforms). Because of their sampled data behavior, SCFs are subject to the problems of aliasing. (See Chapter 14, Sec. 14.1 for a detailed discussion of the sampling process and how it affects signals in the frequency domain.) For SCFs to function properly, the input signals that they process must contain no significant spectral power above one-half the capacitor switching frequency. This requirement is known as the Nyquist criterion.

Basically, an SCF is an active filter using op-amps in which each resistor is replaced by a MOS capacitor which is rapidly and synchronously switched by MOS transistor analog switches driven by a two-phase, non-overlapping clock waveform (see Fig. 9.17[C]). The MOS capacitors are generally small (in the picofarad range) and take little room on a MOS LSI chip. MOS capacitors have low dissipation factors and low tempcos (about 20 ppm/C). Their ratios on a chip can be held to better than 0.1%.

A heuristic analysis of SCFs treats a switched capacitor as a resistor. Referring to Fig. 9.17(B), we see that when clock phase 1 is high, the capacitor charges rapidly to V_1 through the MOS switch, Q_1. Q_1 then opens. Then phase 2 goes high, connecting C_s to V_2 through Q_2, completing the cycle. The net charge transferred through Q_2 is simply

$$\Delta q = C_s(V_1 - V_2) \tag{9.60}$$

The current transferred in one clock cycle (T_c) is thus

$$i_c = \frac{\Delta q}{\Delta t}$$
$$= \frac{C_s(V_1 - V_2)}{T_c} = \frac{V_1 - V_2}{1/f_c C_s} \tag{9.61}$$

Note that the denominator of the last form of Eq. 9.61 has the dimensions of resistance. Hence, in our analysis of SCFs, we can replace the MOS switches and switched capacitor by a simple equivalent resistor of value

$$R_{eq} = \frac{1}{f_c C_s} \tag{9.62}$$

Table 9.1 gives us a feel for the R_{eq} sizes encountered in practice.

Table 9.1 R_{eq} values for various switched-capacitor values (C_s) and clock frequencies (f_c)

C_s (pF)	f_c (Hz)		
	10^4	10^5	10^6
0.1	10^9	10^8	10^7
1	10^8	10^7	10^6
10	10^7	10^6	10^5
100	10^6	10^5	10^4

ohms

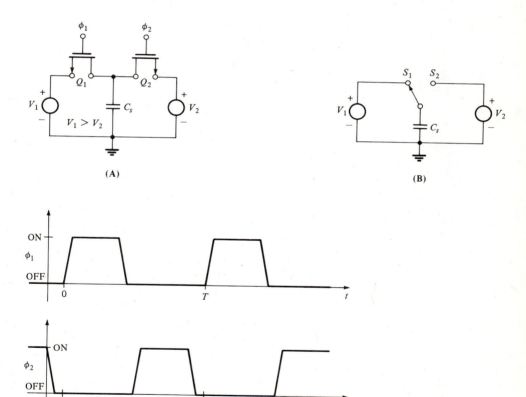

(A)

(B)

(C)

Figure 9.17 (A) MOS transistor switched capacitor. (B) Equivalent circuit for (A). (C) Two-phase clock waveform used to switch the MOS transistors.

A major problem in the design of SCFs is the parasitic (stray) capacitance that exists between the bottom electrode of C_s and the semiconductor substrate (small-signal ground). There is also stray capacitance to ground between the top electrode of C_s and its leads. These parasitic capacitances can be of the order of 5% to 20% of C_s depending on the IC fabrication technique. The negative effects of parasitic capacitances can be mitigated by using a dual SPDT MOS switch geometry (Grebene, 1984) employing four MOS transistors, as shown in Fig. 9.18.

To illustrate how this architecture works, consider the noninverting switched-capacitor integrator shown in Fig. 9.19 with parasitic capacitances C_p and C_p'. When clock phase 1 goes high, Q_1 and Q_4 turn ON, and Q_2 and Q_3 are OFF. C_s and C_p charge rapidly to V_1. C_p' has zero steady-state charge because Q_4 shorts it to ground in this half clock cycle. In the next half clock cycle, phase 1 goes low and phase 2 goes

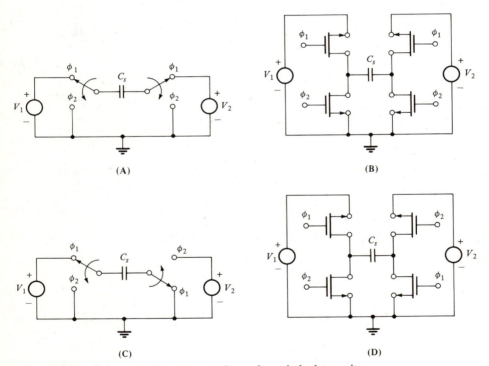

Figure 9.18 Four-switch implementations of a switched capacitor designed to reduce the effects of parasitic capacitance. **(A)** Equivalent circuit of a noninverting switched capacitor. **(B)** MOS switch implementation of (A). **(C)** Equivalent circuit of an inverting switched capacitor. **(D)** MOS switch implementation of (C).

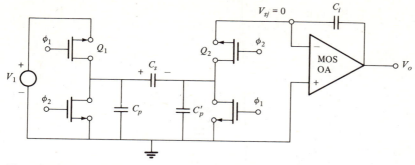

Figure 9.19 Switched-capacitor, noninverting integrator. The four MOS switches and the parasitic capacitors to ground are shown.

high; Q_2 and Q_3 turn ON, shorting the left side of C_s and C_p to ground. C_p' remains uncharged because it is connected to the summing junction, where $V_{sj} = 0$. Hence neither parasitic capacitance affects the charge transferred through C_s; the R_{eq} of C_s is thus not affected by C_p and C_p'. The waveforms seen in the operation of the non-inverting switched-capacitor integrator of Fig. 9.19 are shown in Fig. 9.20. When C_s

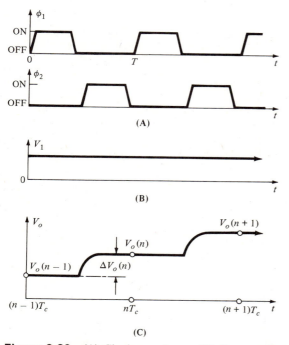

Figure 9.20 **(A)** Clock waveforms. **(B)** Constant input voltage. **(C)** Output voltage of the noninverting switched-capacitor integrator of Fig. 9.19.

is connected to the summing junction, the op-amp holds V_{sj} to zero by driving V_o rapidly positive. This action causes C_i to accumulate charge. The charge is supplied to C_s on the $(n-1)$th clock cycle. That is,

$$q(n) = V_1 C_s = V_o C_i \qquad\qquad 9.63$$

The change in V_o between the $(n-1)$th and nth clock cycle is thus

$$V_o(n) = \frac{+V_1(n-1)C_s}{C_i} \qquad\qquad 9.64$$

Now a complete expression for V_o can be written as a difference equation:

$$V_o(n) = V_o(n-1) + \frac{V_1(n-1)C_s}{C_i} \qquad\qquad 9.65$$

This difference equation can be Fourier transformed to yield

$$\mathbf{V}_o(j\omega) = \mathbf{V}_o(j\omega)e^{-j\omega T_c} + \mathbf{V}_1(j\omega)\left(\frac{C_s}{C_i}\right)e^{-j\omega T_c} \qquad\qquad 9.66$$

where T_c is the clock (sampling) period. The $e^{-j\omega T_c}$ term comes from the delay operation in sampled charge transfer. Equation 9.66 can be rearranged to obtain the integrator's frequency response function:

$$\frac{\mathbf{V}_o}{\mathbf{V}_1}(j\omega) = \frac{\left(\dfrac{C_s}{C_i}\right)e^{-j\omega T_c}}{1 - e^{-j\omega T_c}} \qquad\qquad 9.67$$

If the numerator and denominator of Eq. 9.67 are multiplied by $(2j)e^{+j\omega T_c/2}$, and the Euler identity for $\sin(x)$ is used, we find

$$\frac{\mathbf{V}_o}{\mathbf{V}_1}(j\omega) = \frac{\left(\dfrac{C_s}{C_i}\right)e^{-j\omega T_c/2}}{2j\dfrac{(e^{j\omega T_c/2} - e^{-j\omega T_c/2})}{2j}} = \frac{\left(\dfrac{C_s}{C_i}\right)e^{-j\omega T_c/2}}{2j\sin\left(\dfrac{\omega T_c}{2}\right)} \qquad\qquad 9.68$$

For $\omega \ll 2\pi/T_c$, $\sin(x) = x$ in radians, and Eq. 9.68 reduces to

$$\frac{\mathbf{V}_o}{\mathbf{V}_1}(j\omega) \cong \left(\frac{1}{j\omega}\right)\left(\frac{C_s f_c}{C_i}\right)e^{-j\omega T_c/2} \qquad\qquad 9.69$$

This is simply the frequency response function of a noninverting integrator with gain $C_s f_c / C_i$. At low frequencies, this gain magnitude is the same as that for the traditional op-amp analog integrator shown in Fig. 9.21. Note that the phase lag term in Eq. 9.69,

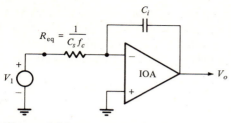

Figure 9.21 Analog integrator equivalent of the switched-capacitor integrator of Fig. 9.22(B) when the highest frequency in V_1 is much less than $f_c/2$.

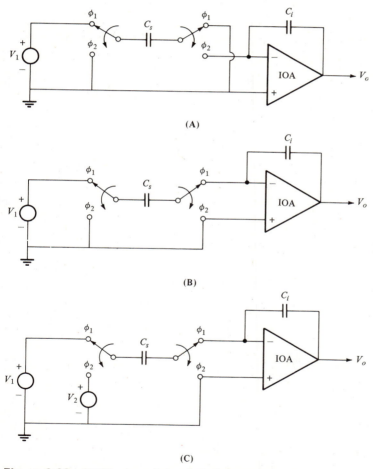

(A)

(B)

(C)

Figure 9.22 (A) Noninverting switched-capacitor integrator. (B) Inverting switched-capacitor integrator. (C) Differential switched-capacitor integrator. $V_o = -(V_1 - V_2)(f_c C_s/C_i)/s$.

although small at low frequencies, can be a problem in designing cascaded SCFs. Techniques exist for compensating for this phase error (Grebene, 1984).

It is also possible to realize inverting and differential input integrators using quad MOS switches and a single switched capacitor. These designs are shown in Fig. 9.22.

We will now analyze a few simple examples of SCFs.

EXAMPLE 9.1 ●

Design of a Noninverting, Real-Pole, Switched-Capacitor, Active Filter

The first example shown in Fig. 9.23(A), is a noninverting, single-real-pole, low-pass filter (LPF). Figure 9.23(B) illustrates its equivalent circuit. By inspection of Fig. 9.23(B), we see that the transfer function of the switched-capacitor LPF is

$$\frac{V_o}{V_1}(s) = \frac{\dfrac{1}{G_b + sC_i}}{R_a}$$

$$= \frac{\dfrac{R_b}{R_a}}{1 + sC_iR_b} \qquad\qquad 9.70$$

$$= \frac{\dfrac{C_a}{C_b}}{1 + s\dfrac{C_i}{f_cC_b}}$$

The filter's dc gain is set by the ratio C_a/C_b, and the break frequency is at

$$\omega_o = \frac{f_cC_b}{C_i} \text{ rad/s} \qquad\qquad 9.71$$

Note that ω_o is clock frequency dependent. ●

EXAMPLE 9.2 ●

Design of a Biquad, Bandpass, Switched-Capacitor, Active Filter

As a second example of an SCF realization of an active filter, consider the basic analog biquad filter illustrated in Fig. 9.9. The bandpass transfer function for this filter is given by Eq. 9.31, and expressions for its ω_n and damping factor are given in Eq. 9.30. Figure 9.24 illustrates an SCF realization of the biquad bandpass filter.

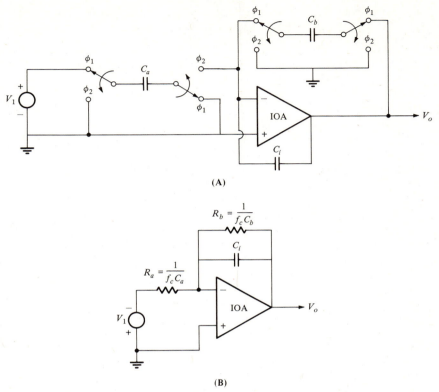

(A)

(B)

Figure 9.23 **(A)** Noninverting, real-pole, switched-capacitor LPF.
(B) Equivalent circuit of the switched-capacitor LPF.

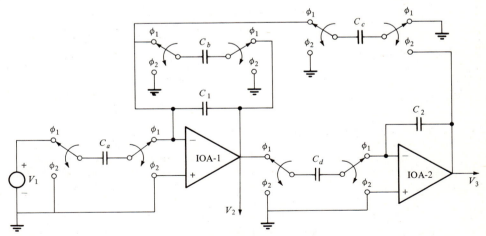

Figure 9.24 SCF realization of a biquad bandpass filter. Notice
that the switching of C_c performs the necessary sign inversion
around the feedback loop, eliminating the need for a third op-amp.

Recognizing that at signal frequencies much less than $2\pi/T_c$, the switched capacitors can be replaced by equivalent resistors according to Eq. 9.62, we can use Mason's rule to write the transfer function, V_2/V_1:

$$\frac{V_2}{V_1}(s) = \frac{-\left[\dfrac{1/(sC_1 + f_cC_b)}{1/f_cC_a}\right]}{1 - \left[\left(\dfrac{-1}{sC_2(1/f_cC_d)}\right)\left(\dfrac{+1/(sC_1 + f_cC_b)}{1/f_cC_c}\right)\right]}$$

9.72

The bandpass filter transfer function of Eq. 9.72 can be put into time-constant format:

$$\frac{V_2}{V_1}(s) = \frac{-s\left(\dfrac{C_2C_a}{C_cC_d}\right)}{s^2\left(\dfrac{C_1C_2}{f_c^2C_cC_d}\right) + s\left(\dfrac{C_2C_b}{f_cC_cC_d}\right) + 1}$$

9.73

where

$$\omega_n = f_c\sqrt{\frac{C_cC_d}{C_1C_2}} \text{ rad/s}$$

9.74

and

$$Q = \frac{\sqrt{C_1C_2C_cC_d}}{C_2C_b} = \frac{1}{2\xi}$$

9.75

The gain at $\omega = \omega_n$ is

$$\frac{V_2}{V_1}(j\omega_n) = -\frac{C_a}{C_b}$$

9.76

Ratios of CMOS capacitor values are seen to be important in determining critical switched-capacitor BPFs. To obtain a feel for the numbers involved in SCF design, let us design a filter where $f_c = 10^5$ Hz, $(V_o/V_1)(j\omega_n) = 10$, $Q = 10$, and $\omega_n = 6.28 \times 10^3$ rad/s. First we pick $C_a = 10$ pF. Therefore, $C_b = 1$ pF by Eq. 9.76. By Eq. 9.74,

$$\sqrt{\frac{C_cC_d}{C_1C_2}} = \frac{6.28 \times 10^3}{10^5} = 6.28 \times 10^{-2}$$

9.77

and, by Eq. 9.75,

$$Q = \frac{\sqrt{C_1 C_2 C_c C_d}}{C_2 C_b} = 10 \qquad\qquad 9.78$$

Equations 9.77 and 9.78 are solved for $C_1 = 159.2$ pF; C_2 is arbitrarily picked to be 10 pF. This gives $C_c C_d = 6.285$ pF, and if C_c is assumed to be equal to C_d, then $C_c = C_d = 2.507$ pF. ●

To summarize, SCFs are versatile, and many filters can be put in a relatively small volume and consume low power. It appears that any filter function realizable by conventional analog active filter techniques can be implemented by SCFs. SCFs are relatively noisy, however, and have limited dynamic range. They are limited to applications ranging from voice band to upwards of hundreds of kilohertz because of limitations on MOS transitor switching speed, aliasing, etc.

9.3 Review of Basic Filter Format Types

Most filters fall into four basic categories: low-pass, high-pass, bandstop, and bandpass. It is possible to design a filter belonging to one of these categories using a filter format type such as Butterworth, Chebychev, Bessel, Thomson, elliptic, or Legendre. In this section, we review and comment on certain of the filter format types, assuming a low-pass category.

9.3.1 Butterworth Filter

This format is also called the maximally flat filter. The frequency response magnitude for this filter has no ripple in the passband; its phase response is nonlinear, so the propagation time through the filter is not constant with frequency. The time domain transient response of this filter exhibits considerable overshoot for filter order n greater than 4. The magnitude response for the Butterworth LPF is given by

$$|\mathbf{H}_B(j\omega)| = \frac{|A_o|}{\sqrt{1 + (\omega/\omega_o)^{2n}}} \qquad\qquad 9.79$$

where

$A_o = $ the dc gain of the filter

$\omega_o = $ the frequency at which the LPF's gain is down by -3 dB (i.e., $|\mathbf{H}_B(j\omega)| = |A_o|/\sqrt{2}$)

The Butterworth LPF's poles are equally spaced at $180°/n$ degrees on a semicircle of radius ω_o in the left-half s-plane. Locations of the complex-conjugate pole pairs on the Butterworth circle can be given in vector form:

$$\mathbf{s}_k = -\sigma_k \pm j\omega_k \qquad (n: k = 0, 1, \ldots, n-1)$$

$$\sigma_k = \omega_o \cos\left(\frac{n\pi + (2k+1)\pi}{2n}\right) \qquad\qquad 9.80$$

$$\omega_k = \omega_o \sin\left(\frac{n\pi + (2k+1)\pi}{2n}\right)$$

As an example, let $n = 3$. This gives a real pole at $\mathbf{s} = -\omega_o$ (odd-n Butterworth filters all have a real pole at $\mathbf{s} = -\omega_o$), and a complex-conjugate pole pair. The LPF's denominator can be written as

$$Q(s) = \left(\frac{s}{\omega_o}\right)^3 + 2\left(\frac{s}{\omega_o}\right)^2 + 2\left(\frac{s}{\omega_o}\right) + 1 \qquad\qquad 9.81$$

which is factored to give

$$Q(s) = \left(1 + \frac{s}{\omega_o}\right)\left[1 + \frac{s}{\omega_o} + \left(\frac{s}{\omega_o}\right)^2\right] \qquad\qquad 9.82$$

From consideration of Eq. 9.82, we see that the LPF's poles are at $\mathbf{s} = -\omega_o$, and at $\omega_o(-1/2 \pm j\sqrt{3}/2)$.

When n is even, all poles occur as complex-conjugate pairs. Butterworth LPF designs are easily realized by cascading biquad LPF stages.

9.3.2 Chebychev Filters

The Chebychev LPF has a passband ripple, but none in the stopband. Its transfer function can be written

$$|H_c(j\omega)| = \frac{|A_o|}{\sqrt{1 + \varepsilon^2 C_n^2(\omega/\omega_o)}} \qquad\qquad 9.83$$

where

A_o = the filter's maximum dc gain in the passband (not necessarily the gain at $\omega = 0$)

ε = a number related to the peak-to-peak ripple of the frequency response magnitude in the passband

$C_n(\omega/\omega_o)$ = a *Chebychev polynomial*

It can be shown that

$$\varepsilon^2 = \frac{1}{\gamma^2} - 1 \qquad\qquad 9.84$$

where γ is the minimum of the passband ripple. The peak-to-peak passband ripple is just $|A_o|(1 - \gamma)$. Some Chebychev polynomials are given below:

Order n	$C_n(\omega/\omega_o)$
0	1
1	(ω/ω_o)
2	$2(\omega/\omega_o)^2 - 1$
3	$4(\omega/\omega_o)^3 - 3(\omega/\omega_o)$
4	$8(\omega/\omega_o)^4 - 8(\omega/\omega_o)^2 + 1$
5	$16(\omega/\omega_o)^5 - 20(\omega/\omega_o)^3 + 5(\omega/\omega_o)$
6	$32(\omega/\omega_o)^6 - 48(\omega/\omega_o)^4 + 18(\omega/\omega_o)^2 - 1$
$n + 1$	$C_{n+1}(\omega/\omega_o) = 2(\omega/\omega_o)C_n(\omega/\omega_o) - C_{n-1}(\omega/\omega_o)$

The poles of an nth-order Chebychev LPF lie on an ellipse in the left-half s-plane. They have equal angular spacing, however, of $180°/n$. Chirlian (1981) gives the pole locations as

$$s_k = -\sigma_k \pm j\omega_k$$

$$\sigma_k = \omega_o\cos\left(\frac{n\pi + (2k + 1)\pi}{2n}\right)\sinh\left[\frac{1}{n}\sinh^{-1}\left(\frac{1}{\varepsilon}\right)\right]$$

$$\omega_k = \omega_o\sin\left(\frac{n\pi + (2k + 1)\pi}{2n}\right)\cosh\left[\frac{1}{n}\sinh^{-1}\left(\frac{1}{\varepsilon}\right)\right]$$

$$\text{9.85}$$

The order, n, of the Chebychev LPF required to give a scalar transmission magnitude of $\beta|A_o|$, where $\beta \ll 1$ at a frequency $(\Delta\omega + \omega_o)$, can be shown to be given by

$$n \geq \frac{\cosh^{-1}\sqrt{\dfrac{1/\beta^2 - 1}{1/\gamma^2 - 1}}}{\cosh^{-1}\left(1 + \dfrac{\Delta\omega}{\omega_o}\right)} \qquad \text{9.86}$$

Generally, a Butterworth LPF having the same slope in the attenuation band as a comparable Chebychev LPF requires a higher-order n. Thus the price for saving biquad active filter stages in realizing the Chebychev LPF is paid for in passband ripple and poorer time domain performance for pulses and other transients.

9.3.3 Bessel and Thomson Filters

This class of filters trades off steep attenuation slope in the transition band to obtain a linear phase response. Because any order of Bessel filter is free from ringing and overshoot to transient inputs, these filters make good LPFs where faithful time do-main signal reproduction is required. The passband of the Bessel LPF slopes down gradually with frequency and has no ripple.

9.3.4 Elliptic Filters

The passband and stopband of this LPF format both have ripple. There is a very steep slope in the transition band. Relatively low order is required to attain the steep transition band slope, compared to other filter formats.

9.3.5 Legendre Filters

The design constraints on this filter form maximize the attenuation rate in the transi-tion band under conditions of no ripple in the pass- and stopbands. It has a steeper attenuation than the Butterworth LPF of the same order.

So far, we have discussed certain LPF formats. It is interesting to note that high-pass and bandpass filters can be designed from basic LPF formats by changing variables in s. To convert a given LPF format to a high-pass filter of the same format and corner frequency, we substitute $1/s$ for s in the LPF denominator polynomial. To convert a LPF to a bandpass filter, we substitute $(s + \omega_b^2/s)$ for s in the LPF's denominator, then put the denominator in time-constant form. It can be shown that the resulting bandpass filter has a center frequency, ω_b, and a bandwidth equal to the $(0, -3 \text{ dB})$ passband of the LPF, ω_o. Thus the Q of the transformed bandpass filter is simply $Q = \omega_o/\omega_b$. If the original LPF is an nth-order Butterworth, then the transformation has the effect of translating the center of the semicircular contour of radius ω_o on which the LPF's n poles lie (which is centered on the origin of the s-plane) to $s = \pm j\omega_b$. The bandpass filter's order is now $2n$, and there are n zeros at the origin of the s-plane. To explore the design of various filters in greater depth, the interested student is urged to consult the bibliography on active filter designs.

SUMMARY

In this chapter, we have seen that there are many types of active filters. Some are more useful than others in offering ease of design, adjustment of transfer function parameters, and parametric stability (low sensitivities). We have described certain active filters that are frequently used in current engineering practice. They include infinite-gain, single-feedback filters; infinite-gain, multiple-feedback filters; controlled-source filters; state variable filters; biquad filters; and generalized impedance con-verter-derived filters.

A major section has been devoted to the analysis and design of switched-capacitor active filters. Although SCFs can be implemented in all of the analog active filter architectures, the biquad SCF is probably the most useful design.

The salient properties of Butterworth, Chebychev, Bessel, Thomson, elliptic, and Legendre filters were reviewed to give the reader an appreciation for how basic analog or switched-capacitor active filters can be combined to build complex, high-order filters to meet various design criteria.

In working the examples in this chapter, we have assumed that all op-amps are ideal, except for the case of capacitor-free active filters. However, certain end-of-chapter problems intended to be solved by an ECAP consider the effects of practical op-amp characteristics on active filter performance at high frequencies. When an ECAP is used, it is easy to vary systematically a parameter such as op-amp dc gain and see how it affects the active filter's transfer function.

PROBLEMS

9.1 An infinite-gain, multiple-feedback active filter is shown in Fig. P9.1. Assume an ideal op-amp.

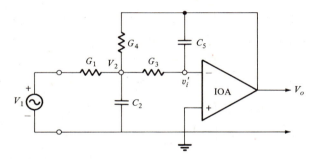

Figure P9.1

(a) Find an expression for the filter's transfer function in time-constant form.
(b) Find expressions for ω_n, ξ, and the mid-band gain, A_{vo}, in terms of circuit parameters.
(c) Sketch and dimension a Bode magnitude plot for the filter.

9.2 (a) Find an expression for $(V_o/V_1)(s)$ in time-constant form for the Sallen and Key high-pass filter shown in Fig. P9.2(a). Assume the op-amp is ideal.

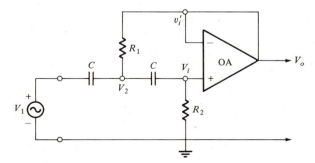

Figure P9.2(a)

(b) Find expressions for ω_n and ξ in terms of R_1, R_2, and C.

(c) Now assume the op-amp is nonideal, so

$$V_o = \frac{K_v}{\tau_a s + 1}(v_i - v_i')$$

Repeat parts (a) and (b).

9.3 Design a Sallen and Key low-pass filter such that $f_n = 500$ Hz and $\xi = 0.5$. Assume the op-amp is ideal, and keep $10^3 \ \Omega < R < 10^6 \ \Omega$. Write the frequency response function in time-constant format.

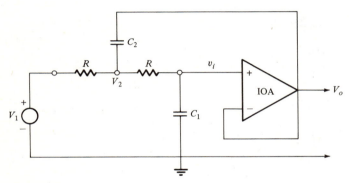

Figure P9.3

9.4 Use an ECAP such as Micro-Cap II to verify your design of a Sallen and Key high-pass filter with $f_n = 5$ kHz, $\xi = 0.6$, and a high-frequency -3 dB point of greater than 3 MHz. A practical op-amp must be chosen having an adequate f_T to meet these specifications. Verify your design with a Bode plot.

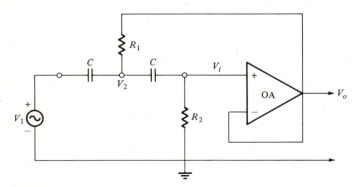

Figure P9.4

9.5   Derive an expression for V_o/V_1 in time-constant form. Give expressions for ω_n and the damping factor, ξ. Assume an ideal op-amp.

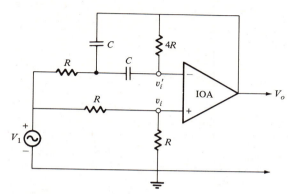

Figure P9.5

9.6   Use the GIC circuit of Fig. 9.15; assume ideal op-amps. Let $Z_1 = Z_3 = Z_4 = Z_5 = R = 10^4\ \Omega$, and $Z_2 = 1/sC$, where $C = 10^{-6}$ F.

(a) Find an expression for $\mathbf{Z}_{in}(j\omega) = V_1/I_1$. What is the value of the equivalent component seen as $\mathbf{Z}_{in}$?

(b) The GIC is connected as shown in Fig. P9.6(b). Derive an expression for V_1/V_s in time-constant form. Make a dimensioned Bode magnitude plot for this transfer function. Note that this plot may easily be done with an ECAP.

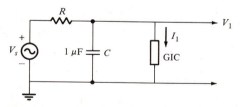

Figure P9.6(b)

9.7   This problem is easily done with the aid of an ECAP such as Micro-Cap II. The GIC circuit of Fig. 9.15 is built with $Z_1 = Z_3 = 1/sC$ and $Z_2 = Z_4 = Z_5 = R = 10^4\ \Omega$. $C = 10^{-6}$ F.

(a) The GIC circuit just described is connected in a simple voltage divider with a 10 kΩ resistor as shown in Fig. P9.7(a). Find an expression for V_1/V_s in time-constant form. Make a dimensioned Bode magnitude plot for this transfer function.

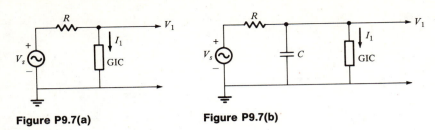

Figure P9.7(a)   **Figure P9.7(b)**

(b) In Fig. P9.7(b), the GIC low-pass filter of part (a) is given damping by adding a parallel capacitor, $C = 10^{-6}$ F. Find an expression for V_1/V_s in time-constant form. Make a dimensioned Bode magnitude plot for this transfer function.

9.8 Use an ECAP such as Micro-Cap II to make Bode magnitude and phase plots for the GIC-derived notch filter shown in Fig. P9.8. Plot from 1 kHz to 1 MHz. Assume LM741A op-amps; let $R = 10^4 \ \Omega$ and $C = 5 \times 10^{-10}$ F.

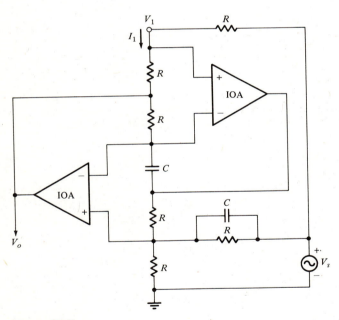

Figure P9.8

9.9 A GIC is used to make an all-pass filter. Use an ECAP to make Bode magnitude and phase plots for this system. Let $C = 5 \times 10^{-10}$ F and $R = 10^4 \ \Omega$. Assume LM741A op-amps.

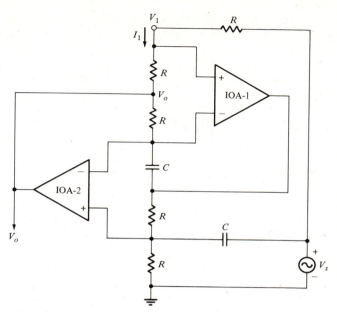

Figure P9.9

9.10 Two op-amps are connected to make a capacitorless active filter. Assume

$$\frac{V_o}{V_2}(s) = \frac{10^5}{10^{-2}s + 1} \qquad \frac{V_2}{V_i'}(s) = \frac{-10^5}{10^{-2}s + 1}$$

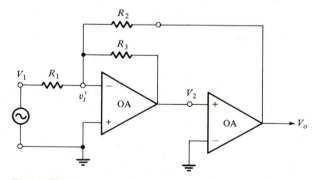

Figure P9.10

(a) Find an expression for V_o/V_1 in time-constant form.

(b) Let $G_1 = G_2 = G_3 = 10^{-3}$ S. Evaluate ω_n, V_o/V_1 at dc, and the damping factor, ξ, for the active filter.

9.11 A bandpass SCF is to be analyzed. The op-amp is assumed to be ideal.

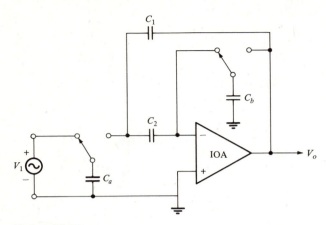

Figure P9.11

(a) Derive an expression for V_o/V_1 in time-constant form in terms of circuit parameters and the switch clock period, T_c. Also find expressions for the filter's ω_n, Q, and peak gain (at $\omega = \omega_n$).

(b) Let $C_1 = C_2 = 100$ pF, $T_c = 5 \times 10^{-6}$ s, $\omega_n = 2\pi \times 10^3$ rad/s, and $Q = 5$. Find the required values of C_a and C_b, and the peak response of the filter, that is, $|V_o/V_1|$ at $\omega = \omega_n$.

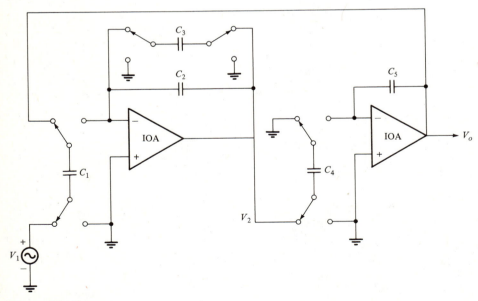

Figure P9.12

9.12 The circuit is shown in Fig. P9.12 is an SCF low-pass filter. Assume that the op-amps are ideal and that the clock period is T_c.

   (a) Derive an expression for V_o/V_2 with the loop open.
   (b) Derive an expression for V_o/V_1 in time-constant form with the loop closed. Give expressions for ω_n, ξ, and the dc gain.

9.13 Derive an expression for V_2/V_1 in time-constant form for the biquad, switched-capacitor, bandpass filter shown in Fig. P9.13. Give expressions for ω_n, Q, and V_2/V_1 at ω_n.

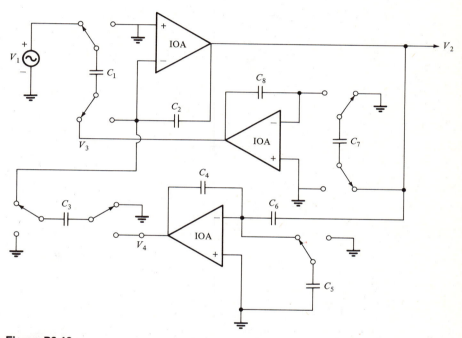

Figure P9.13

9.14 Derive an expression for V_o/V_1 in time-constant form for the switched-capacitor notch filter shown in Fig. P9.14. Give expressions for the dc gain, damping factor, and ω_n. Let $C_2 = C_3 = C$.

9.15 A basic biquad SCF is shown in Fig. 9.22

   (a) Derive an expression for V_3/V_1 in time-constant form. Give expressions for the filter's dc gain, ω_n, and damping factor.
   (b) Let the clock frequency be 10^5 pps and $C_2 = C_6 = 50$ pF. Find the values of C_1, C_3, C_4, and C_5 to make a LPF with $\omega_n = 2\pi(5 \times 10^3)$, unity dc gain, and $\xi = 0.5$.

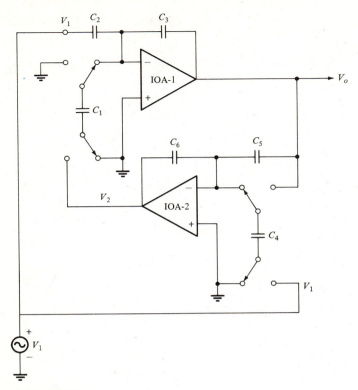

Figure P9.14

9.16 Design a switched-capacitor version of the GIC filter of Problem 9.6(b). That is, implement a GIC low-pass filter with switched-capacitor elements for R_2, R_4, and R_5. Make $f_n = 100$ Hz, $\xi = 0.5$, and $T_c = 10^{-5}$ s. Assume ideal op-amps.

9.17 **(a)** For the biquad active filter shown in Fig. P9.17(a) write the transfer function V_4/V_1 in time-constant form.

   (b) Write the transfer function V_2/V_1 for the active filter in time-constant form. Give expressions for ω_n, Q, and V_2/V_1 at ω_n.

   (c) Write an expression for the transfer function V_5/V_1 in time-constant form. Use an ECAP to make a Bode plot of V_5/V_1. Let $R = 1$ kΩ, $R_1 = 10$ kΩ, $R_4 = 1$ kΩ, and $C = 10$ nF. Assume LM741A op-amps are used.

9.18 Find an expression for V_4/V_1 in time-constant form for the biquad active filter shown in Fig. P9.18. Give expressions for ω_n, Q, and $|V_4/V_1|$ at ω_n. Assume ideal op-amps.

9.19 **(a)** Draw the equivalent analog active filter that will mimic the switched-capacitor filter shown in Fig. P9.19(a). Note that all switched capacitors C_1 are switched with clock

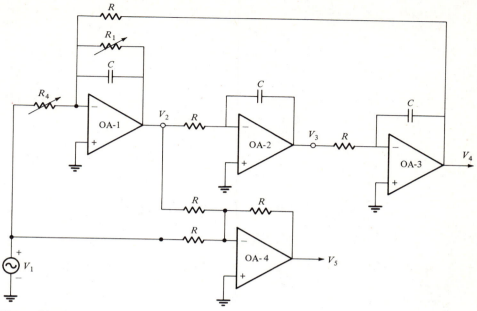

Figure P9.17(a)

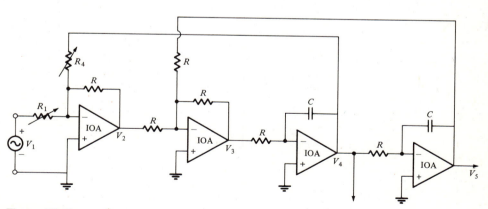

Figure P9.18

period T_1, all C_2's are switched with period T_2, and all C_3's are fixed. Note that the bottom C_1 inverts V_3.

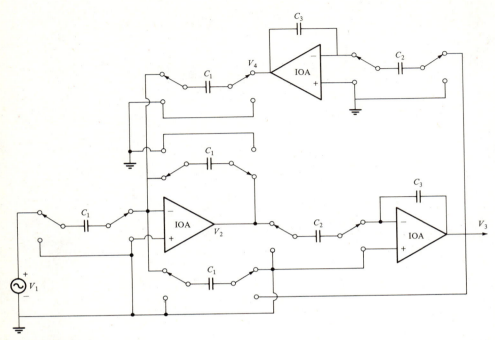

Figure P9.19(a)

(b) Find an expression for V_4/V_1 in time-constant form. (*Hint:* Use Mason's rule.)

(c) Give expressions for ω_n and the damping factor for this filter. How can ω_n be easily varied?

Chapter *10*

Tracking and Adaptive Active Filters

A tracking active filter is defined in this text as an active filter in which one or more filter parameters (i.e., ω_n or ξ) can be varied electrically in response to control signals. An adaptive active filter is a tracking active filter in which one or more parameters are automatically adjusted in response to control signals derived from the filter's input or output signals. In some adaptive active filters, such as the well-known Dolby B system used for noise reduction in high-fidelity systems, the criterion for filter adjustment is determined by the instantaneous level of high-frequency energy in the signal plus noise power density spectrum. Other criteria for tracking active filter parameter adjustment are determined by the purpose of the filter. For example, the cutoff frequency of a series of cascaded, quadratic low-pass filters used for anti-aliasing can be set digitally in response to setting the sampling frequency of an analog-to-digital conversion interface.

At the heart of all tracking and adaptive active filters is one or more electronically settable analog variable gain elements (VGEs). VGEs include such ICs as operational transconductance amplifiers (OTAs), analog multipliers, certain multiplying digital-to-analog converters (MDACs), and programmable gain amplifiers (PGAs) generally having digital control inputs. The following section describes some of the modern VGEs available to the designer of tracking and adaptive active filters.

10.1 Variable Gain Elements

10.1.1 Operational Transconductance Amplifiers

The operational transconductance amplifier (OTA) is an IC, two-quadrant analog multiplier in which a differential analog input voltage is converted to an output current with a transconductance that is proportional to a control current, I_{ABC}, which can range over three decades. OTAs are manufactured by RCA (CA-3080, CA-3280, CA-3060, and CA-3094) and by Exar (XR-13600); the various models cover a wide range of output slew rates, bandwidths, and output characteristics. The equivalent circuit of the RCA CA-3280 OTA is shown in Fig. 10.1. The OTA's output is from a VCCS, the transconductance of which is given as

$$g_m = 16.67 I_{ABC} \qquad \text{10.1}$$

g_m is in microsiemens if I_{ABC} is in microamperes; it is seen to vary linearly for I_{ABC} in the range from 0.1 μA to 1000 μA. The maximum I_{ABC} that can be sunk by the CA-3280 OTA is 10 mA; the slew rate is 125 V/μs, and the Norton output resistance is 63 MΩ (in parallel with the controlled current source).

Figure 10.2 illustrates the use of an OTA as a two-quadrant multiplier VGE in which

$$V_o = -I_o R_4 = +V_1 g_m R_4 = V_1 (16.67) I_{ABC} R_4 \qquad \text{10.2}$$

I_{ABC} can be set by an op-amp BJT voltage-current converter. Exact analysis of this circuit with a nonideal op-amp is complex. The approximate circuit behavior can be described easily if we assume the op-amp to be ideal; hence $V_{sj} = 0$, and we assume

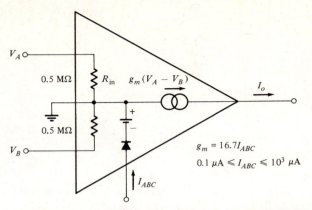

Figure 10.1   Equivalent circuit of an RCA CA-3280 OTA. Note that this IC has two OTAs in the package.

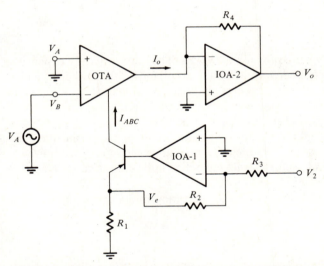

Figure 10.2   Two-quadrant multiplier using an OTA and two op-amps.

that the BJT's $h_{fe} > 100$ and $R_2 \gg R_1$. Thus we can write

$$V_e \cong -I_{ABC}R_1 \qquad\qquad 10.3$$

Because the summing junction is at virtual ground, the input current through R_3 must be equal to the current through R_2 to the v_e node, so

$$V_2G_3 = -V_eG_2 \qquad\qquad 10.4$$

Substitution of Eq. 10.3 for v_e into Eq. 10.4 yields

$$I_{ABC} = \frac{V_2 R_2}{R_3 R_1}$$

10.5

The OTA, two-quadrant multiplier's transfer function is found by substituting the expression for I_{ABC} into Eq. 10.2:

$$V_o = V_1 V_2 \left(\frac{16.67 R_4 R_2}{R_3 R_1} \right)$$

10.6

Note that the control voltage, V_2, must be positive and be held within a range determined by the constants in Eq. 10.5 so that I_{ABC} will not exceed the desired maximum value for the OTA used. V_2 is generally a dc or slowly varying control signal, whereas V_1 is an ac signal whose amplitude is being conditioned.

Examples of tracking active filters with OTAs as control elements are given in the problems for this chapter.

10.1.2 Analog Multipliers

Analog multipliers are ideally suited for use as VGEs. They are generally expensive, however, and there is a trade-off between accuracy and bandwidth, neither of which is cheap. Relevant analog multiplier response characteristics include multiplication accuracy versus frequency, tempco, total harmonic distortion (related to nonlinearity), bandwidth, output noise, and slew rate. Analog multiplier ICs are made by a number of companies, including Analog Devices, Burr-Brown, Intronics, Motorola, and Raytheon. It is impossible to review the characteristics of all of these devices; however, two devices that we have found useful in tracking active filter designs are the Analog Devices AD-532 and AD-539.

The AD-532 has two differential inputs, 1% accuracy, and a frequency response to above 800 kHz. Its output is given by

$$V_o = 0.1(X_1 - X_2)(Y_1 - Y_2)$$

10.7

Note that a scaling constant of 0.1 is used by the AD-532, and also by many other four-quadrant, IC analog multipliers. This scaling constant value allows the differential inputs to be as large as ± 10 V for an output maximum of ± 10 V.

The AD-539 is a two-channel, two-quadrant analog multiplier ideally suited for tuning biquad tracking active filters. The AD-539 is a 5 MHz device with 0.5% or 1% accuracy. It has single-ended bipolar inputs that are limited to ± 2 V. The control input, V_x, which is common to both multipliers on the chip, can range from 0 to 3 V. The AD-539's outputs are described by the equations

$$V_{o(1)} = V_{y(1)} V_x$$
$$V_{o(2)} = V_{y(2)} V_x$$

10.8

The basis for most IC analog multiplier designs is the Gilbert multiplier circuit, which is described in Chapter 13.

10.1.3 Multiplying Digital-to-Analog Converters as Variable Gain Elements

The multiplying digital-to-analog converter (MDAC) has both a digital input and an analog reference input which can be varied. The conversion scale factor of the MDAC can be varied by changing the reference input, producing two-quadrant, and in some systems, four-quadrant, multiplication. When used in tracking or adaptive active filters, the high-frequency analog signal is applied to the MDAC's reference input, and the digital input is used for the gain control. Good examples of MDACs suitable for VGEs are the Datel DAC-HA10B and the Analog Devices AD-7520. These 10-bit devices use the R-$2R$ inverted ladder architecture (see Chapter 14). Ten CMOS digital switches are used to switch the laser-trimmed, nichrome, thin-film resistors to vary the MDAC's output current. The MDAC output current is converted to a voltage with an op-amp transresistor. The input reference voltage "sees" a 10 kΩ load and can vary ± 10 V in a dc to 200 kHz bandwidth. The output settling time for digital inputs is 500 ns.

An example of a two-quadrant, MDAC VGE is shown in Fig. 10.3. The analog output is given by

$$V_o(t) = \frac{-V_R(t)}{2^N} \sum_{k=1}^{N} B_k 2^{k-1} \qquad (N = 10)$$

10.9

Note that -10 V $\leq V_R \leq +10$ V, and the digital scale factor can range from 0/2 (all $B_k = 0$) to $(2-1)/2$ (all $B_k = 1$).

Many manufacturers market MDACs; these devices are generally more expensive than OTAs or analog multipliers when used as VGEs in active filter designs.

10.1.4 Digitally Programmable Gain Amplifiers

Digitally programmable gain amplifiers (DPGAs) are based on the MDAC design. Designs are available that offer linear or logarithmic attenuation controlled by a digital input. The Analog Devices AD-7525 is in the linear attenuation category. It is a $3\frac{1}{2}$-digit, BCD-controlled attenuator using CMOS technology. This 18-pin IC has 10 mV resolution with a 10 V input, and a gain function that behaves according to Table 10.1. Note that 13 digital input lines are used to control the gain of the AD-7525.

The Analog Devices AD-7110, AD-7111, and AD-7118 are examples of digitally controlled logarithmic attenuators. The AD-7118 is a CMOS MDAC that has attenuation ranging from 0 dB (000000 input) to -88.5 dB (111011 input) in 1.5 dB increments. Its frequency response is flat to 100 kHz. The AD-7110 offers the same attenuation features as the AD-7118 but also includes three MOS "loudness" switches which can be used to enable additional bass and treble boost at low volume settings when this chip is used for audio attenuator applications. The AD-7111 is similar to

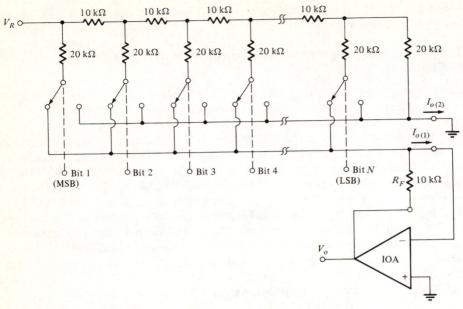

Figure 10.3   Unipolar binary operation of an inverted ladder, *R*-2*R* MDAC such as the AD-7520 or DAC-HA10B. The system is a two-quadrant multiplier in which the positive control input is an *N*-bit digital word, and the bipolar, time-varying signal is applied to the reference input terminal.

Table 10.1   BCD digitally selected gain factors for the AD-7525 digitally controlled gain element

BCD Input				Equivalent Decimal Input	V_o/V_i
1.0	0.1	0.01	0.001		
1	1001	1001	1001	1.999	−1.999
1	0000	0000	0001	1.001	−1.001
1	0000	0000	0000	1.000	−1.000
0	1001	1001	1001	0.999	−0.999
0	0101	0000	0001	0.501	−0.501
0	0000	0000	0000	0.000	−0.000

the two logarithmic VGEs described above, except that it has an 8-bit input and uses 0.375 dB steps to go from 0 dB to -88.5 dB attenuation. It has zero transmission for inputs of 1111,0000 and larger.

The final DPGA we will consider is the Micro Networks MN-2020 (or Hybrid Systems Corp. HS-2020). This amplifier is a hybrid IC whose gain can be set to 1, 2, 4, 8, 16, 32, 64, or 128 (6 dB steps) using a 3-bit digital input word. The MN-2020 small-signal bandwidth ranges from 5 MHz at $K_v = 1$ to 40 kHz at $K_v = 128$. The device's slew rate is 12 V/μs.

As we will see in the next section, all of the VGEs mentioned thus far can be used to design electronically adjustable tracking and adaptive active filters.

10.2 Examples of Tracking Active Filter Design

EXAMPLE 10.1

Tracking Low-Pass Active Filter

The circuit for this filter is shown in Fig. 10.4. We assume a four-quadrant analog multiplier (AM) is used in which the output, z, is $xy/10$. We can write this filter's transfer function by inspection, using Mason's rule on each transmission subunit of the circuit:

$$\frac{V_o}{V_1} = \frac{\left(-\dfrac{R_2}{R_1}\right)\left(\dfrac{-|y|}{10}\right)\left(\dfrac{-1}{R_4Cs}\right)}{1 - \left[\left(\dfrac{-R_2}{R_1}\right)\left(\dfrac{-|y|}{10}\right)\left(\dfrac{-1}{R_4Cs}\right)\right]} \qquad 10.10$$

In this system, we see that the sign of the net loop gain must be negative for stability. Hence the control input to the multiplier, y, must be a negative voltage. If Eq. 10.10

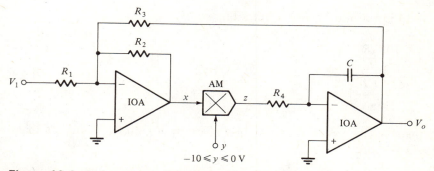

Figure 10.4 Voltage-controlled low-pass filter. y is the negative control voltage.

is written in time-constant form, we have

$$\frac{V_o}{V_1} = \frac{-\dfrac{R_3}{R_1}}{s\left(\dfrac{10R_3R_4C}{|y|R_2}\right) + 1} \qquad\qquad 10.11$$

in which the low-frequency gain is simply

$$K_v = -\frac{R_3}{R_1} \qquad\qquad 10.12$$

and the break frequency, ω_o, is

$$\omega_o = |y|\frac{R_2}{10R_3R_4C} \text{ rad/s} \qquad\qquad 10.13$$

Under the assumption that y can effectively range from -10 V to -0.01 V, ω_o can range from R_2/R_3R_4C rad/s to $0.001R_2/R_3R_4C$ rad/s.

EXAMPLE 10.2

Tracking High-Pass Active Filter

A high-pass tracking active filter design is shown in Fig. 10.5. Again we can write the transfer function by inspection with the aid of Mason's rule:

$$\frac{V_o}{V_1} = \frac{-\dfrac{R_2}{R_1}}{1 - \left[\left(\dfrac{-|y|}{10}\right)\left(\dfrac{-R_2}{R_3}\right)\left(\dfrac{-1}{R_4Cs}\right)\right]} \qquad\qquad 10.14$$

When Eq. 10.14 is put into time-constant form, we have

$$\frac{V_o}{V_1} = \frac{\left(-\dfrac{10R_3R_4C}{|y|R_1}\right)s}{s\left(\dfrac{10R_3R_4C}{|y|R_2}\right) + 1} \qquad\qquad 10.15$$

The break frequency, ω_o, is the same as that given by Eq. 10.13; the filter's high-frequency gain is simply

$$K_v = -\frac{R_2}{R_1} \qquad\qquad 10.16$$

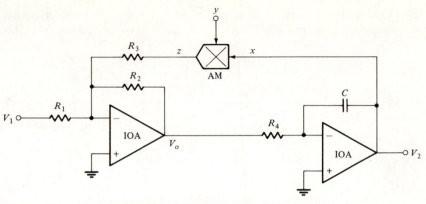

Figure 10.5 Voltage-controlled high-pass filter with one real pole. The control voltage, y, must be negative for filter stability.

EXAMPLE 10.3

Tracking Biquad Bandpass Active Filter

In this example, we consider the design of a quadratic bandpass filter, illustrated in Fig. 10.6, that has independent control of Q and the undamped natural frequency, ω_n. In this case, we assume that AD-7118 CMOS digitally controlled attenuators are used as the VGEs. Their attenuation can range from 0 dB to -88.5 dB in 1.5 dB steps. This corresponds to a maximum attenuation of 3.758×10^{-5} or gain factor steps

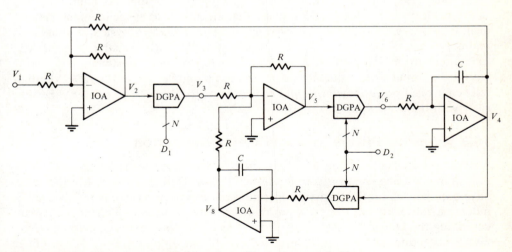

Figure 10.6 Digitally controlled bandpass filter. D_1 controls filter Q; D_2 controls the center frequency, ω_n.

of 0.8414. This active filter is seen to have two touching feedback loops (in signal flow graph terminology), so its transfer function, written by inspection using Mason's rule, is

$$\frac{V_4}{V_1} = \frac{(-1)(N_1)(-1)(N_2)\left(\dfrac{-1}{RCs}\right)}{1 - \left[(-1)(N_1)(-1)(N_2)\left(\dfrac{-1}{RCs}\right) + (-1)(N_2)\left(\dfrac{-1}{RCs}\right)(N_2)\left(\dfrac{-1}{RCs}\right)\right]} \qquad 10.17$$

Simplification of this transfer function to time-constant form yields

$$\frac{V_4}{V_1} = \frac{\left(-\dfrac{N_1}{N_2} RC\right)s}{s^2\left(\dfrac{RC}{N_2}\right)^2 + s\left(\dfrac{N_1}{N_2} RC\right) + 1} \qquad 10.18$$

From Eq. 10.18, we can see that the filter's parameters are

$$K_v = -1 \quad \text{at } \omega = \omega_n$$

$$\omega_n = \frac{N_2}{RC} \text{ rad/s} \qquad 10.19$$

$$Q = \frac{1}{N_1}$$

Since the attenuations N_1 and N_2 are independently digitally controlled and can range from 1 to 3.758×10^{-5}, we see that an enormous range of system adjustment is theoretically possible. However, practical limitations on system stability (N_1) and noise (N_1 and N_2) preclude using the full dynamic range available with the AD-7118 attenuators. Note that the AD-7118 allows logarithmic adjustment of ω_n and Q with digital input word value. If linear variation is sought, use of the AD-7525 digitally controlled potentiometer should be contemplated, because its output varies linearly with the control input, as shown in Table 10.1.   ●

10.3 Adaptive Filters in Audio Noise Reduction

An important example of an adaptive active filter applied to high-frequency noise reduction in audio systems is seen in the well-known Dolby B system. Many articles have been written on the theory of operation of this system, which is widely used in modern high-fidelity recording and reproduction equipment. A general block diagram describing the Dolby B system is shown in Fig. 10.7. Note that this adaptive active filter architecture utilizes a control signal derived from within the loop to adjust filter parameters. Depending on the *Dolby level* (DL) at point Z, an automatic modification of the filters' high-frequency response is obtained. The encoding filter, used in record-

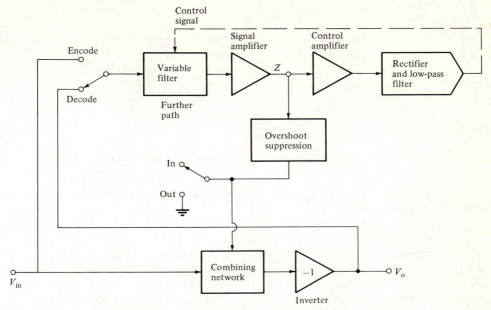

Figure 10.7 Dolby B adaptive active filter organization (after Berkovitz and Gundry, 1973).

ing sound on magnetic tape, provides variable high-frequency preemphasis dependent on the Dolby level. The decoding, or playback, Dolby B filter provides a variable high-frequency attenuation which is greatest when little high-frequency control signal power is available in the tape playback signal.

Inspection of the encoder characteristics shows that they have a fixed zero at 500 Hz and a pole that moves between 500 Hz and 2 kHz in response to the control signal (see Fig. 10.8). Its transfer function is of the form

$$\mathbf{E}(j\omega) = \frac{V_o}{V_1} = \frac{j\omega \dfrac{1}{2\pi(500)} + 1}{j\omega \dfrac{1}{2\pi f_p} + 1} \tag{10.20}$$

Similarly, the decoder has a fixed pole at 500 Hz and an adjustable zero which can range from 500 Hz to 2 kHz. Its transfer function is

$$\mathbf{D}(j\omega) = \frac{V_o}{V_1} = \frac{j\omega \dfrac{1}{2\pi f_o} + 1}{j\omega \dfrac{1}{2\pi(500)} + 1} \tag{10.21}$$

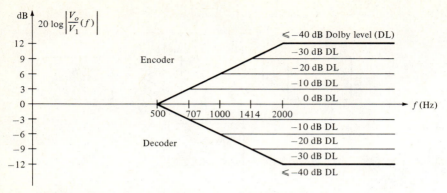

Figure 10.8   Frequency response of the Dolby B system. Bode plot asymptotes are shown.

Examination of the Dolby B encoder schematic as given in U.S. Patent #3,729,693, "Compressor/Expander Switching Methods and Apparatus," April 24, 1973 (inventor R. M. Dolby), shows that the encoder can be broken down into a system block diagram as shown in Fig. 10.9. The blocks labeled HWR and LPF are, respectively, a half-wave rectifier and a low-pass filter used to develop a control signal, V_c, from V_2. The time constant τ_p in the further path filter is made to vary from 2×10^{-3} s when V_2 is at 0 dB Dolby level, to 5×10^{-4} s when V_2 is at -40 dB Dolby level (100 times smaller). In other words, the time constant τ_p varies as a logarithmic function of the rectified and low-pass filtered output V_c of the further path filter. The overall encoder transfer function is seen to be

$$V_{oe} = -V_1 + V_1 \frac{s(\tau_p - \tau_o)}{s\tau_p + 1} = -V_1 \frac{s\tau_o + 1}{s\tau_p + 1} \qquad 10.22$$

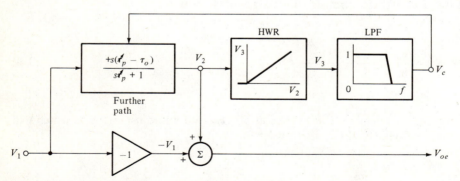

Figure 10.9   Block diagram of the Dolby B encoder system. The rectified and low-pass filtered output of the further path filter is used to adjust this filter's pole and direct current gain.

which is of the desired form given in Eq. 10.20. Note that a parametric control law must be derived to make the encoder's pole move according to the law suggested by Fig. 10.8. A 100-fold decrease in V_2 causes a 4-fold decrease in τ_p. This law is summarized in Table 10.2. Note that V_2 is the output of the further path filter and is given by

$$\frac{V_2}{V_1} = -\frac{j\omega(\tau_o - \tau_p)}{j\omega\tau_p + 1} \qquad\qquad 10.23$$

where $\tau_p < \tau_o$. The high-frequency gain of this filter is

$$K_v = 20 \log \left| \left(\frac{\tau_o}{\tau_p} - 1\right) \right| \text{dB} \qquad\qquad 10.24$$

An interesting property of the Dolby B system is that it is *output adaptive;* that is, its parameters are a function of the output voltage of the further path filter whose parameters are being varied to adjust the desired decoder or encoder transfer function. Figure 10.10 illustrates how the frequency response of the further path filter varies with Dolby level.

If the power density spectrum of the input signal contains significant spectral energy above 500 Hz, this will be sensed as an increased $|V_2|$. A high $|V_2|$, in turn, will cause f_p to approach 500 Hz and give a flat encoder characteristic. However, as $f_p \to$ 500 Hz, the gain of the further path filter will decrease and prevent $|V_2|$ from becoming large enough to cause an additional decrease in the further path filter's gain. Hence, it appears in practice that $|V_2|$ can never reach an amplitude that will give flat encoder frequency response, because a flat encoder frequency response requires $f_p = 500$ Hz, which requires a large $|V_2|$. However, the gain of the further path filter approaches $-\infty$ dB as f_p approaches 500 Hz, as shown in Table 10.2.

Table 10.2 Relation between the Dolby B encoder's movable pole at f_p and the level of the average rectified output of the further path filter, V_c

f_p (Hz)	τ_p (ms)	$20 \log\left(\frac{\|V_2\|}{\text{DL}}\right)$	V_c	K_v of Further Path Filter
500	2.000	0 dB	DL × 1	$-\infty$ dB
707	1.414	−10 dB	DL × 0.316	−10.28 dB
1000	0.100	−20 dB	DL × 0.1	0 dB
1414	0.707	−30 dB	DL × 0.0316	5.24 dB
2000	0.500	−40 dB	DL × 0.01	9.54 dB

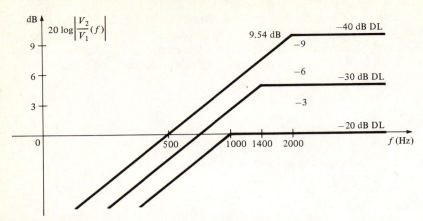

Figure 10.10   Frequency response of the further path filter.

The Dolby B decoder (used for tape playback) is also an output adaptive system. This decoder is shown in block diagram form in Fig. 10.11. Inspection of this diagram shows that the decoder's output can be written

$$V_{od} = -V_1 + V_1 \frac{s(\tau_p - \tau_o)}{s\tau_p + 1} = V_1 \frac{s\tau_p + 1}{s\tau_o + 1} \qquad 10.25$$

which has a frequency response function given by the bottom portion of Fig. 10.8. Here the pole is fixed at 500 Hz, and the zero varies with the Dolby level from 500 Hz (0 dB DL) to 2000 Hz (-40 dB DL).

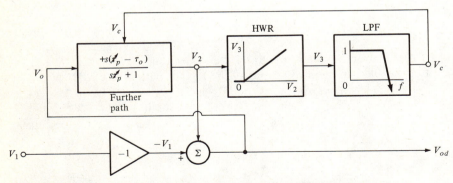

Figure 10.11   Block diagram for the Dolby B decoder system. Note that it is the output of the further path filter that is processed and used to control the time constant, τ_p.

The output of the decoder's further path filter is given by the following transfer function. Note that in contrast to Eq. 10.23, this filter has a *fixed* pole at 500 Hz.

$$\frac{V_2}{V_1}(j\omega) = \frac{j\omega(\tau_o - \tau_p)}{j\omega\tau_o + 1}$$

10.26

The high-frequency gain for this filter is

$$K_v = 20 \log\left|\left(\frac{\tau_o - \tau_p}{\tau_o}\right)\right| \text{ dB}$$

10.27

and ranges from $-\infty$ dB when $\tau_p = \tau_o$ (0 dB DL) to -2.5 dB (-40 dB DL). The same problem occurs in determining $|V_2|$ for the decoder as with the encoder: As the magnitude of $|V_2|$ increases because of a high level of high-frequency energy in the power density spectrum of the input, V_1, the transmission of the further path filter decreases, causing $|V_2|$ to be self-limiting, and the zero never to reach 500 Hz to give a flat decoder filter characteristic.

One way to avoid this output dependence of V_2 is to derive the adaptive control signal for τ_p from a fixed high-pass filter as shown in Fig. 10.12. We also observe that the architecture of the Dolby B filter system is awkward because it is constrained by the need to use the same variable further path filter in both encoder and decoder systems.

In the preceding discussion of the Dolby B system, we saw how a function of the instantaneous spectral power (average rectified voltage) from a region of a signal's power density spectrum can be used to adjust a filter's bandpass characteristics. The Dolby B system uses simple passive diode rectification and R-C averaging to derive a control signal that can be used to adjust the range of a movable pole (in the encoder) or a moveable zero (in the decoder). There is no reason why other criteria might not be used to derive the control signal, such as the true rms value of a spectral region, or average peak values of a rectified signal.

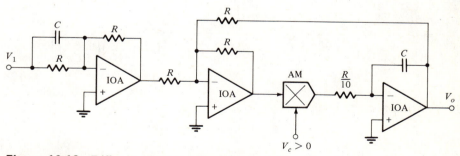

Figure 10.12   Different adaptive active filter encoder realization using op-amps and an analog multiplier as a VGE. The control signal is derived from a fixed high-pass filter.

An important property of any Dolby-type adaptive active filter used in audio applications is the ability of the filter to respond quickly to changes in the high-frequency energy of the controlling signal's spectrum. Thus, under conditions of low high-frequency signal power entering the decoder, the decoder will give maximum high-frequency attenuation (-12 dB above 2000 Hz). This attenuation minimizes high-frequency background noise from the recording tape. If a large, suddenly applied burst of 5 kHz signal occurs on the tape, the decoder must sense this and readjust its zero to give a flat response. This adjustment must be done rapidly so that minimum attenuation of the leading edge of the tone burst occurs. Some delay time is required for the rectifier and LPF to respond to the onset of the tone burst. This lag time will ordinarily be of the order of two or three time constants of the LPF. If an encoder is used in recording the tone burst on tape, the response lag in the encoder pole adjustment gives a transient, $+12$ dB boost in the leading edge of the tone burst at the encoder's output, which is recorded on the tape. When a Dolby B decoder is used to play back a Dolby B encoded tape, the natural lag in the decoder's response acts to cancel the leading edge, high-frequency boost caused by the control lag in the encoder. It is evident, then, that sound recorded with a Dolby B encoder sounds best when played back with a Dolby B decoder. If a Dolby B decoder is used to play back sound recorded with ordinary means, it will lack high-frequency brilliance.

10.4 An Automatic Tracking Active Bandpass Filter

In this section, we outline the design of a quadratic bandpass filter that automatically adjusts its center frequency, ω_n, to track the frequency of the coherent input signal, under constant-Q conditions. This tracking active filter finds applications where the amplitude information of the input signal is to be retained, while output signal-to-noise ratio is improved through the elimination of broadband noise.

The basis for the automatic tracking bandpass filter is shown in Fig. 10.13. This system is a tracking bandpass filter in which all op-amps are assumed to be ideal. The desired bandpass transfer function is found by applying Mason's rule to the transfer function subunits of each op-amp in Fig. 10.13:

$$\frac{V_4}{V_1} = \frac{(-1)(-\eta_1)\left(\frac{V_c}{10}\right)\left(\frac{-10}{RCs}\right)}{1 - \left[(-1)(-\eta_1)\left(\frac{V_c}{10}\right)\left(\frac{-10}{RCs}\right) + \left(\frac{V_c}{10}\right)\left(\frac{-10}{RCs}\right)(-1)\left(\frac{V_c}{10}\right)\left(\frac{-10}{RCs}\right)\right]} \qquad 10.28$$

This expression is easily reduced to

$$\frac{V_4}{V_1} = \frac{\left(\frac{-\eta_1 V_c}{RC}\right)s}{s^2 + s\left(\frac{\eta_1 V_c}{RC}\right) + \left(\frac{V_c}{RC}\right)^2} \qquad 10.29$$

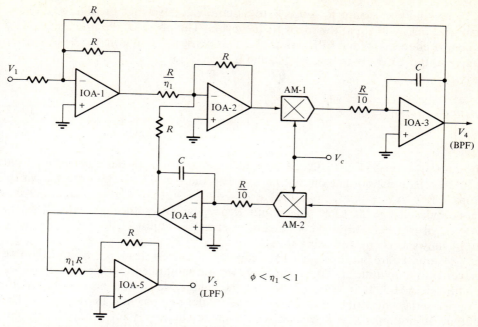

Figure 10.13   Tracking biquad active filter. The positive control voltage, V_c, sets the filter's center frequency, ω_n, under conditions of constant Q and gain.

in which it is easily seen that at $\omega = \omega_n$,

$$K_v = -1 \tag{10.30a}$$

$$\omega_n = \frac{V_c}{RC} \qquad (0 < V_c \le 10 \text{ V}) \tag{10.30b}$$

$$Q = \frac{1}{\eta_1} \tag{10.30c}$$

Using a similar analysis, we find the transfer function for the low-pass output of this tracking active filter:

$$\frac{V_5}{V_1} = \frac{-\left(\dfrac{V_c}{RC}\right)^2}{s^2 + s\left(\dfrac{\eta_1 V_c}{RC}\right) + \left(\dfrac{V_c}{RC}\right)^2} \tag{10.31}$$

The low-frequency gain for this transfer function is seen to be $K_v = -1$.

It is now necessary to cause V_c to assume values so that the filter's resonant frequency, ω_n, will equal the input frequency, ω. The phase of the input signal, V_1, is compared to that of the LPF output, V_5. This phase is found from the LPF transfer function, Eq. 10.31, by letting $s = j\omega$, and is

$$\psi = -180° - \tan^{-1}\left[\frac{\omega\left(\dfrac{\eta_1 V_c}{RC}\right)}{\left(\dfrac{V_c}{RC}\right)^2 - \omega^2}\right] \qquad 10.32$$

The $-180°$ term in Eq. 10.32 comes from the minus sign in the numerator of the LPF transfer function. The output, V_5, thus lags the input, V_1, by ψ degrees. ψ varies from $-180°$ at $\omega = 0$, to $-(3/2)180°$ at $\omega = \omega_n$, to $-360°$ as $\omega \to \infty$. Because this monotonic change in LPF phase with frequency passes through $-(3/2)180°$ at $\omega = \omega_n$, it can be used to derive a control signal to adjust ω_n to equal ω. Figure 10.14 shows a scheme for doing this.

V_1 and V_5 are put through infinite clippers to convert them to zero-mean square waves with amplitude A. These square waves are multiplied together by an analog multiplier, AM-3. It is well known that the square waves can be represented by Fourier series consisting of odd harmonics of the fundamental frequency, ω. The output of multiplier AM-3 will have an average value only for products of input harmonics of the same frequency; cross-products have no average value. The dominant average value is a result of the product of the fundamental frequency terms. Thus,

$$z = \frac{x_1 y_1}{10} = 0.1\left\{\left[A\frac{4}{\pi}\sin(\omega t)\right]\left(A\frac{4}{\pi}\sin(\omega t + \psi)\right)\right\} \qquad 10.33$$

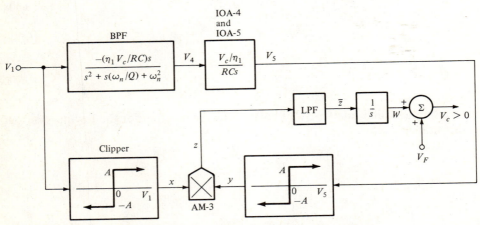

Figure 10.14 Control scheme for tracking the active bandpass filter center frequency, $\omega_n = V_c/RC$.

which, by trigonometric identity, becomes

$$z = 0.1 \left(A \frac{4}{\pi} \right)^2 \frac{1}{2} \left[\cos(-\psi) - \sin(2\omega t + \psi) \right] \tag{10.34}$$

The LPF averages the multiplier output, z, giving

$$\bar{z} = \frac{16}{20} \left(\frac{A}{\pi} \right)^2 \cos \left\{ \pi + \tan^{-1} \left[\frac{\omega \left(\frac{\eta_1 V_c}{RC} \right)}{\left(\frac{V_c}{RC} \right)^2 - \omega^2} \right] \right\} \tag{10.35}$$

which is the same as

$$\bar{z} = -\frac{4}{5} \left(\frac{A}{\pi} \right)^2 \cos \left\{ \tan^{-1} \left[\frac{\omega \left(\frac{\eta_1 V_c}{RC} \right)}{\left(\frac{V_c}{RC} \right)^2 - \omega^2} \right] \right\} \tag{10.36}$$

The cosine of the arctangent can be written

$$\bar{z} = -\frac{4}{5} \left(\frac{A}{\pi} \right)^2 \frac{\left(\frac{V_c}{RC} \right)^2 - \omega^2}{\sqrt{\left[\left(\frac{V_c}{RC} \right)^2 - \omega^2 \right]^2 + \left(\omega \frac{\eta_1 V_c}{RC} \right)^2}} \tag{10.37}$$

This function is sketched in Fig. 10.15. Note that $\bar{z}$ is positive for $\omega > \omega_n$.

$\bar{z}$ is integrated and added to a constant, V_f, to derive the positive control voltage, V_c, which sets ω_n. Assume that the system is at equilibrium; that is, $\omega = \omega_n$, $\bar{z} = 0$,

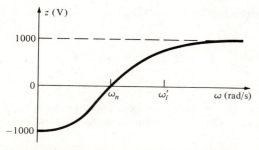

Figure 10.15 Relation between the average multiplier output, $\bar{z}$, and the input frequency, ω, to the tracking active filter. $K = (A^2)16/20(\pi^2)$.

and $V_c = \omega RC$. Now let the input frequency, ω, abruptly increase to a value $\omega' > \omega_n$. This increase causes a positive $\bar{z}$ to appear at the LPF output, which is integrated, causing W to increase; hence V_c increases, causing $\omega_n \to \omega'$. This action makes $\bar{z}$ decrease. The process continues until a new equilibrium, $V'_c = \omega' RC$, is developed. Since $\omega' = \omega'_n$ in the steady state, $\bar{z} = 0$, and there is no further change in the integrator output, W.

In this tracking active filter control system, the analog multiplier, AM-3, is used as a phase detector to derive the control signal, $\bar{z}$, and its integral, W. The inputs to AM-3 are independent of signal amplitude; this characteristic is important because $\bar{z}$ must be a function of the tracking active filter phase only in order for ω_n to track ω. This system is *not* a phase lock loop; the controlled variable is the filter's center frequency, ω_n, and not the voltage-controlled oscillator's phase.

Note that the study of the behavior of this tracking active filter for noisy inputs, and for transient changes in input frequency, is beyond the scope of this chapter. This system should make the basis for interesting experimental studies, however.

SUMMARY

In this chapter, we have introduced variable gain elements as a means of electronically varying an active filter's transfer function parameters. Variable gain elements were shown to be useful in the design of tracking and adaptive active filters which can be tuned by a control voltage, a current, or a digital word. Independent control of a biquad filter's ω_n and damping factor was shown to be possible.

A major use of parametrically tuned active filters was illustrated through the analysis of the well-known Dolby B audio noise reduction system. This filter was shown to be an output adaptive system where the signal that changes the filter's frequency response is taken at the output of the variable filter.

An example of a tracking adaptive filter was given in the analysis of an automatically tracking bandpass filter in which the center frequency follows changes in the input frequency. Biquad active filter architecture was used in this system.

The following problems illustrate the utility of using variable gain elements to realize electronically variable tracking and adaptive active filters.

PROBLEMS

10.1 An operational transconductance amplifier is used to make a current-programmed active filter. Assume the op-amp is ideal, and the OTA's output current is given by

$$I_2 = V_i(16.7)I_{ABC}$$

The Norton output conductance of the OTA is assumed to be zero.

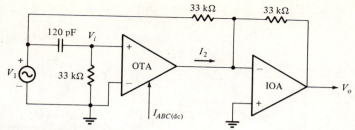

Figure P10.1

(a) Find $(V_o/V_1)(s, I_{ABC})$ is time-constant form.
(b) Sketch a Bode magnitude plot of V_o/V_1 to scale. Show corner frequencies for $I_{ABC} = 10^{-6}, 10^{-5}, 10^{-4},$ and 10^{-3} A.

10.2 (a) Find an expression for V_o/V_1 in time-constant form for the Dolby B decoder active filter shown. Assume the op-amps are ideal, and

$$V_4 = V_2 V_3/10$$

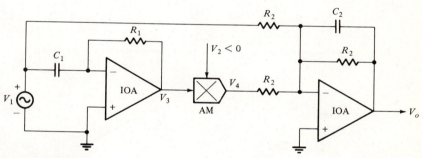

Figure P10.2(a)

(b) Assume the frequency of the fixed pole is 500 Hz. V_2 must range between 1 V (corresponding to the filter zero at 2 kHz) and 4 V (corresponding to the filter zero at 500 Hz). Choose reasonable values for $R_1, C_1, C_2,$ and R_2. Note that a 0 to −40 dB range of control signal, V_c, must cause V_2 to swing through +4 V to +1 V. Write an approximate equation for $V_2 = f(V_c)$ and plot this relation.

10.3 Find an expression for $(V_o/V_1)(s, V_c)$ in time-constant form for the Dolby B encoder active filter shown in Fig. P10.3. Assume

$$\frac{V_c R_4}{10 R_3} \gg 1$$

$$\frac{V_c V_3}{10} = V_o$$

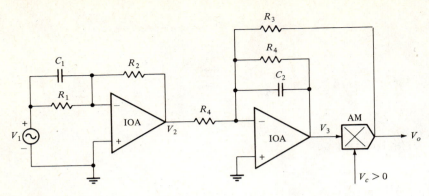

Figure P10.3

10.4 A voltage-tunable active filter is shown in Fig. P10.4. The op-amps are ideal, and

$$V_2 = \frac{V_o V_y}{10}$$

where $V_y < 0$.

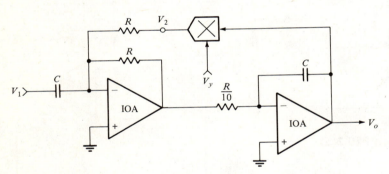

Figure P10.4

(a) Find $(V_o/V_1)(s, V_y)$ in time-constant form.
(b) Sketch and dimension $20 \log|(V_o/V_1)(j\omega)|$ versus ω. Show break frequencies and 0 dB/octave gains for $V_y = -0.1, -1$, and -10 V.

10.5 A voltage-tunable active filter is shown in Fig. P10.5. Assume the op-amps are ideal. Assume that the analog multiplier output is given by

$$V_z = \frac{V_x V_y}{10}$$

where -10 V $< V_y < -0.1$ V.

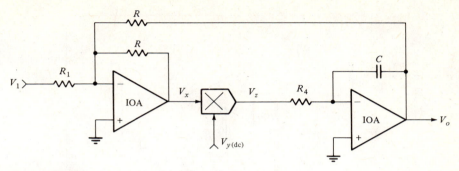

Figure P10.5

(a) Find an expression for $(V_o/V_1)(s, V_y)$ in time-constant form.
(b) Write expressions for the dc gain, A_o, and break frequency, ω_b.
(c) Specify values for R_1, R, R_4, C, and the range of negative V_y such that $A_o = -10$ and 0.1 rad/s $\leq \omega_b \leq 100$ rad/s.
(d) What are the consequences of $V_y > 0$?

10.6 Two MN-2020 programmable gain amplifiers are used to make a wide-range tunable active filter. $K_1 = V_3/V_2 = 2^{D_1}$, $K_2 = V_4/V_3 = 2^{D_2}$, $D_1 = (A_2, A_1, A_0)$, and $D_2 = (A_5, A_4, A_3)$. Find an expression for $(V_o/V_1)(s, K_1, K_2)$ in time-constant form.

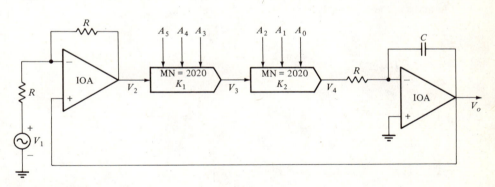

Figure P10.6

10.7 The circuit shown in Fig. P10.7 is an electronically tuned all-pass filter. Op-amps are ideal, and the OTA current source is

$$I_2 = -V_2(16.7)I_{ABC}$$

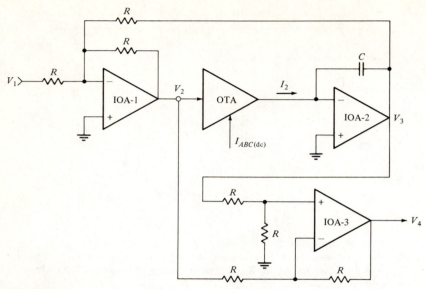

Figure P10.7

(a) Find an expression for $(V_3/V_1)(s)$ in time-constant form.
(b) Find an expression for $(V_4/V_1)(s)$ in time-constant form.
(c) Write an expression for the phase of $(V_4/V_1)(j\omega)$ versus I_{ABC}.

10.8 Two MN-2020 variable gain amplifiers are used to make a digitally controlled LPF. Op-amps are ideal. The MN-2020 is a VCVS with gain

$$K_{mn} = 2^{D_{10}}$$

where $0 \le D_{10} \le 7$.

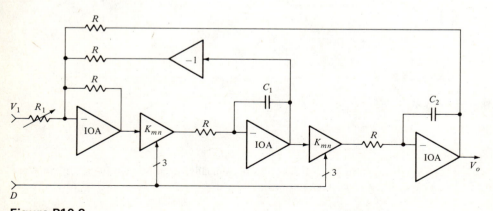

Figure P10.8

(a) Find an expression for $(V_0/V_1)(s, D_{10})$ in time-constant form.

(b) Find expressions for the filter's dc gain, ω_n, and ξ.

(c) Design the filter such that $A_o = -1$, $\xi = 0.707$, and 10 rad/s $\leq \omega_n \leq$ 1280 rad/s (specify R_1, R, C_1, and C_2; note that K_{mn} ranges from 2^0 to 2^7 in eight steps).

10.9 Three analog multipliers are used to make a biquad, voltage-tunable bandpass filter with independent control of ω_n and $Q = 1/2\xi$. Ideal op-amps are assumed. Also assume that

$$V_3 = \frac{V_2 V_{y(1)}}{10}$$

$$V_6 = \frac{V_5 V_{y(2)}}{10}$$

$$V_7 = \frac{V_4 V_{y(2)}}{10}$$

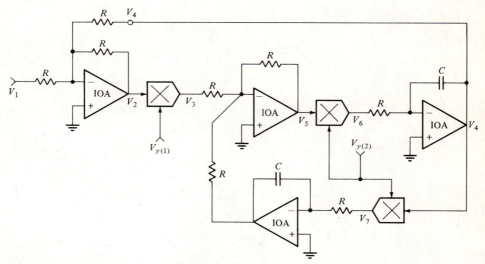

Figure P10.9

(a) Find an expression for $(V_4/V_1)(s, V_{y(1)}, V_{y(2)})$ in time-constant form. Should $V_{y(1)}$ and $V_{y(2)}$ be greater or less than zero?

(b) Find expressions for the filter's ω_n, Q, and $(V_4/V_1)(j\omega_n) = K_{vpk}$.

10.10 Three OTAs are used to make a current-tunable, bandpass, biquad active filter. Assume the OTA transconductances are given by $g_m = 16.7 I_{ABC}$. Find an expression for $(V_0/V_1)(s, I_{ABC(1)}, I_{ABC(2)}, I_{ABC(3)})$ in time-constant form. Also give expressions for ω_n, Q, and V_0/V_1 at ω_n.

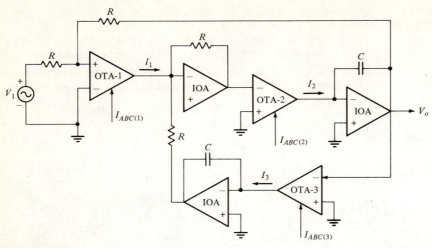

Figure P10.10

10.11 The circuit shown is a biquad, digitally tuned notch filter. A 13-bit BCD word is used to set K_p in the range 0.000 to 1.999. Ideal op-amps are assumed.

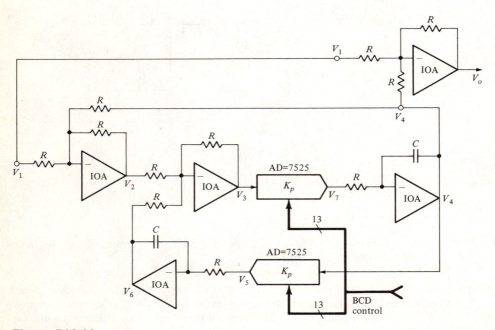

Figure P10.11

(a) Find an expression for $(V_4/V_1)(s)$ in time-constant form.
(b) Find expressions for ω_n, ξ, and $(V_4/V_1)(j\omega_n) = K_{vpk}$.
(c) Find an expression for $(V_o/V_1)(s)$ in time-constant form.
(d) Plot and dimension $20 \log_{10}|(V_o/V_1)(j\omega)|$ for $\omega_n = 1$ rad/s, $\xi = 0.707$, and $K_p = 1.000$.

10.12 The circuit shown is a voltage-tuned oscillator: Ideal op-amps are assumed. Also assume the following:

$$V_c < 0$$

$$V_3 = \frac{V_c V_2}{10}$$

$$V_6 = \frac{V_c V_1'}{10}$$

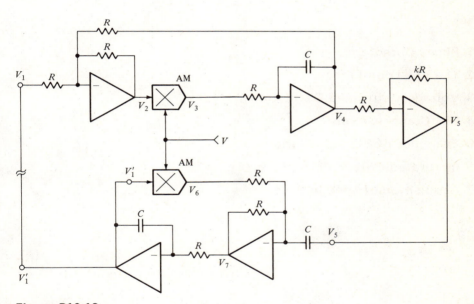

Figure P10.12

(a) Isolate the $V_1 \rightarrow V_4$ circuit. Find $(V_4/V_1)(s)$ in time-constant form.
(b) Isolate the $V_4 \rightarrow V_5$ circuit. Find V_5/V_4.
(c) Isolate the $V_5 \rightarrow V_1'$ circuit. Find $(V_1'/V_5)(s)$ in time-constant form.
(d) Find $A_L(s)$ in Laplace format.
(e) Draw the system's root-locus to scale at $V_c = -10$ V.
(f) Find an expression for $\omega_o = f(V_c, R, C, K)$ and find the minimum K for the system to oscillate; use root-locus.

Chapter 11

Introduction to Phase-Lock Loops

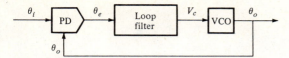

Figure 11.1   Basic phase-lock loop. PD = phase detector.

A phase-lock loop (PLL) is a closed-loop (feedback) system in which the phase of the PLL's periodic output is made to follow, under steady-state conditions, the phase of a periodic input signal. Response time constants to a transient phase change in the input signal are, in general, several orders of magnitude larger than the period of the input signal.

Phase-lock loops have many applications in communications, control systems, and instrumentation systems. They can be used to modulate and demodulate narrow-band FM signals, demodulate AM signals, demodulate double-sideband/suppressed-carrier signals, synthesize frequencies used in digital communications equipment, decode telephone status signals, control motor speed, measure motor speed (ta-chometer), measure Doppler frequency shift (speed), generate controlled phase shifts, and so on. Selected examples of PLL applications are presented in this chapter and in the problems section.

The block diagram of a basic PLL is shown in Fig. 11.1. Note that the input variable is the phase θ_i of the periodic input signal. The input carrier typically is a sinusoid or periodic logic waveform. The periodic PLL output can also be either a sinusoid or logic format and, in the steady state, has the same frequency as the input.

Note that there are three major subsystems in a PLL. The phase detector has an average output voltage proportional to a function of the difference between the input phase and the output phase. It can be a four-quadrant analog multiplier operating on sinusoidal signals or a sequential logic system to detect the phase difference between two digital signals. The loop filter introduces appropriate dynamic compensation to give the closed-loop PLL desired transient response properties to changes in input phase. The voltage-controlled oscillator (VCO) is driven from the loop filter output; it produces a periodic output signal whose phase is made to track that of the input signal.

In the following sections, we will discuss the electronic systems used to implement phase detection, loop filtering, and VCOs. We will also give some applications of PLLs in communications, control, and measurement. We will also discuss problems associated with a PLL locking to its input signal (capture) and remaining locked when there are transient changes in input frequency.

11.1 Phase Detectors

The first, and most basic, form of phase detector is the four-quadrant analog multiplier. Its operation may be described as follows. Assume the PLL input is a sine wave, described by $x(t) = A \sin(\omega_i t + \theta_i)$. The fed-back signal to the analog multiplier phase detector is typically the PLL VCO output. Under lock conditions, this output

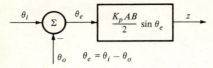

$\theta_e = \theta_i - \theta_o$

Figure 11.2 System equivalent to an analog multiplier phase detector at lock.

is $y(t) = B\cos(\omega_o t + \theta_o)$. The output of the analog multiplier is simply $z(t) = x(t)y(t)K_m$, where K_m is a constant peculiar to the multiplier. Thus $z(t)$ can be written

$$z(t) = \left(\frac{K_m AB}{2}\right)\{\sin[(\omega_i - \omega_o)t + (\theta_i - \theta_o)] + \sin[(\omega_i + \omega_o)t + (\theta_i + \theta_o)]\} \quad \text{11.1}$$

When the PLL is locked and tracking input phase variations, $\theta_i(t)$, the VCO output frequency, ω_o, equals the input frequency, ω_i. Hence the phase detector output becomes

$$z(t) = \left(\frac{K_m AB}{2}\right)\{\sin(\theta_i - \theta_o) + \sin[2\omega_i t + (\theta_i + \theta_o)]\} \quad \text{11.2}$$

Because the phase detector is generally followed by a low-pass loop filter, the second term in Eq. 11.2 is attenuated severely and does not affect loop dynamics; hence it may be dropped. Figure 11.2 shows that at lock, the analog multiplier phase detector can be represented by a simple differencer followed by a $\sin \theta_e$ operation. When the magnitude of the phase difference θ_e is less than 15°, then the $\sin \theta_e$ nonlinearity is negligible and the analog multiplier phase detector may be further reduced to the system shown in Fig. 11.3. Note that in this system, the phase error output depends not only on the phase difference, $(\theta_i - \theta_o)$, but also on the amplitudes of the sinusoids entering the phase detector. Also, the VCO output must be in quadrature (a cosine wave) with the sinusoidal input to obtain a proper lock.

A large step change in the input signal frequency, ω_i, can cause a PLL with a multiplier phase detector to loose lock. That is, the VCO output frequency, ω_o, no

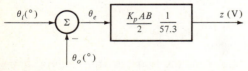

Figure 11.3 System equivalent to an analog multiplier phase detector at lock when the phase error magnitude is less than 15°.

$x(t) = A \sin(\omega_i t + \theta_i)$

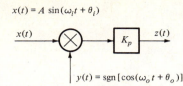

$y(t) = \text{sgn}[\cos(\omega_o t + \theta_o)]$

Figure 11.4   Switching phase detector. K_p is the phase detector conversion gain.

longer tracks ω_i. Instead, the VCO generates several transient cycles in which ω_o approaches ω_i. Lock is finally attained, by definition, when $\omega_i = \omega_o$.

Switching phase detectors are often used in IC PLL designs instead of true four-quadrant analog multiplier phase detectors. One input to a switching phase detector can be a sine wave, the other a square wave switching function. The latter input is usually a logic signal. The effective operation of a switching phase detector is shown in Fig. 11.4. Note that the signum function, sgn(x), generates a square wave from a periodic analog signal, $x(t)$, according to the rule

$$\text{sgn}(x) = 1 \quad \text{for } x(t) > 0$$
$$\text{sgn}(x) = -1 \quad \text{for } x(t) < 0$$

11.3

The square wave $y(t)$ has the argument of the generating cosine wave and can be expressed as a Fourier series:

$$y(t) = 2 \sum_{\substack{n=1 \\ (n \text{ odd})}}^{\infty} \text{sinc}\left(\frac{n}{2}\right) \cos(\omega_o nt + n\theta_o)$$

11.4

Note that $\text{sinc}(x) = \sin(\pi x)/\pi x$.

The effective output of the switching phase detector is the product of the input sinusoid, $x(t)$, times all the terms of the Fourier series. If the PLL is in lock, $\omega_o = \omega_i$, and the dominant, low-frequency output term from the switching phase detector is

$$z(t) = K_p A \sin(\omega_i t + \theta_i)\left(\frac{4}{\pi}\right)\cos(\omega_i t + \theta_o) = \left(\frac{K_p A2}{\pi}\right)\sin(\theta_i - \theta_o)$$

11.5

This is basically the same result obtained when a four-quadrant analog multiplier phase detector operates on two sinusoids (see Eq. 11.2 and Fig. 11.2).

A schematic of one type of switching phase detector is shown in Fig. 11.5. Fast op-amps or video amps are used to invert or buffer the analog signals, and a fast BIFET analog switch pair, such as the Precision Monolithics SW-01-04N, controlled

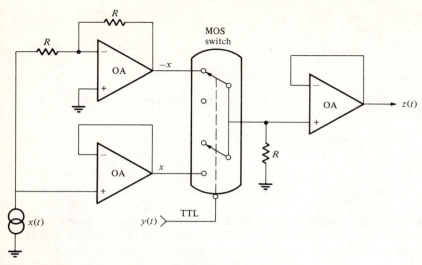

Figure 11.5   Switched-amplifier phase detector.

by the TTL square wave from the PLL's VCO, is used to commute the positive or negative analog signal. BIFET analog switches have useful rates limited by their turn-on and turn-off times, typically 200 ns to 300 ns. Faster switching phase detectors can be constructed from high-speed (Schottky) diodes in the form of a balanced-bridge or ring demodulator.

One form of a diode-switched ring demodulator is shown in Fig. 11.6. If we assume that $V_1 = +V_c$ and $V_2 = -V_c$, then gate A is ON (if V_c is large enough). These same control voltages turn gate B OFF, so $v'_i = 0$. It is not difficult to show that

$$v_i = \frac{v_x R_1 R_c}{R_1 R_c + R_1 R_1 + R_1 R_c/2} \tag{11.6}$$

Note that because of circuit symmetry, terms in $\pm V_c$ and the diode forward drops, V_d, cancel out of the expression for v_i/v_x. v'_i/v_x is given by the same expression when gate B is conducting.

A switched phase detector may also be built using an FET differential amplifier with BJT switch drain loads. The design of such a phase detector is shown in Fig. 11.7. Waveforms found in the switching phase detector of Fig. 11.7 are shown in Fig. 11.8.

The switched phase detectors of Figs. 11.5 through 11.7 all require a 90° phase shift between the analog input, v_x, and the square wave switch control, v_y, to obtain a zero-average output for zero dynamic phase difference. The transfer characteristic

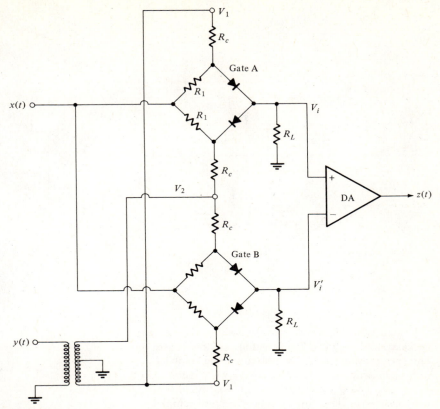

Figure 11.6 Switching phase detector consisting of two, two-diode bridge transmission gates connected to a differential amplifier. The square-wave input $y(t)$ generates a symmetrical switching waveform, $\pm V_c$, which connects an attenuated input sinusoid, $x(t)$, to either the noninverting or the inverting DA inputs, depending on the relative input phase between $x(t)$ and $y(t)$.

for this type of phase detector is given by Eq. 11.5 for $v_x = A \sin(\omega_i t + \theta_i)$, when $v_y = \text{sgn}[\cos(\omega_o t + \theta_o)]$.

We continue the discussion of PLL phase detectors by considering digital phase detectors that are used when both detector inputs are logic waveforms. These sequential logic circuits can be as simple as an exclusive-OR gate. More complex gate arrays are possible, including systems known as phase/frequency detectors. This latter class of digital phase detector has poor noise immunity and should not be used where extra or missing zero crossings are anticipated.

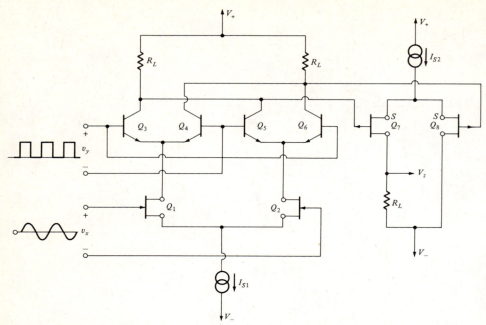

Figure 11.7 Transistor differential amplifier switched phase detector. The sine wave v_x drives an FET differential amplifier consisting of Q_1, Q_2, and current source $I_{S(1)}$. The polarity of the amplifier's differential output is switched 180° by the square wave impressed on the BJT's Q_3, Q_4, Q_5, and Q_6. The differential output is converted to a single-ended phase error signal by a second differential amplifier, consisting of Q_7, Q_8, and current source $I_{S(2)}$.

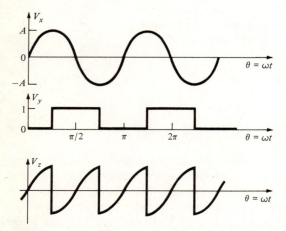

Figure 11.8 Phase relationships in a switching phase detector. $\overline{V}_z = 0$ is shown.

One simple sequential digital phase detector is the *RS* flip-flop. A simple NAND gate realization of this detector is illustrated with waveforms, output, and transfer function in Fig. 11.9. The inputs to this phase detector must be narrow, complementary pulses generated at the positive or negative zero crossings of the PLL input and VCO output. This phase detector transfer function has a positive slope, sawtooth shape which improves relock speed when noise is present. Zero-average output occurs when $\omega_i = \omega_o$ and there is a fixed, 180° phase difference (θ_o lags θ_i by 180°).

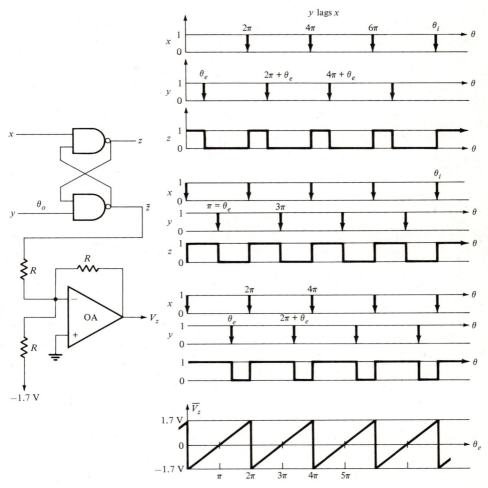

Figure 11.9 *RS* flip-flop sequential (digital) phase detector. Note that the TTL output, $\bar{z}$, is conditioned by an op-amp, so that $\bar{V}_z$ has zero average value when the VCO phase lags the input phase by 180°.

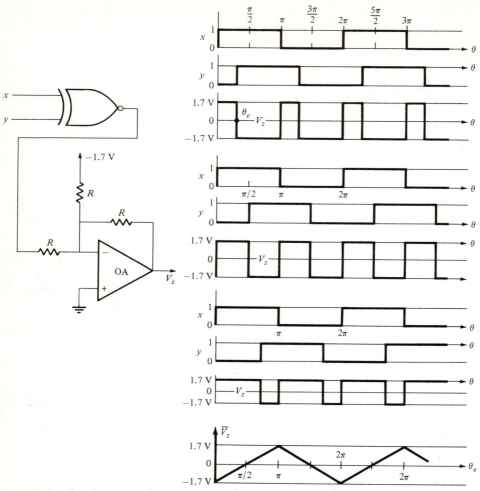

Figure 11.10   Exclusive-NOR phase detector.

An exclusive-NOR gate digital phase detector is shown in Fig. 11.10 with relevant waveforms and transfer function. Note that this digital phase detector has a narrower linear range than does the *RS* flip-flop phase detector, and also alternating unstable regions of negative slope. It requires 50% duty cycle square waves to operate.

A more complicated phase/frequency detector can be constructed that will signal average frequency differences when the PLL is out of lock as well as provide an output proportional to that when the system is locked and tracking θ_i. This system uses a 7400 quad, dual-NAND-gate IC and a Fairchild 9300 (or 7495) bidirectional shift register to detect the conditions: lock ($Q_0 = 0$, $Q_2 = 1$, and $Q_1 = \theta_e$); $\omega_i > \omega_o$

$(Q_0 = 0, Q_1 = 0,$ and $Q_2 = \omega_i - \omega_o)$; and $\omega_o > \omega_i$ $(Q_1 = 1, Q_2 = 1,$ and $Q_0 = \omega_o - \omega_i)$. This detector is illustrated in Fig. 11.11. It is seen that $V_z = 0$ when $\theta_e = 180°$, and the phase detector function is linear over $0° < \theta_e < 360°$ but is not periodic in θ_e. That is, Q_1 remains high (or $V_z = 1.7$ V) for $\theta_e > 360°$ and $\omega_o > \omega_i$, or Q_1 remains low ($V_z = -1.7$ V) for $\theta_e < 0$ and $\omega_o < \omega_i$. Because this phase detector has *memory*, it has relatively poor performance in noise.

Another commercially available sequential logic phase detector is the Motorola MC-4044 phase/frequency detector. This system, too, has a memory which can contribute to poor performance in the presence of phase noise. It uses a *charge pump* to generate a V_z proportional to θ_e. V_z is linear over $\pm 360°$; also, $V_z(0) = 0$. The MC-4044 is ideally suited to generate an (averaged) analog output signal proportional to the phase difference over a $\pm 360°$ range, using low-noise inputs.

The sample-and-hold phase detector is still another design that yields an analog output voltage proportional to the phase difference between two input signals. This phase detector provides a relatively smooth, ripple-free output when the PLL is in lock. In this system, the phase detector reference input waveform is used to generate a triangle wave with positive slope. The triangle wave is reset to its starting value and begins to rise on the $(n - 1)$th positive zero crossing (or the rising edge of a logic waveform) of the reference input; it is again reset to its starting level and begins to rise on the nth positive zero crossing of the reference wave; and so on. The nth positive zero crossing of the compared waveform causes a sample-and-hold circuit to sample and output the level of the nth triangle wave. This sampled voltage is held until the $(n + 1)$th positive zero crossing of the compared waveform, when the sample-and-hold process repeats, causing resampling of the triangle wave in its next, $(n + 1)$th, cycle.

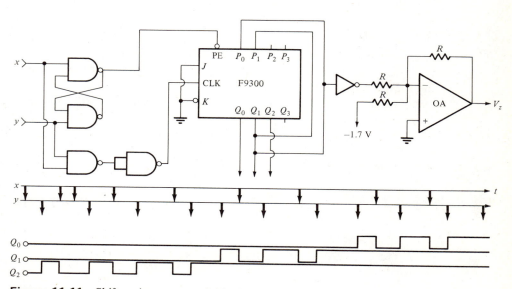

Figure 11.11 Shift register, sequential logic phase detector.

If the triangle wave has zero mean value, then a 180° phase difference between phase detector reference waveforms results in a 0 V (average) phase detector output. Phase differences of 0° to 180° cause a negative (average) output, and phase differences from 180° to 360° cause a positive (average) output. Thus the linear range of the sample-and-hold phase detector is from 0° to 360° phase difference, with 180° being the equilibrium value. The detector output as a function of phase error has a sawtooth characteristic given by the equation

$$V_o = A\left(\frac{\theta_e}{180°} - 1\right)$$

where

θ_e = the phase difference in degrees

A = the peak height of the sampled triangle wave

V_o = the phase detector output, held over one period of the compared waveform

Although the sample-and-hold phase detector appears linear, note that if the period of the reference input changes, so will the peak height of the triangle wave, A. Hence the phase detector gain will decrease with increasing input frequency.

We have seen that the outputs of all phase detectors, whatever their type, have an average value that follows the relatively slow changes in phase difference between the input signal, v_x, and the VCO output, v_y, under conditions of lock. High-frequency phase detector output components are generally attenuated to negligible levels by the low-pass filter elements (loop filter and VCO) following the phase detector in the PLL.

11.2 The PLL Loop Filter

The loop filter is generally a low-pass filter with a transfer function designed to give the PLL a desired characteristic in responding to transient changes in input phase. The loop filter, $F(s)$, is usually of the form

$$F(s) = \frac{K_f(s + a)}{s} \qquad\qquad 11.7$$

but sometimes lead-lag or simple low-pass filters are used, such as

$$F(s) = \frac{K_f(s + a)}{s + b} \qquad\qquad 11.8$$

where $a > b$, or

$$F(s) = \frac{K_f}{s + b} \qquad\qquad 11.9$$

Loop filters are generally realized using passive or active RC filters. More will be said about the design and application of loop filters in Sec. 11.4 on PLL compensation.

11.3 Voltage-Controlled Oscillators

A PLL VCO may have a logic (square-wave) output, a sinusoidal output, or both. A VCO generally has a frequency range (f_{max}, f_{min}) over which its input will linearly control its output oscillations. In the absence of an input signal $(V_c = 0)$, the VCO can be designed to oscillate at a center frequency, f_c, or at a minimum frequency, f_{min}.

Desirable properties of a VCO depend on the PLL's application; in general, they are:

1. Linearity (f_o versus V_c)
2. Wide tuning range ($[f_{max} - f_{min}]/[f_{max} + f_{min}]$)
3. Low phase noise and drift with temperature
4. Spectral purity with sinusoidal output
5. Low cost

The architecture of VCOs depends on the frequency range over which they are intended to work. Many VCOs used on IC PLLs employ simple RC transistor multivibrators in which the capacitor charging current is varied in response to the control voltage. This type of VCO can output a logic square wave or, with some extra circuitry, a triangle wave or sinusoid. The triangle and square wave outputs can be obtained in some cases at frequencies in excess of 100 MHz.

An emitter-coupled multivibrator VCO is illustrated in Fig. 11.12(A). Each lower current source (I_c) is equal to $V_c h_{fe}/h_{ie}$ of the lower transistors, Q_4 and Q_5. The frequency of the emitter-coupled multivibrator is found by assuming that the multivibrator is running in the steady state and that Q_1 has just turned off and Q_2 has turned on. Inspection of Fig. 11.12(B) shows that under these conditions, the emitter current of Q_2, and hence its collector current (assuming $h_{fe} \gg 1$), is $2I_c$. This current is large enough to cause D_2 to conduct heavily. Hence $V_{C(2)} = V_{B(4)} = (V_{CC} - 0.7)$ V, and $V_{B(1)} = V_{E(4)} = (V_{CC} - 0.7)$ V. Little current enters the base of Q_3, so $V_{B(3)} = V_{CC}$, and $V_{E(3)} = V_{B(2)} = (V_{CC} - 0.7)$ V. $V_{E(2)}$ is one base-emitter diode drop below Q_2's base; hence $V_{E(2)}$ is at a low-impedance point and is held at $(V_{CC} - 1.4)$ V. As the capacitor charges from right to left, $V_{E(2)}$ remains at $(V_{CC} - 1.4)$ V, and $V_{E(2)}$ drops linearly with time until it is low enough to turn Q_1 on. Q_1 turns on when $V_{E(1)} = (V_{CC} - 2.1)$ V. Switching is fast because of the positive feedback between Q_1 and Q_2 during the transition time. When Q_1 turns on at $T/2$, $V_{E(1)}$ becomes a low-impedance voltage source held at $(V_{CC} - 1.4)$ V. Consequently, since $V_c(t)$ cannot change abruptly (no impulses of current flow in the capacitor), $V_{E(2)}$ jumps to a value at $t = T/2$ given by Kirchhoff's voltage law, $(V_{CC} - 0.7)$ V. Immediately following the switching at $t = T/2$, the emitter of Q_2 appears to be a high-impedance point, and its

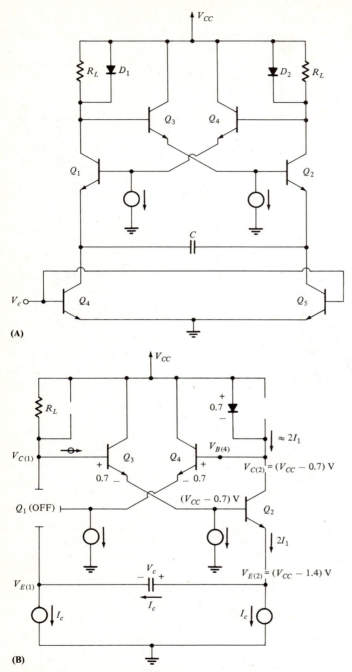

(A)

(B)

Figure 11.12 **(A)** Emitter-coupled, voltage-controlled, multivibrator circuit. **(B)** Emitter-coupled multivibrator at the moment Q_2 turns ON and Q_1 turns OFF.

voltage is free to change as C charges from left to right as the multivibrator oscillation cycle continues. When $V_{E(2)} = (V_{CC} - 2.1)$ V at $t = T$, Q_2 again turns on and Q_1 turns off. The steady-state waveforms for $V_{E(1)}$, $V_{E(2)}$, and $V_c = (V_{E(2)} - V_{E(1)})$ are shown in Fig. 11.13. Note that the voltage swings are small, and that the transistors do not saturate. Thus this type of VCO can work at frequencies in excess of 100 MHz.

The frequency of the emitter-coupled VCO is easy to calculate. In one half-period, a current $I_C = V_c h_{fe}/h_{ie}$ charges the capacitor linearly from -0.7 V to $+0.7$ V. Hence

$$\Delta V = 1.4 = \frac{1}{C}\int_0^{-T/2} I_C\, dt = \left(\frac{V_c h_{fe}}{Ch_{ie}}\right)\left(\frac{T}{2}\right) \qquad 11.10$$

from which we can write

$$f_o = \frac{V_c h_{fe}}{2Ch_{ie}(1.4)} = \frac{I_B h_{fe}}{2C(1.4)} \qquad (11.11)$$

This analysis assumes the discharge cycle has the same duration as the charge cycle. I_B is the base current of Q_4 or Q_5; h_{fe} is their current gain.

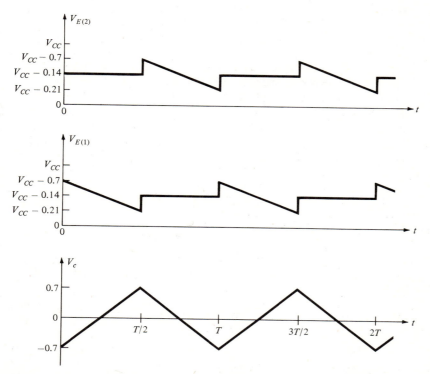

Figure 11.13 Waveforms in the emitter-coupled multivibrator VCO. Note that there are no steps in the $V_c(t)$ waveform.

VCO designs for VHF and UHF PLLs often make use of varactor diode tuning. Hartley, Clapp, and tunnel diode free-running oscillators can be tuned by application of a low-frequency control signal across a back-biased, pn junction diode, varying its capacitance according to the rule

$$C(V_c) = \frac{K}{(\psi_o - V_d)^{1/n}} \qquad\qquad 11.12$$

where

$K =$ a constant that depends on physical properties of the diode

$\psi_o =$ the contact potential of the pn junction determined by the carrier concentrations in the p- and n-materials

ψ_o ranges from 0.2 V to 0.9 V; it is typically about 0.75 V at room temperature for a silicon diode. The constant n which ranges from 1/3 to 3, depends on the doping profile near the pn junction; $n = 1/2$ for an abrupt junction. Since the frequency of a tuned circuit is proportional to $1/\sqrt{LC}$, choice of $n = 2$ makes the change in the tuned frequency of oscillation proportional to ΔV_d, the change in voltage across the varactor diode. Normally, $V_d < 0$ because the diode is back-biased.

In addition to LC oscillators whose sinusoidal output frequency is tunable by a varactor diode, crystal oscillators can also be deviated around their center frequency by the use of varactor diodes. Such variable crystal-controlled oscillators (VCXOs) usually operate in the range of a few megahertz to 20 or 30 MHz. Frequency multiplication using VCXO harmonics and tuned filters is used to obtain high-frequency outputs. A typical center frequency might be 100 MHz, around which the VCXO must deviate ± 10 kHz. If a 20 MHz, AT-cut quartz crystal is used along with $5 \times$ frequency multiplication, the actual deviation of the oscillator needs to be ± 2 kHz. This is a $\pm 0.01\%$ deviation, which is easily obtained. Maximum deviations as large as 0.25% to 0.5% of the VCXO center frequency have been reported. Further discussion of the design and application of VCOs and VCXOs can be found in the texts by Blanchard (1976) and Gardner (1979).

11.4 PLL Compensation

A block diagram representing a basic PLL as a negative feedback system is shown in Fig. 11.14(A). The input, or reference variable, is considered to be changes in the phase of the input signal other than the phase argument generating the sine wave. That is, for the input signal,

$$x(t) = A \sin[\omega_i t + \theta_i(t)]$$

θ_i is treated as the phase input and is considered to be a variable whose maximum rate of change is much less than ω_i. The VCO output, here assumed to be cosinusoidal, also has a phase variable $\theta_o(t)$.

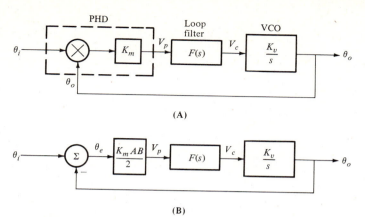

Figure 11.14 **(A)** PLL using an analog multiplier phase detector. **(B)** Linearized PLL. Note that A is the input sine wave peak amplitude; B is the VCO sine wave peak amplitude. $\theta_e \cong \theta_i - \theta_o$; $|\theta_e| < 15°$ for a valid approximation.

The VCO output under lock conditions is a cosine wave of the same frequency as the input carrier but, in general, with a different phase $\theta_o(t)$. Thus $\theta_o(t)$ is the effective output of the PLL which tries to make $\theta_e = (\theta_i - \theta_o) \rightarrow 0$ under steady-state conditions. The analog multiplier phase detector shown in Fig. 11.14(A) reduces, for $\theta_e < 0.25R$, to the simple summer followed by a gain of $K_pAB/2$ (see Figs. 11.2 and 11.3).

The VCO generates an output frequency of K_v rad/s/V in response to its input, $V_c(t)$. Because the VCO output phase changes are considered in the PLL dynamic model, we must integrate the VCO output frequency to obtain $\theta_o(t)$ in radians. Hence the PLL has a loop gain with a pole at the origin supplied by the VCO.

The dynamics of the closed-loop PLL relate how changes in the VCO output phase respond to changes in the input carrier's phase under lock conditions. Let us first consider the loop filter to be a simple real-pole, low-pass filter. Also for simplicity, let $K_mAB/2 = K_p$, the phase detector conversion constant. The loop gain for this PLL configuration is

$$A_L(s) = -\frac{K_pK_fK_v}{(s+b)s} \qquad 11.13$$

The root-locus diagram for this system shows all possible locations for the closed-loop PLL's poles as a function of the gain product $K_pK_fK_v$. It is given in Fig. 11.15.

Often we desire the closed-loop PLL to have a critically damped response, such as that given by $\xi = 0.707$. This value corresponds to closed-loop system complex-conjugate poles lying on lines drawn from the origin at angles of $\pm\gamma = \cos^{-1}(\xi) = 45°$ from the negative real axis in the s-plane. The root-locus of the PLL with a simple real-pole loop filter intersects the $\gamma = 45°$ line ($\xi = 0.707$) at points P and P' in Fig.

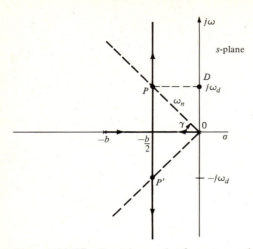

Figure 11.15 Root-locus plot for a two-pole, loop gain PLL. A gain product $K_pK_fK_v$ is sought that will give the closed-loop PLL a suitable damping constant, $\xi = \cos\gamma$. Note that the distance OP (and OP') is the undamped natural frequency, ω_n, of the closed-loop PLL's complex-conjugate pole pair. The distance OD is the damped natural frequency, ω_d, of the system. It may be shown that $\omega_d = \omega_n\sqrt{1-\xi^2}$.

11.15. The gain required for this pole location is found from the root-locus magnitude criterion:

$$|A_L(\mathbf{P})| = 1 = \frac{K_pK_fK_v}{\left(\frac{b}{2}\sqrt{2}\right)^2} \qquad 11.14$$

This gives

$$K_pK_fK_v = \frac{b^2}{2} \qquad 11.15$$

for $\xi = 0.707$ in the closed-loop system's quadratic response.

We next consider the case where the loop filter is of the form

$$F(s) = \frac{Kf(s+a)}{s} \qquad 11.16$$

This loop filter produces a PLL loop gain of

$$A_L(s) = -\frac{K_pK_fK_v(s+a)}{s^2} \qquad 11.17$$

The root-locus diagram for this system is shown in Fig. 11.16. Note that the curved part of the loci can be shown to be a circle of radius a rad/s. For this root-locus diagram, we note that for a closed-loop system damping factor of $\xi = 0.707$, the system poles will be at points P and P'. The gain required to obtain this desired damping factor is again found from the root-locus magnitude criterion:

$$|A_L(P)| = 1 = \frac{K_p K_f K_v a}{(a\sqrt{2})(a\sqrt{2})} \qquad 11.18$$

so

$$K_p K_f K_v = 2a \qquad 11.19$$

As a final example of a loop filter, let $F(s)$ be a lag-lead function:

$$F(s) = \frac{Kf(s + a)}{s + b} \qquad 11.20$$

The PLL's root-locus for this filter also contains a circle, as shown in Fig. 11.17. The circle's radius in this case can be shown to be the geometric mean distance from the two real-axis poles to the zero. That is,

$$R = \sqrt{a(a - b)} \qquad 11.21$$

Because the circle loci do not pass through the origin in this case, there is no simple algebraic solution for the $K_p K_f K_v$ product that will give a closed-loop PLL damping factor of 0.707. Note that the root-locus circle intersects the 45°, $\xi = 0.707$ line at

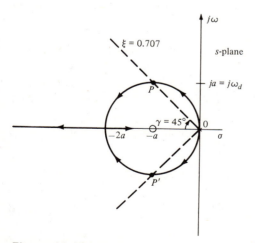

Figure 11.16 Root-locus plot for a PLL with a loop filter of the form $F(s) = K_f(s + a)/s$. The closed-loop PLL has two poles.

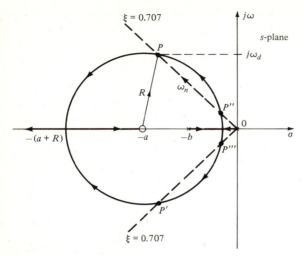

Figure 11.17   Root-locus diagram for a PLL with a lag-lead loop filter.

two points in Fig. 11.17, P and P''. Generally, the gain product that will put the closed-loop PLL's poles at points P and P' is best, because it will result in a faster system rise time in response to step changes in θ_i than will the lower gain which sets the poles at P'' and P'''.

11.5 Examples of PLL Applications

EXAMPLE 11.1

Design of a PLL Motor Speed Control System

The first PLL system that we will examine is a motor speed control. This system can be used as a speed regulator for a tape recorder drive or a phonograph turntable, or any application that requires constant motor speed under varying torque loads. The example we consider is a direct-drive turntable system. A block diagram is shown in Fig. 11.18.

   To analyze this system, we first note that the dc, permanent-magnet (PM) servomotor has a torque-speed characteristic given by

$$T_m = I_m K_T - K_c \dot{\theta}_m \qquad\qquad 11.22$$

This torque-speed behavior is shown in Fig. 11.19, along with the equivalent circuit for the motor's armature. From Kirchhoff's voltage law, we can write

$$I_m = \frac{V_A - K_T \dot{\theta}_m}{R_A} \qquad\qquad 11.23$$

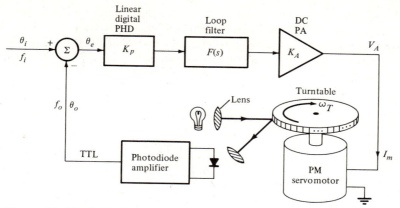

Figure 11.18 PLL phonograph motor speed control. The turntable
edge has alternating black white markings which generate an output
frequency, f_o, as the turntable rotates at its desired speed $\dot{\theta}_m$. A light
and a photodiode are used to sense the moving edge markings.

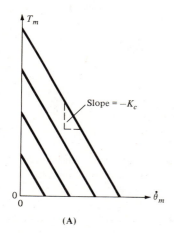

(A)

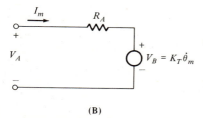

(B)

Figure 11.19 **(A)** dc servomotor torque-speed curve. **(B)** Equivalent
circuit for a dc servomotor's armature. V_B is the back-emf.

Now the torque generated by the motor is used to accelerate (or decelerate) the inertia, J, of the turntable, to overcome viscous friction, B, and to compensate for any external load torque, T_L (as produced by wiping the record while the turntable is running). This torque may be expressed mathematically as

$$T_m = J\ddot{\theta}_m + B\dot{\theta}_m + T_L \qquad 11.24$$

If Eqs. 11.22 and 11.23 are substituted into Eq. 11.24, we can write

$$\frac{V_A K_T}{R_A} = J\ddot{\theta}_m + \dot{\theta}_m\left(B + K_c + \frac{K_T^2}{R_A}\right) + T_L \qquad 11.25$$

If this differential equation is Laplace transformed, we find that the motor speed is related to V_A and T_L by the transfer function

$$\dot{\theta}_m(s) = \frac{V_A(s)\left(\dfrac{K_T}{R_A J}\right) - T_L(s)\left(\dfrac{1}{J}\right)}{\left(s + \dfrac{B + K_c + K_T^2/R_A}{J}\right)} \qquad 11.26$$

which can be written in simpler form as

$$\dot{\theta}_m(s) = \frac{V_A(s)K_m - T_L(s)K_2}{s + b} \qquad 11.27$$

where the parameters K_m, K_2, and b have obvious correspondence in Eq. 11.26. The units of $\dot{\theta}_m$ are assumed to be radians per second, so to convert $\dot{\theta}_m$ to revolutions per second of the turntable, we must divide $\dot{\theta}_m$ by 2π.

The turntable is assumed to be 12 inches in diameter, so its circumference is $12\pi = 37.7$ inches. The turntable edge is engraved with one cycle of black-white square wave every 0.1 inch, so 377 square wave cycles appear on the edge of the turntable. A photoelectric sensor is used to detect the motion of the turntable edge. Thus 377 pulses occur at the output of the photodiode circuit for every revolution of the turntable. Let us assume that the desired turntable speed is 33.33 rev/min, or 0.5555 rev/s. This speed will give a phototransistor output frequency of $f_o = (0.5555)(377) = 209.4$ pps in the steady state. Note that the input frequency to the speed regulator PLL must also be 209.4 pps if the turntable is to run at 33.33 rev/min in the steady state. A block diagram summarizing this design is shown in Fig. 11.20.

To adjust the dynamic response of the closed-loop motor speed control, we must choose a loop filter format. Using op-amp active filter design, we can synthesize a proportional-plus-derivative filter, that is, a loop filter having a real zero. Thus,

$$F(s) = Kf(s + a) \qquad 11.28$$

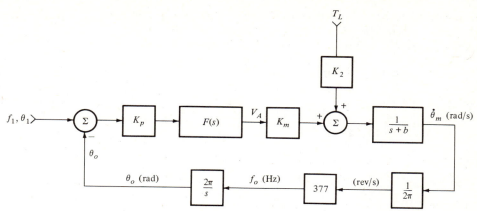

Figure 11.20 PLL motor speed control showing load torque. Turntable speed in revolutions per second must be multiplied by the number of black-white cycles on the turntable edge to give the phototransistor output frequency, f_o, in pulses per second. Integration of f_o and multiplication by 2π gives the phase of the tachometer output in radians.

The system loop gain is then

$$A_L(s) = -\frac{K_p K_f(s + a)K_m(377)}{[s(s + b)]}$$

11.29

The root-locus diagram for this system is similar to that shown in Fig. 11.17. Design for closed-loop system poles with maximum bandwidth and a damping factor of 0.707 is best carried out graphically, with recognition that the circle part of the root-locus diagram has a radius $R = \sqrt{a(a - b)}$ and is centered at $s = -a$.

The closed-loop system's velocity response to changes in load torque may be written from Fig. 11.20:

$$\frac{\dot{\theta}_m(s)}{T_L(s)} = \frac{-\left(\dfrac{K_2}{s + b}\right)}{1 + \dfrac{K_p K_f(s + a)K_m(377)}{s + b}}$$

11.30

This transfer function reduces to

$$\frac{\dot{\theta}_m(s)}{T_L(s)} = \frac{-sK_2}{s^2 + s(1 + 377K_p K_f K_m) + a(377)K_p K_f K_m}$$

11.31

From this velocity/load torque relation, we see, using the Laplace final value theorem, that there is no net change in turntable velocity in the steady state given a step change in load torque. This characteristic is illustrated qualitatively in Fig. 11.21.   ●

Next we will show some examples of how PLLs are used in FM communications systems. Angle modulation is a process in which a carrier, $f_c(t) = A \sin(\omega_c t)$, is modulated so that

$$f_{cm}(t) = A \sin[\theta_m(t)] \qquad\qquad 11.32$$

where $\theta_m(t)$ is the radian phase argument of the carrier sinusoid. The instantaneous frequency of the carrier, r_i, is defined as

$$r_i = \dot{\theta}_m(t) \qquad\qquad 11.33$$

In *phase modulation* (PM), a modulating signal $x_m(t)$ is applied so that the carrier phase is given by

$$\theta_m(t) = \omega_c t + \phi_c + K_1 x_m(t) \qquad\qquad 11.34$$

and the instantaneous frequency of the PM carrier is

$$r_i = \omega_c + K_1 \dot{x}_m(t) \text{ rad/s} \qquad\qquad 11.35$$

In *frequency modulation* (FM),

$$\theta_m(t) = \omega_c t + \phi_c + K_2 \int_0^t x_m(t)\, dt \text{ rad/s} \qquad\qquad 11.36$$

and the instantaneous frequency of the FM carrier is

$$r_i = \omega_c + K_2 x_m(t) \text{ rad/s} \qquad\qquad 11.37$$

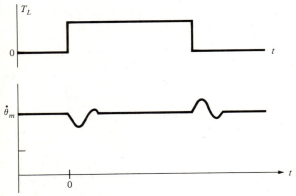

Figure 11.21   Turntable velocity response to a step of load torque.

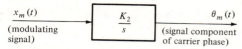

Figure 11.22   Relation of the FM carrier phase to the modulating signal.

Thus, from Eq. 11.36, we see that there is a simple relation between the information content of the FM carrier phase and the modulating signal. This relationship is shown in Fig. 11.22.

EXAMPLE 11.2

Narrowband FM Generation by a PLL

We now examine the conditions under which a PLL may generate FM. An example of one such scheme is shown in Fig. 11.23. The loop gain of this PLL is easily seen to be

$$A_L(s) = \frac{-K_p K_f K_v (s + a)}{s^2}$$

11.38

Good closed-loop pole placement gives the system a damping factor of $\xi = 0.707$. This value occurs for

$$|A_L(s = P)| = 1 = \frac{K_p K_f K_v a}{(a\sqrt{2})^2}$$

11.39

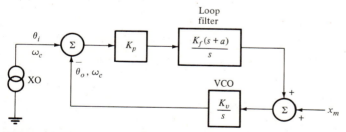

Figure 11.23   PLL, narrowband FM generation system. The low-frequency modulating signal, $x_m(t)$, is added to the output of the loop filter. At lock, the PLL tries to maintain zero phase error to the perturbations caused by x_m. Narrowband FM is widely used in police, military, and fire voice radio systems. XO = crystal oscillator.

from which we see

$$K_p K_f K_v = 2a \tag{11.40}$$

must be satisfied. The root-locus plot for this PLL is shown in Fig. 11.24. From the geometry, the undamped natural frequency of the closed-loop system is seen to be $\omega_n = 2a$ rad/s. Since the VCO output is the phase of the FM carrier (see Eq. 11.36), we can write

$$\frac{\theta_o(s)}{X_m(s)} = \frac{\dfrac{K_v}{s}}{1 + \dfrac{K_p K_f K_v(s + a)}{s^2}} = \frac{sK_v}{s^2 + s(2a) + 2a^2} \tag{11.41}$$

using Eq. 11.40.

Examining the frequency response of this transfer function, where ω is the frequency of the modulating signal, $x_m(t)$, we find that θ_o/x_m behaves like an integrator, generating FM from $2\omega_n < \omega < \infty$. That is, in this range of frequencies,

$$\frac{\theta_o}{x_m} \cong \frac{K_v}{s} \tag{11.42}$$

and FM is generated at the PLL VCO output. Note that the PLL FM modulator closed-loop ω_n should be about one-half the lowest frequency expected in the modulating signal, $x_m(t)$. ●

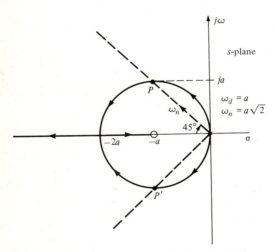

Figure 11.24 Root-locus plot for the FM generator PLL showing the positions for the closed-loop poles at P and P' desired for good (critically damped) dynamic response.

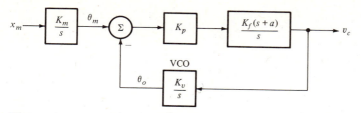

Figure 11.25 PLL used to demodulate an FM carrier. v_c is proportional to x_m from dc to a rad/s.

Other PLL configurations can also be used to generate an FM carrier from a bandwidth-limited, ac-coupled modulating signal. See the problem set at the end of this chapter for examples.

Since the phase of an FM carrier has been shown to be related to the modulating signal by an integral relationship, demodulation of an FM signal by a PLL must reverse this process. A PLL system to demodulate FM is shown in Fig. 11.25. The modulation process is shown as an integrator with gain K_m. The gain factor $K_pK_fK_v$ is set to give the closed-loop, complex-conjugate PLL poles a good damping factor, generally set to be 0.707. Under this condition, it is easy to show that $K_pK_fK_v = 2a$, and the transfer function between loop filter output, V_c, and the low-frequency modulating signal, x_m, is seen to be

$$\frac{V_c(s)}{X_m(s)} = \frac{K_mK_pK_f(s + a)}{s^2 + s(2a) + 2a^2} \tag{11.43}$$

Hence, for frequencies $x_m < a$ rad/s,

$$\frac{V_c}{X_m}(j\omega) \cong \frac{K_mK_pK_fa}{aK_pK_fK_v} = \frac{K_m}{K_v} \tag{11.44}$$

Equation 11.44 means that a low-frequency output voltage, V_c, proportional to the modulating signal, x_m, will appear at the output of the PLL's loop filter. Demodulation occurs from dc up to the damped natural frequency of the PLL. Thus, for effective operation, the ω_n of the closed-loop PLL FM demodulator should be about twice the highest frequency expected in the modulating signal, $x_m(t)$.

Amplitude-modulated (AM) carriers can also be detected using PLL design. An AM carrier is characterized by the expression

$$x_c(t) = A[1 + m(t)]\sin(\omega_c t + \theta_c) \tag{11.45}$$

where

$m(t)$ = the (normalized) low-frequency modulating signal, with $\max|m(t)| < 1$
ω_c = the carrier frequency
A and θ_c = constants

If the AM carrier, $x_c(t)$, is multiplied by a VCO output signal, $x_o(t)$, given by

$$x_o(t) = B \sin(\omega_o t + \theta_o)$$

11.46

it is easily shown that the product, $x_p(t) = x_o(t)x_c(t)$, is given by

$$x_p(t) = \left(\frac{AB}{2}\right)[1 + m(t)]\{\cos[(\omega_c - \omega_o)t + \theta_c - \theta_o]$$
$$- \cos[(\omega_c + \omega_o)t + \theta_c + \theta_o]\}$$

11.47

Under steady-state lock conditions, $\omega_c = \omega_o$, and $\theta_o = \theta_c$. Thus Eq. 11.47 reduces to

$$x_p(t) = \left(\frac{AB}{2}\right)[1 + m(t)][1 - \cos(2\omega_c t + \theta_o + \theta_c)]$$

11.48

If $x_p(t)$ is passed through a bandpass filter that blocks dc and frequencies above the highest spectral component in $m(t)$, then $x_p(t)$ becomes $x_f(t)$, where

$$x_f(t) = K_f\left(\frac{AB}{2}\right)m(t)$$

11.49

The dc term due to the carrier drops out, as does the high-frequency component at $2\omega_c$. A PLL system to implement this demodulation of AM is shown in Fig. 11.26. The PLL is assumed to use a multiplier-type phase detector, so the VCO output will be in quadrature with the input. The PLL locks on the carrier.

The output of the PLL phase detector, $x_e(t)$, may be written

$$x_e(t) = x_c(t)x_o(t)$$
$$= K_p\left(\frac{AB}{2}\right)[1 + m(t)]\{\sin[(\omega_c - \omega_o)t + \theta_c - \theta_o]$$
$$+ \sin[(\omega_c + \omega_o)t + \theta_c + \theta_o]\}$$

11.50

At lock, when the loop is tracking the input phase, $\omega_o = \omega_c$, and

$$x_e(t) = \left(\frac{AB}{2}\right)[1 + m(t)][\sin(\theta_c - \theta_o) + \sin(2\omega_c t + \theta_c + \theta_o)]$$

11.51

The double-frequency term is filtered out, and if the closed-loop PLL has a very low frequency pair of poles at ω_n, variations in $m(t)$ are also not responded to, so the output of multiplier M-1 reduces to

$$x_e(t) = K_p\left(\frac{AB}{2}\right)\sin(\theta_c - \theta_o) \cong K_p\left(\frac{AB}{2}\right)(\theta_c - \theta_o)$$

11.52

where $(\theta_c - \theta_o)$ is in radians. Note that if the $\sin(\square)$ argument is small, the $\sin(\theta_c - \theta_o)$ term can be replaced by its argument, giving effective linear phase detector operation.

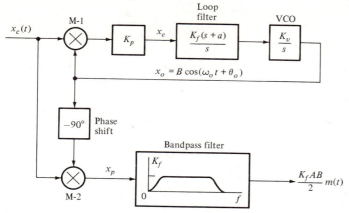

Figure 11.26 PLL used to demodulate an AM carrier. Two mixers (analog multipliers) are used. A 90° phase shift is required to recover $m(t)$.

Note that the 90° phase shift is required to enable the AM detector mixer, M-2, to operate with even phase error. The PLL phase detector, M-1, must generate an odd error signal to have a stable PLL; hence x_c and x_o must be in quadrature, so the error x_e will be in the form of Eq. 11.52.

Note that the AM detector PLL of Fig. 11.26 can be modified to act as a lock detector to sense the presence of a carrier. In this system, the bandpass filter shown in Fig. 11.26 is replaced with a low-pass filter, and its output is processed by a threshold detector (comparator). A PLL lock detector is shown in Fig. 11.27. A high average output, $\overline{x_p}$, occurs when the system is in lock and tracking.

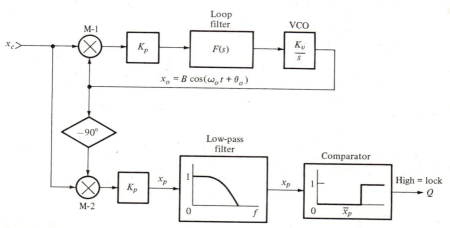

Figure 11.27 Lock detector. The average value of mixer M-2 output remains near zero until $\omega_o = \omega_c$ and $\theta_o \to \theta_c$. Then x_p rises, and the threshold detector output goes high, indicating lock.

Still another PLL application is the demodulation of double-sideband/suppressed carrier (DSBSC) signals. A special type of PLL, called the Costas loop, is used for this application. DSBSC signals are found in many control and measurement systems. A DSBSC signal may be written

$$x_1(t) = A x_m(t) \cos(\omega_c t + \theta_c)$$ 11.53

A Costas PLL is shown in Fig. 11.28. Its successful operation requires that the modulating signal, $x_m(t)$, be nonzero, or zero for only short intervals, so that the Costas loop will not lose lock. A heuristic analysis of how $x_m(t)$ is recovered by the Costas loop follows. Note that this PLL uses three mixers and three low-pass filters.

The output of the VCO, x_6, is seen to be

$$x_6 = B \cos(\omega_o t + \theta_o)$$ 11.54

and the output of mixer M-1, x_2, is found to be

$$x_2(t) = x_1(t) x_6(t) = (ABx_m)\cos(\omega_c t + \theta_c)\cos(\omega_o t + \theta_o)$$ 11.55

or

$$x_2(t) = \left(\frac{AB}{2}\right) x_m \{\cos[(\omega_c - \omega_o)t + \theta_c - \theta_o] + \cos[(\omega_c + \omega_o)t + \theta_c + \theta_o]\}$$

11.56

Again, at lock, $\omega_c = \omega_o$ and $\theta_o \to \theta_c$, so LPF-1 acting on x_2 produces

$$x_3(t) = \left(\frac{AB}{2}\right) x_m(t)$$ 11.57

which is the desired demodulated output.

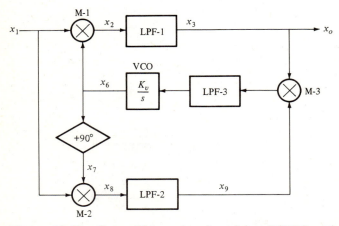

Figure 11.28 Costas PLL used to demodulate DSBSC-modulated carriers.

We must still examine other signals in the Costas loop to see how it locks. We note that, by trigonometric identity,

$$x_7(t) = -B \sin(\omega_o t + \theta_o) \qquad (11.58)$$

The output of mixer M-2 is then

$$x_8(t) = x_1(t)x_7(t) = (-ABx_m)\cos(\omega_c t + \theta_c)\cos(\omega_o t + \theta_o) \qquad 11.59$$

which reduces to

$$x_8(t) = \left(\frac{-AB}{2}\right)x_m\{\sin[(\omega_c + \omega_o)t + \theta_c + \theta_o] + \sin[(\omega_o - \omega_c)t + \theta_o - \theta_c]\}$$

$$11.60$$

$x_8(t)$ is operated on by LPF-2. Assuming lock ($w_c = w_o$), we have at the filter's output

$$x_9(t) = \left(\frac{-AB}{2}\right)x_m\sin(\theta_c - \theta_o) \qquad 11.61$$

Hence the output of mixer M-3, $x_4(t)$, is given by

$$x_4(t) = x_3(t)x_9(t) = \left(\frac{ABx_m}{2}\right)^2 \sin(\theta_c - \theta_o)\cos(\theta_c - \theta_o) \qquad 11.62$$

By trigonometric identity, this reduces to

$$x_4(t) = \left[\frac{(ABx_m)^2}{8}\right]\{\sin[2(\theta_c - \theta_o)] + \sin(0)\} \qquad 11.63$$

or, for low phase error at lock,

$$x_4(t) \cong \left[\frac{(ABx_m)^2}{4}\right](\theta_c - \theta_o) \qquad 11.64$$

Note that the odd phase error term, x_4, drives the VCO, and that the loop gain (and lock) depends on the mean-squared value of the modulating signal. Thus a Costas loop is highly nonlinear. If $x_m(t) \to 0$ for a significant time, the input, $x_1(t) \to 0$, and the Costas loop will lose lock.

Another wide area of application for PLLs is the synthesis of a number of uniformly closely-spaced, precision frequencies from one or more master crystal oscillator inputs. PLL frequency synthesizers are widely used in modern two-way radio equipment operating on assigned, fixed channels, such as citizen's band or marine radiotelephone systems. PLL synthesizers are used to generate the carrier frequency for the transmitter, as well as one or two local oscillator frequencies used in each channel of the superheterodyne receiver.

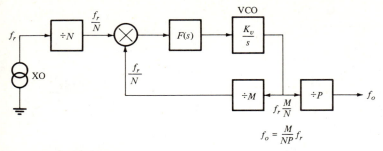

Figure 11.29   Divided-output synthesizer.

A wide variety of synthesizer architectures have been developed. We will consider two here to illustrate some of the techniques used in synthesizer design.

Synthesizers generally use digital, variable modulus counters as frequency division elements, and may use Schottky TTL or ECL logic for speed. Figure 11.29 shows a divided-output synthesizer. The input is at frequency f_r from a reference crystal oscillator (XO). f_r is reduced by an integer amount N at the PLL input; the loop VCO oscillates at frequency $f_r(M/N)$ because the VCO output frequency is divided by M, and the phase detector input frequencies must be equal at lock. The VCO output is further divided at the output by integer P. Although M, N, and P are integers, the output frequency, f_o, is obviously not related to f_r by an integer. Adjustment of f_o is accomplished by changing M, N, and P. Additional flexibility in the design of divided-output synthesizers can be obtained by using synchronous binary rate multipliers, such as the 7497. The 7497 gives a frequency output of

$$f_o = \left(\frac{M}{64}\right) f_i \tag{11.65}$$

where

$$M = D_0(1) + D_1(2) + D_2(4) + D_3(8) + D_4(16) + D_5(32) \tag{11.66}$$

and $D_0, \ldots, D_5$ are the TTL rate programming inputs to the rate multiplier. (Inspection of relations 11.65 and 11.66 shows that this IC is really a rate divider in its action. We also note that two or three 7497s can be cascaded to obtain 12- or 18-bit rate multiplication.)

A second type of PLL frequency synthesizer, the vernier loop synthesizer, is illustrated in Fig. 11.30. In this system, frequencies f_o and f_1 are mixed at M-3, and the lower (difference) product is selected; this is $(K_1/N)f_r$. A little algebra, left as an exercise for the reader, reveals that the output frequency at lock, f_o, is given by

$$f_o = f_r\left(\frac{K_1 + K_2 + K_1/N}{N + 1}\right) \tag{11.67}$$

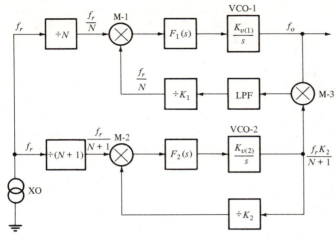

Figure 11.30 Vernier loop synthesizer. The LPF selects the lower (difference) frequency output of mixer M-3. It is easily shown that $f_o = [f_r/(N + 1)](K_1 + K_2 + K_1/N)$.

Adjustment of the integral dividers K_1 and K_2 allows f_o to be varied in a systematic manner.

11.6 Integrated Circuit PLLs

We have seen that a PLL is a relatively complex electronic feedback system containing subsystems such as a voltage-controlled oscillator, a phase detector, and a loop filter. A number of manufacturers have designed and marketed complete PLL systems as single-chip ICs. We will examine several examples of IC PLLs.

The Signetics NE564 is a radio-frequency PLL, operating up to 50 MHz. It finds use in narrowband FM communications systems, frequency shift keying systems, frequency synthesizers, and high-speed modem equipment. On the chip is a limiter to condition the input signal, a phase comparator/loop filter module, a VCO, a dc retriever, and a Schmitt trigger with TTL output. This IC is easily configured as an FM demodulator, FM modulator, or frequency synthesizer. The NE564 VCO output is a TTL waveform, and the VCO input to the phase detector is also TTL. A sine wave can be used as the NE564 reference input, however.

The postdetection processor section of the NE564 PLL consists of a comparator and a unity-gain transconductance amplifier. The comparator has variable hysteresis so phase jitter in the output can be reduced. The transconductance amplifier can be used to demodulate FSK signals, or as a postdetection filter for linear FM demodulation.

The Signetics NE/SE567 tone decoder PLL is another versatile IC PLL system. It can operate from 0.01 Hz to 500 kHz. Its primary use is in decoding touch-tone signals used in modern telephone systems. All dialed numbers and status signals (dial tone, busy signal, ringing) are composed of sums of pairs of standard, coherent, audio-frequency signals. To decode numbers and other telephone signals, combinations of NE567 PLLs can be used to sense the simultaneous occurrence of certain tone pairs. The output of this PLL is a TTL signal which goes low when the loop locks on the desired input frequency; simple NORing of the outputs of a pair of NE567s will detect the simultaneous presence of the desired pair of tones. The NE567 requires only four external components (three capacitors and one resistor) to set bandwidth, center frequency, and output delay independently. Frequency stability of the NE567 is ± 140 ppm/°C, or 1%/V supply. This stability is adequate under most conditions to demodulate standard touch tones, which have a 10% spacing.

Other applications of the versatile NE567 include decoding paging signals transmitted on radio-frequency carriers, and enabling emergency communications receivers (tone monitors) used by fire departments and ambulance corps. Tone decoder PLLs are also used in air-transmitted ultrasonic systems used for remote control of TV receivers, VCRs, and garage door openers. The maximum rate at which the NE567 can detect a keyed (ON-OFF) tone is about $f_o/10$ baud, where f_o is the frequency of the keyed carrier.

The 4046B is a CMOS, integrated circuit PLL system that consumes low power and operates up to 2.3 MHz. On this chip are a low-power digital VCO, a source-follower buffer amplifier, a 7 V zener diode, and two phase detectors. One phase detector, PHD-1, is an exclusive-OR logic gate. For maximum lock range with PHD-1, both logic inputs should have 50% duty cycles. Note that the inputs of an exclusive-OR phase detector are in quadrature at lock. PHD-2 is an edge-triggered digital memory circuit with four flip-flop stages and associated control and output gates; its operation appears similar to that of the Motorola MC-4044 TTL phase detector IC. At lock, PHD-2 has zero phase error between its input signals. It also has an *in-lock* status line output. Loop filters for the 4046B PLL are constructed externally from a capacitor and one or two resistors. The 4046B is a generally useful PLL IC with many applications in communications, instrumentation, and control.

Other specialized PLLs can be assembled from commercially available IC modules. VCOs, phase frequency detectors, logic rate multipliers (frequency dividers), active filters, and other devices can be used to create custom PLLs with special output waveforms, lock characteristics, and so on.

11.7 Acquisition of Lock in PLLs

In the preceding sections of this chapter, we have assumed the PLLs to be in lock; that is, the average frequency of the VCO is equal to the average frequency of the input signal ($\omega_o = \omega_i$). This assumption was made in order that linear systems analysis techniques could be used in the analysis and design of PLLs. However, it should be noted that all PLLs given a *cold start* are initially out of lock and therefore must gain lock through a process known as acquisition.

Acquisition is generally a nonlinear process and is difficult to model analytically. The dynamics of the process depend on the type of phase detector in the PLL, the type of loop filter, and the frequency error at start-up. PLL acquisition can be simulated using such programs as CSMP. However, we will examine acquisition in first-and second-order PLLs using the heuristic approach of Gardner (1979). (Note that an nth-order PLL has an s^n term in the denominator of its loop gain expression.)

First we will examine a simple first-order PLL that uses an analog multiplier phase detector to compare the phase of two sinusoidal signals, the reference input, V_i, and the VCO output, V_o. This PLL is shown in Fig. 11.31. A simple low-pass loop filter is assumed of the form

$$F(s) = \frac{K_f}{s/a + 1} \qquad\qquad 11.68$$

The output voltage of the analog multiplier phase detector is given by Eq. 11.1, and if we assume that the loop filter attenuates the high-frequency term to a negligible magnitude, we can express the multiplier phase detector output as

$$v_d = K_d \sin(\theta_e) \qquad\qquad 11.69$$

where $\theta_e = \theta_i - \theta_o$. v_d is multiplied by the low-frequency gain of $F(s)$, K_f, and then drives the VCO. The phase at the VCO output can be written

$$\theta_o(t) = \omega_o t + K_d K_f K_v \int_0^t \sin(\theta_e)\, dt + \theta_o(0) \qquad\qquad 11.70$$

The phase error can thus be expressed as

$$\theta_e(t) = \omega_i t - \omega_o t - K_d K_f K_v \int_0^t \sin(\theta_e)\, dt - \theta_o(0) \qquad\qquad 11.71$$

If Eq. 11.71 is differentiated, we obtain the simple nonlinear differential equation

$$\dot{\theta}_e = \Delta\omega - K_d K_f K_v \sin(\theta_e) \qquad\qquad 11.72$$

where $\Delta\omega = \omega_i - \omega_o$. The loop is locked if $\dot{\theta}_e = 0$. Examination of Eq. 11.72 under lock conditions leads to the observation that the loop can lock only if

$$|\Delta\omega| < K_d K_f K_v \qquad\qquad 11.73$$

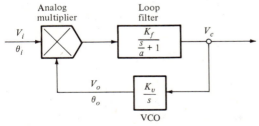

Figure 11.31   A first-order PLL with an analog multiplier phase detector.

Because every cycle in the nonlinear equation 11.72 has a stable null (with positive slope), θ_e cannot change by more than one cycle (360°) before the first-order PLL locks. The time required to reach lock, of course, depends on the initial values of frequency and phase error; Gardner gives the time to lock as approximately $3/K_d K_f K_v$ for the first-order PLL.

If $\theta_e(0)$ is close to an unstable null in Eq. 11.72, that is, where $\dot{\theta}_e/K_d K_f K_v$ crosses zero with negative slope, then the phase error can remain near the null for an extended time; this is called the hang-up effect. The hang-up effect can also occur in higher-order PLLs.

In examining the lock acquisition processes for a second-order PLL, we again assume an analog multiplier phase detector. The loop filter in this case is assumed to be of the form

$$F(s) = \frac{K_f(s + a)}{s} \qquad\qquad 11.74$$

Before the PLL locks, ω_o is close to ω_i, and the output of the analog multiplier phase detector, given by Eq. 11.1, contains a high-frequency term and a more significant, low-frequency, *beat frequency* term. Inspection of the phase detector output as the PLL pulls in to lock shows that it contains an average or dc component which is, in fact, the signal that drives the VCO to reach $\omega_o = \omega_i$ at lock. Gardner argues that there is a critical value of this average voltage required for lock, called the pull-in voltage, v_p, which is approximated by

$$v_p = K_d \left[\frac{\Delta\omega}{K} - \sqrt{\left(\frac{\Delta\omega}{K}\right)^2 - 1} \right] \qquad\qquad 11.75$$

for $|\Delta\omega| > K$, where $K \triangleq K_d K_f K_v$.

The instantaneous frequency difference between the VCO output and the input can be written for this second-order PLL as

$$\Delta\omega = \Delta\omega(0) - (K_f a K_v) \int_0^t v_p(t)\, dt \qquad\qquad 11.76$$

where $\Delta\omega(0)$ is the frequency difference at $t = 0$, and the low-frequency gain of the loop filter, $K_f a$, is used on v_p. If Eq. 11.76 is differentiated with respect to time, and Eq. 11.75 for v_p is substituted, we can write

$$dt = -\frac{d(\Delta\omega)}{aK\left[\dfrac{\Delta\omega}{K} - \sqrt{\left(\dfrac{\Delta\omega}{K}\right)^2 - 1}\right]} \qquad\qquad 11.77$$

Equation 11.77 can be integrated to find the time T_p required for the second-order PLL to go from some large initial $\Delta\omega(0)$ to the lock limit, $|\Delta\omega| = K$. Thus we integrate the right side of Eq. 11.77 between $\Delta\omega(0)$ and K to find

$$T_p = \left(\frac{1}{a}\right)\left[\frac{\Delta\omega(0)}{K}\right]^2 \qquad\qquad 11.78$$

Gardner cautions that this formula for T_p is based on approximations and should not be used if $\Delta\omega(0)$ is small (near $|\Delta\omega| = K$).

If the loop filter integrator is imperfect, the PLL becomes first-order, albeit with a large but finite dc gain. In this case, the loop filter transfer function is of the lag-lead form and is given by

$$F(s) = \frac{K_f(s + a)}{s + b} \qquad\qquad 11.79$$

Now the instantaneous frequency difference between input and VCO can be written

$$\Delta\omega = \Delta\omega(0) - K_v\left(\frac{K_f a}{b}\right)v_p \qquad\qquad 11.80$$

When relation 11.75 for v_p is substituted into Eq. 11.80, we can solve for the maximum $\Delta\omega(0)$, ω_p, for which the PLL will never acquire lock. This quadratic solution is based on the assumption (for a real root) that

$$\Delta\omega(0) > K_d K_f K_v \sqrt{\frac{2a}{b} - 1} \qquad\qquad 11.81$$

where $a/b \gg 1$, and is simply

$$\Delta\omega_p = K_d K_f K_v \sqrt{\frac{2a}{b}} \qquad\qquad 11.82$$

Finally, Gardner cites a derivation for the lock time of a third-order loop. If the loop gain is

$$A_L(s) = -\frac{K_d K_f K_v(s + a)^2}{s^3} \qquad\qquad 11.83$$

the lock time can be shown to be approximately

$$T_p = \left(\frac{\sqrt{\pi}}{a}\right)\left(\frac{\Delta\omega(0)}{K_p K_v K_f}\right) \qquad\qquad 11.84$$

In the discussion and developments just given, we have considered self-acquisition of lock in certain types of PLLs. Often, faster and certain frequency acquisition can be obtained by operating a phase detector in the phase frequency mode. The VCO input, v_c, is set to a maximum (or minimum) value when the phase frequency detector signals that $\omega_i \neq \omega_o$, thus slewing ω_o toward ω_i at a maximum rate. When the phase frequency detector signals lock, the VCO input is reconnected to the output of the loop filter. Another way of expediting acquisition is to sweep the VCO output signal over its range when no lock is signaled, in effect searching for ω_i. When lock is obtained, the sweep mechanism is disabled. It is important that the sweep rate be kept less than ω_n^2 in order for lock to occur (ω_n is the natural frequency of the closed-loop PLL). Other adaptive and clever means have been described to insure that PLLs lock rapidly and maintain lock in the presence of noise and transients in ω. These descriptions can be found in the reference section at the back of the book.

SUMMARY

In this chapter on the analysis and applications of phase-lock loops, we have first described the component subsystems that comprise an analog PLL. Phase detector designs were first described and analyzed; these included the analog multiplier, the sample-and-hold phase detector, and various logic circuit phase detectors that operate on logic signal waveforms. The transfer functions of various loop filters commonly used in PLLs were described, and certain commonly used voltage-controlled oscillator designs were given.

Dynamic compensation of PLLs was approached using the root-locus technique, assuming linear loop operation at lock. Root-locus was used to set a closed-loop PLL's damping factor and to find its natural frequency at a given loop gain.

Several examples of PLL applications were given, including a motor speed control system, modulators and demodulators used in various communications systems, and frequency synthesizers.

The problems associated with PLL acquisition of lock were discussed. The large-signal behavior of PLLs was seen to be quite nonlinear and to defy simple analysis. Approximate formulas were derived defining relations between PLL loop gain and acquisition time, and the maximum initial frequency difference for pull-in as a function of PLL parameters.

PROBLEMS

11.1   A PLL is used to generate a variable phase shift.

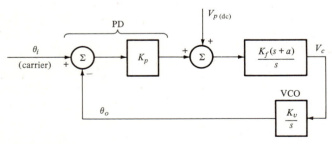

Figure P11.1

(a) Use root-locus to find $K_p K_f K_v$ to give the closed-loop PLL a ξ of 0.707.
(b) Find the phase difference, $\theta_i - \theta_o$, in the steady state for $V_p > 0$.
(c) If V_p is a low-frequency sinusoid, what form of modulation is seen at the VCO output?

11.2 A 4046 CMOS integrated circuit PLL is used as a tachometer. Assume $K_p K_v$ such that the closed-loop ξ is 0.707.

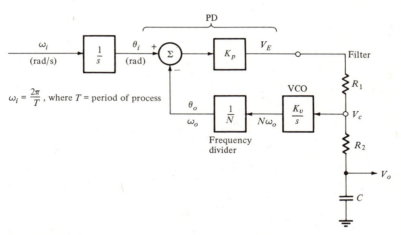

$$\omega_i = \frac{2\pi}{T}, \text{ where } T = \text{period of process}$$

Figure P11.2

(a) Write an expression for $(V_o/f_i)(s)$ in time-constant form.
(b) Sketch and dimension $V_o(t)$, given $f_i = f_{io}$ for $t < 0$ and $f_i = 0$ for $t \geq 0$.

11.3 The PLL in Fig. P11.3 is used as a phase modulator. A PM carrier has the form

$$f_{cm}(t) = A \sin[\omega_c t + K_m X_m(t)]$$

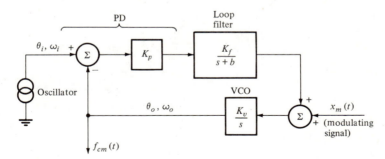

Figure P11.3

(a) Use root-locus to find the scalar gain product $K_p K_f K_v$ required to give the closed-loop PLL a ξ of 0.707.

(b) Find $(\theta_o/X_m)(s)$ in time-constant form.
(c) Make a dimensioned Bode plot of $(\theta_o/X_m)(j\omega)$. Use the $K_pK_fK_v$ value found in part (a).
(d) Show the ω region where the VCO output is phase-modulated by $X_m(t)$.
(e) Find an expression for K_m, the PM coefficient in $f_m(t)$.

11.4   The PLL in Fig. P11.4 is used to generate narrowband FM.

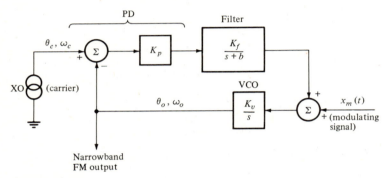

Figure P11.4

(a) Write $A_L(s)$ for the PLL.
(b) Sketch the PLL's root-locus diagram.
(c) Find $(\theta_o/X_m)(s)$ in time-constant form.
(d) Make a Bode magnitude plot of $(\theta_o/X_m)(j\omega)$ versus ω.
       (*Note:* ω values are of X_m.)
(e) Show the range of ω where true narrowband FM is generated at the VCO output.

11.5   The carrier phase in a narrowband signal is related to the modulating signal by an integration.

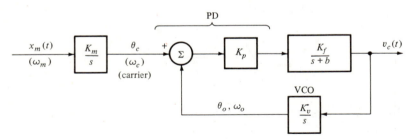

Figure P11.5

(a) Write the system's $A_L(s)$.
(b) Use root-locus to find an expression for the gain product $K_pK_fK_v$ to give the closed-loop PLL complex-conjugate poles with $\xi = 0.5$.
(c) Find an expression for $(V_c/X_m)(s)$ in time-constant form.
(d) Make a Bode magnitude plot of $|(V_c/X_m)(j\omega)|$ versus ω. Use $\xi = 0.5$ and ω_n found in part (b).

(e) Specify a numerical value for $K_p K_f K_v$ that will give the demodulator a corner frequency, f_n, at 5000 Hz.

11.6 A PLL tone decoder (Fig. P11.6) is to be used to detect the musical note A_{440} (assume sinusoidal) to within $\pm 1\%$. This means the loop must capture and lock to the input signal in a capture range $\Delta f_c = 8.8$ Hz centered on $f_1 = 440$ Hz. In a conventional PLL, the capture range is given by

$$\Delta\omega_c = 2\pi\,\Delta f_c = K_T|F(j\,\Delta\omega_c)|$$

where $F(j\,\Delta\omega_c)$ is the low-pass (loop) filter's magnitude response at $\omega = \Delta\omega_c$ rad/s and $K_T = K_p K_v F(0)$.

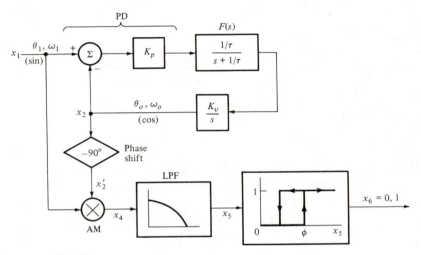

Figure P11.6

(a) Find the filter time constant, τ, and the gain factor $K_p K_v$ required for the lock specifications indicated, and to give the PLL a ξ of 0.707.

(b) Find the PLL's ω_n and its lock range, $\Delta\omega_L = K_T$.

11.7 A *long-loop*, superheterodyne, coherent AM, PLL radio receiver is shown in Fig. P11.7. Assume the following:

$$X_1(t) = A[1 + m(t)]\sin(\omega_1 t + \theta_1) \qquad \text{(AM input)}$$

$m(t) = $ modulating signal; maximum $m(t) < 1$

At lock: $\omega_5 = \omega_4 = (\omega_1 - \omega_2)$ rad/s

$\qquad\qquad (\theta_1 - \theta_5 - \theta_2) \to 0$

$$X_2(t) = B\cos(\omega_2 t + \theta_2)$$

$$X_5(t) = C\cos(\omega_5 t + \theta_5)$$

Mixers are ideal multipliers; i.e., $z = xy$

Find expressions for $X_3(t)$, $X_4(t)$, $X_5'(t)$, $X_6(t)$, $X_7(t)$, $X_8(t)$, and $X_9(t)$ at lock.

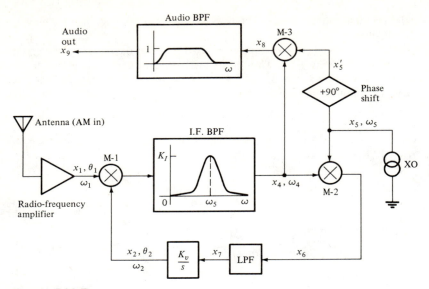

Figure P11.7

11.8 A PLL is used to generate narrowband FM. The modulating signal, $X_m(t)$, acts on a carrier of frequency ω_c to generate an output,

$$v_o(t) = B \sin[\omega_c t + \underbrace{(\psi_o + K_m \int X_m(t)\, dt)}_{\theta_o}], \qquad \omega_o = \omega_c \text{ at lock.}$$

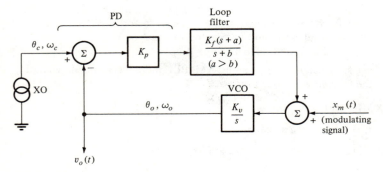

Figure P11.8

(a) Write $A_L(s)$ for the PLL in Laplace format.
(b) Sketch the system's root-locus diagram to scale.

(c) Find the gain product $K_p K_f K_v$ to give the modulator $\xi = 0.707$. Find ω_n for this condition.

(d) Write $(\theta_o/X_m)(s)$ in time-constant form for the condition of part (c).

(e) Sketch a Bode magnitude plot of $(\theta_o/X_m)(j\omega)$. Show the ω range over which FM is generated.

11.9 The PLL in Figure P11.9 is used to generate an angle-modulated carrier.

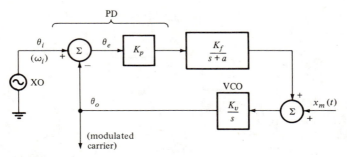

Figure P11.9

(a) Use root-locus to find an expression for $K_p K_f K_v$ to give the closed-loop modulator a ξ of 0.5.

(b) Find $(\theta_o/X_m)(j\omega_m)$ in time-constant form.

(c) Do a detailed dimensioned Bode magnitude plot of $(\theta_o/X_m)(j\omega_m)$. Show the ω_m regions where FM is generated, and where the PM carrier is generated.

11.10 The PLL in Fig. P11.10 is used to generate narrowband FM.

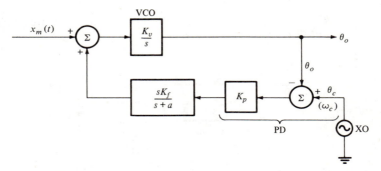

Figure P11.10

(a) Write the system's loop gain, $A_L(s)$, in Laplace format.

(b) Sketch and dimension the system's root-locus diagram to scale.

(c) Find $(\theta_o/X_m)(s)$ in time-constant form.

(d) Sketch a dimensioned Bode magnitude plot for $(\theta_o/X_m)(j\omega)$. Show the regions of ω_m where narrowband FM is produced, and also where PM occurs (if at all).

11.11 A PLL used to demodulate a narrowband FM signal with phase θ_c has a noisy oscillator which injects *phase noise*, θ_n, into the loop. θ_n is effectively added to the VCO's output phase, θ_o. It has a white power density spectrum (units = mean-squared radians per hertz), $S_{\theta_n} = \eta$. Using methods from Chapter 8, write an expression for the mean-squared SNR at v_c, given

$$x_m(t) = X_m \sin(\omega_m t)$$

where $\omega_m = a$. Assume $\omega_o \to \omega_c$ at lock; set $\xi = 0.5$ for the closed-loop PLL.

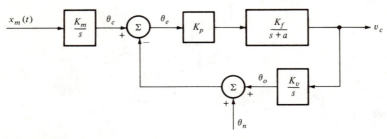

Figure P11.11

11.12 The PLL in Fig. P11.12 is run at a center frequency of 10^5 rad/s; $K_d = 1$ V/rad, $K_f = 1$, and $K_v = 10^2$ rad/s/V. A first-order loop filter is used, where

$$F(s) = \frac{K_f}{sT + 1}$$

Use root-locus techniques to find the filter time constant, T, required to give the closed-loop PLL a ξ of 0.707. Find ω_n for this condition.

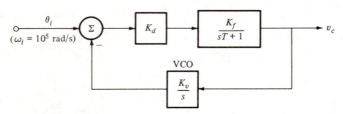

Figure P11.12

11.13 A 4046B CMOS PLL is used to produce a voltage proportional to heart rate (*NASA Tech Briefs* 7(3): 1983, MSC-20078). Actual circuit and system parameters are shown in Fig. P11.13.

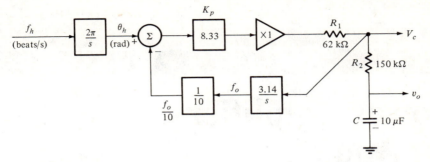

Figure P11.13

(a) Write an expression for the system's loop gain.

(b) Sketch the system's root-locus diagram to scale.

(c) Find the location of the closed-loop system's poles; give ω_n and ξ. (Note that this can be done by trial and error, or by a computer-generated root-locus plot.)

(d) Make a dimensioned Bode magnitude plot of $(V_o/f_h)(j\omega)$.

Chapter 12

Power Amplifiers for Audio and Control Applications

A power amplifier (PA) is intended to deliver electric energy, usually at a rate above 0.5 W, to a load. This power can be used to drive an output transducer such as a loudspeaker, ultrasound transducer array, or dc servomotor. In the analysis and design of PAs, we need to consider not only the small-signal behavior of the circuit, but also the effects of the output semiconductor devices' power dissipation and temperature rise on quiescent biasing, the large-signal transfer nonlinearities which cause distortion, the output impedance of the PA, the load impedance, and power supply regulation.

PAs, like small-signal amplifiers, can be divided into two major classes: those that are reactively coupled and those that are direct-coupled to their loads. The typical organization of PAs of either class consists of a low-power input stage in which there is some voltage amplification, then a driver stage which boosts signal and power levels to drive the high-powered output stage. In this chapter, we will focus on PA power output stages.

A power output stage can have a variety of geometries, as will be seen, and generally uses power BJTs, power Darlington BJTs, or power MOSFETs (VMOSFETs and lateral double-diffused MOSFETs). Older PA designs made excellent use of vacuum tubes (triodes, pentodes) as the output devices. Extensive use was also made of expensive, high-quality audio power transformers to couple low-impedance loads to the power output stages for maximum PA efficiency. The modern trend, however, is either to use direct coupling to the load or to couple to the load through a large electrolytic capacitor if it is not necessary to pass dc. Many modern PAs are of all IC design. IC power amplifier packages are built to be thermally coupled to heatsinks to dissipate PA internal power and reduce operating temperature.

Of special interest are power op-amps, which consist of a regular op-amp coupled to a DC power output stage, all in one package. Power op-amps have all the versatility of conventional op-amps but can deliver appreciable power to their low-impedance loads. They make excellent audio PAs.

12.1 Power Output Devices

The power BJT is probably the most widely used device for PA output stage designs. Power BJTs generally have lower h_{fe}'s than do small-signal BJTs; they also can pass considerably more collector current than a small-signal transistor. Each power output device (BJT, MOSFET, tube) has a permissible operating region defined on its volt-ampere curves. The permissible operating region for a power BJT in the grounded-emitter configuration is shown in Fig. 12.1. Note that the region's boundaries are defined by various physical properties of the BJT. Maximum average BJT power dissipation is given by the hyperbola

$$P_{\max} = I_C V_{CE} \qquad\qquad 12.1$$

$P_{\max}$ is determined by the effectiveness of the transistor's heatsink in keeping its junction temperature below the maximum allowable junction temperature for that device. It is usually specified by the manufacturer for a given set of heatsink conditions. The hyperbolic segment set by $P_{\max}$ is designated by P in Fig. 12.1. The boundary V in

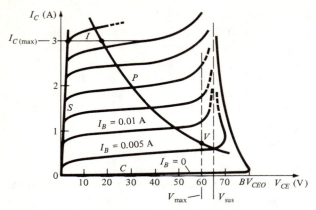

Figure 12.1   Permissible linear, large-signal operating region of a typical power BJT.

Fig. 12.1 is set by the manufacturer-specified maximum collector voltage. V_{max} is just below V_{sus}. At V_{max}, the normally reverse-biased collector-base diode junction begins to break down (avalanche), and the $I_C = f(V_{CE}, I_B)$ curves begin to tip up sharply because of the avalanche multiplication process, causing nonlinearity. The voltage V_{sus} is defined as V_{CE} where transistor small-signal $h_{fe} \to \infty$. The C region is obviously where the BJT is cut off ($I_B = 0$). Only a small, thermally dependent collector leakage current, I_{CEO}, flows in this region. BJT saturation occurs on the S boundary. Here, further increase in I_B causes no further drop in V_{CE}. Also in saturation, the collector-base diode is forward biased. Finally, the I region is set by the maximum continuous collector current specified by the manufacturer. The Ebers-Moll large-signal model of the BJT introduced in Chapter 1, Sec. 1.2 describes these boundaries mathematically. Note that a second type of avalanche breakdown can occur (not shown in Fig. 12.1) when the base-emitter junction is reverse biased (the transistor is cut off) if V_{BE} exceeds the avalanche limit for that junction. This breakdown voltage is typically in the range of -6 V for an npn BJT.

Darlington BJT power transistors have volt-ampere curves similar to those shown in Fig. 12.1, except that Darlingtons have higher net h_{fe}'s than do single power BJTs of comparable P_{max}, V_{sus}, and so on. Most power Darlingtons also include a power diode across the collector-emitter terminals. This diode is reverse biased under normal operating conditions and protects the device from negative voltage transients when it is switching an inductive load.

Power MOSFET transistors also have permissible operating regions, set by their physical properties, as shown in Fig. 12.2. The I, P, V, and C regions are analogous to those of the power BJT shown in Fig. 12.1. However, the S region boundary for the power MOSFET separates the desired, saturated channel region of operation, where r_d is large and I_D is constant, from the unsaturated channel region, where r_d decreases rapidly with V_{DS} and also varies with V_{GS}. The maximum I_D line may, in some devices cross above the intersection of the saturation S line and the maximum

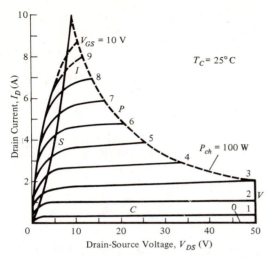

Figure 12.2 Permissible linear, large-signal operating region for a 2SK134 n-channel power MOSFET. Desired operation is under saturated drain conditions to the right of the S line.

power hyperbola, P. The vertical V line indicates the maximum permissible V_{DS}. At V_{DS} values above BV_{DSX}, there is a catastrophic breakdown of the drain-substrate pn junction. BV_{DSX} typically occurs in the range of 50 to 70 V for most power MOSFETs, although some devices can work up to $V_{DS} = 400$ V. The C boundary is the 1 mA drain current line which defines the gate-source threshold voltage, V_T or $V_{GS(th)}$, for the MOSFET.

Three power transistor packages frequently encountered in audio, control, and ultrasound applications are illustrated in Fig. 12.3. Heat is conducted away from the

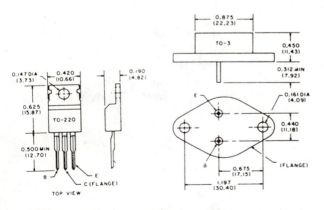
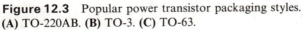
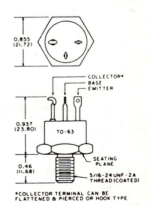

Figure 12.3 Popular power transistor packaging styles. **(A)** TO-220AB. **(B)** TO-3. **(C)** TO-63.

device to a heatsink, which in turn dissipates its thermal load by radiation and convection. High-density power systems frequently use forced-air cooling of heatsinks to ensure adequate heat dissipation under worst-case environmental conditions. Frequently, the device being heatsunk must be electrically insulated from the heatsink, which often is at ground potential. (For example, the cases of many TO-3-mounted power BJTs are connected to the collectors.) To electrically insulate the transistor's case from the heatsink while preserving heat conduction, a thin mica or silicone rubber spacer is used between the transistor and the heatsink. Mounting screws must also be insulated, with plastic bushings, or be plastic. A heat-conducting grease (heatsink compound) is applied in a thin layer between the spacer and the transistor, and between the spacer and the heatsink, to improve heat flow. Heatsink compound is also used even when electrical insulation is not required.

12.2 Heatsinking

One common cause of electronic equipment failure is loss of a power transistor or its associated circuit because the transistor overheated and circuit bias conditions were altered radically under high operating temperatures. A conservative, worst-case approach to heatsink design would eliminate many heat-related equipment failure problems.

To approach the problem of heatsink design for power transistors and IC power amplifiers, we introduce the concept of thermal resistance, Θ. The units of Θ are degrees Celsius per watt. Heat flow or production in watts (joules per second) is treated analogously to current flow (coulombs per second) in a dc circuit. Hence a transistor, SCR, or diode dissipating an average power, P_D, is treated as a current source. Temperatuure in degrees Celsius is analogous to voltage, and of course thermal resistance is modeled by electrical resistance.

Manufacturers' data sheets for power devices generally give the maximum operating junction temperature, $T_{j(max)}$, and the inherent thermal resistance from the junction to the case, Θ_{jc}. Also, the thermal resistance seen in coupling the case to the heatsink, Θ_{ch}, is relatively constant for a given transistor case type; Θ_{ch} is determined by the insulating spacer coated with heatsink compound. The thermal resistance of the heatsink itself is quite variable and depends on the material, size, shape, and orientation of the heatsink, and whether or not there is forced-air cooling.

The equivalent, steady-state analog circuit for a power transistor mounted on a heatsink is shown in Fig. 12.4. Good thermal design involves choosing a heatsink with Θ_h small enough to satisfy the equation

$$T_{j(max)} \leq P_D(\Theta_{jc} + \Theta_{ch} + \Theta_h) + T_a \qquad\qquad 12.2$$

or, equivalently,

$$\Theta_h \leq \frac{T_{j(max)} - T_a - P_D(\Theta_{jc} + \Theta_{ch})}{P_D} \qquad\qquad 12.3$$

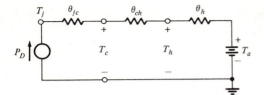

Figure 12.4 Thermal analog circuit for a power device coupled to a heatsink. T_a = the ambient air temperature at the heatsink, T_h = the heatsink temperature (assumed uniform), T_c = the case temperature of the device, and T_j = the (internal) junction temperature of the device. PD is the worst-case (maximum) device power dissipated.

Frequently, power device manufacturers supply a power derating curve. This curve relates maximum permissible average device power dissipation to steady-state device case temperature. A *generic* derating curve is shown in Fig. 12.5. Note that when the case temperature equals $T_{j(max)}$, the device is totally derated; in other words, it should be turned off! Often temperature-sensing switches are used to interrupt power in audio PAs when the heatsink temperature, T_h, reaches a preset high limit. Some modern IC power op-amps also have automatic shutdown when $T_j = T_{j(max)}$.

A simultaneous solution of the derating curve $\bar{P}_D = f(T_c)$ with the temperature difference $(T_c - T_a)$ from Fig. 12.4 can be written

$$\bar{P}_D = \frac{T_c}{\Theta_h + \Theta_{ch}} - \frac{T_a}{\Theta_h + \Theta_{ch}} = f(T_c) \qquad 12.4$$

This solution is shown graphically in Fig. 12.6. Note that the more efficient the heatsink, the lower the value of $(\Theta_h + \Theta_{ch})$ will be, the lower T_c will be, and the higher P_D can be. By inspection of Fig. 12.6 and Eq. 12.4, we can find an expression for the maximum $(\Theta_h + \Theta_{ch})$ value so that $P_D = P_{D(max)}$, given any $T_a < T_{co}$. This expression is

$$(\Theta_h + \Theta_{ch}) \le \frac{T_{co} - T_a}{P_{D(max)}} \qquad 12.5$$

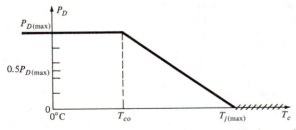

Figure 12.5 Power device derating curve. T_{co} is usually near 25°C, $T_{j(max)}$ is near 150°C, and T_c is the case temperature.

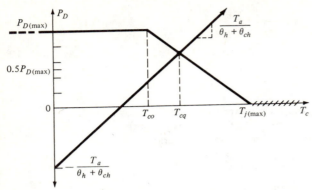

Figure 12.6   Graphical solution to the operating device's case temperature, T_{cq}, given the device's derating curve, the ambient air temperature, T_a, and the combined thermal resistance of the heatsink and thermal coupling ($\theta_h + \theta_{ch}$).

Thus, the thermal resistance of the heatsink and the thermal coupling hardware must be smaller than the value given by the right side of Eq. 12.5 in order that the device can be run at its maximum power rating, $P_{D(max)}$.

12.3 Classification of Power Amplifiers

Traditionally, PAs have been classified according to the fraction of a cycle over which a power device conducts current, given sinusoidal excitation. In class A amplifiers, the device conducts current over the entire cycle; it is never cut off. In class AB the device is cut off for less than one-half cycle; in class B, it conducts current for one-half cycle; and in class C, current flows for less than one-half cycle. Classes A, AB, and B are used in various broadband linear PA designs; class C is generally used at radio frequencies with tuned (RLC) loads. We will not consider tuned class C power amplifiers in this chapter.

A PA may be further classified by the way in which it is coupled to its load. Direct coupling generally requires that $V_o = 0$ when $V_{in} = 0$, so that no quiescent dc current flows in the load. Reactive coupling can be by a large, dc-blocking electrolytic capacitor, C_c, whose value is large enough to ensure that there is little attenuation of the lowest frequency to be delivered to the load. That is,

$$C_c > \frac{1}{2\pi f_{min} RL}$$

12.6

For example, if we wish the low-frequency pole set by C_c to be at 10 Hz, and the load is a nominal 8 Ω loudspeaker, C_c must be 2000 μF.

A second means of reactive coupling is by power transformer. Transformers obviously do not pass dc; they have a rather complex, multicomponent equivalent

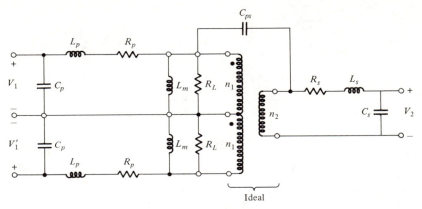

Figure 12.7 Equivalent circuit for an audio power output transformer with a center-tapped primary winding. C_p = the (equivalent)
lumped primary winding capacitance, L_p = the primary leakage
inductance, R_p = the primary winding resistance, L_m = the primary
magnetizing inductance ($L_m \gg L_p$), R_L = the equivalent core loss
resistance (from hysteresis and eddy currents), n_1 = the number of
primary turns (on one side), and n_2 = the number of secondary turns.
The s-subscripted parameters are the secondary winding resistance,
leakage inductance, and capacitance. C_{ps} is the primary-to-secondary
coupling capacitance. An ideal transformer couples the primary to
the secondary circuit.

circuit that describes their behavior at high, mid-, and low frequencies, as shown in
Fig. 12.7. We will consider the transformer here only at mid-frequencies, where it
can best be approximated by an ideal transformer. Power transformers are heavy,
expensive (especially if wide bandwidth = high fidelity is required), and create stability
problems at low and high frequencies when negative feedback is used. The advantages
of transformers are that they allow the load to be matched to the PA's output devices
for maximum power transfer and also block unwanted dc from the load. Transformer
coupling of the power stage to the load is also useful when the ac voltage swing
across the load at full power needs to be much smaller than the maximum linear voltage swing across the transistors, or the ac load swing must be larger than the amplifier's
dc supply voltage.

In an ideal transformer, the volt-ampere relations are

$$V_2 = V_1 \left(\frac{n_2}{n_1}\right) \qquad I_2 = I_1 \left(\frac{n_1}{n_2}\right) \qquad\qquad 12.7$$

When a load, R_L, is attached to the transformer's secondary winding, as shown in
Fig. 12.8, we can use relations 12.7 to find the apparent resistance seen looking into

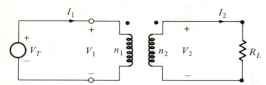

Figure 12.8   Ideal transformer with load R_L on its secondary. The source $V_T = V_1$ is used to find the resistance looking into the primary winding, R'_L.

the primary winding:

$$R'_L = \frac{V_T}{I_1} = \frac{V_1}{I_1} = \frac{V_2(n_1/n_2)}{I_2(n_2/n_1)} \qquad \qquad 12.8$$

Because R_L is equal to V_2/I_2, Eq. 12.8 reduces to

$$R'_L = R_L\left(\frac{n_1}{n_2}\right)^2 \qquad \qquad 12.9$$

A more realistic mid-frequency approximation to the transformer with the secondary load R_L includes primary winding resistance R_1, core loss equivalent resistance R_c, and secondary winding resistance R_2. Thus the PA "sees"

$$R'_L = R_1 + R_c + \left(\frac{n_1}{n_2}\right)^2 (R_L + R_2) \qquad \qquad 12.10$$

Therefore, by controlling n_1/n_2, we can present the devices driving the load with an R'_L that will allow optimum power transfer to R_L.

12.4 Properties of Class A Power Amplifiers

Consider first a single BJT class A power amplifier, as shown in Fig. 12.9(A). This amplifier is biased at point Q on its I_C-versus-V_{CE} curves. Q must lie on the load line halfway between the cutoff point, C, and the saturation point, S. As long as the BJT is not driven beyond the S-C segment of the load line, the collector current and V_{CE} are assumed to be sinusoidal in response to a sinusoidal signal driving I_B.

The power conversion efficiency, η, of the PA is defined as the ratio of the average ac power delivered to the load, P_{ac}, to the average power delivered by the V_{CC} supply, P_{dc}. P_{dc} is independent of the signal level in a class A amplifier; hence, to obtain maximum η, we must have maximum P_{ac}. Maximum P_{ac} occurs when the peak-to-peak swing in V_{ce} and I_c is a maximum. Referring to Fig. 12.9(B), we see that if $V_{CE(\text{sat})}$ is neglected,

$$i_c(t) = I_{CQ}(1 + \sin \omega t) \qquad \qquad 12.11$$

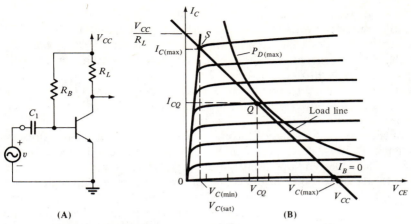

Figure 12.9 (A) Simple class A, oc, BJT power amplifier. (B) BJT volt-ampere curves with the load line plotted. S = the saturation point, Q = the quiescent bias point, and C = the cutoff point. Note $I_B = (V_{CC} - 0.7)/R_B$.

where

$$I_{CQ} = \frac{V_{CC}}{2R_L} \qquad\qquad 12.12$$

$$v_c(t) = \left(\frac{V_{CC}}{2}\right)(1 - \sin \omega t) \qquad\qquad 12.13$$

Note that the collector voltage is 180° out of phase with the collector current. Hence $\bar{P}_{ac}$, the average power in the load, can be found from the following two equations:

$$\bar{P}_{ac} = \frac{1}{2\pi} \int_{-\pi}^{\pi} \frac{V_{CC}}{2R_L}(1 + \sin \omega t)\frac{V_{CC}}{2}(1 - \sin \omega t)\, d\omega t \qquad\qquad 12.14$$

$$\bar{P}_{ac} = \frac{1}{2\pi}\frac{V_{CC}^2}{4R_L}\int_{-\pi}^{\pi}(1 - \sin^2\omega t)\, d\omega t = \frac{V_{CC}^2}{8R_L} \qquad\qquad 12.15$$

Hence the maximum efficiency for the class A BJT amplifier is, from the preceding definition,

$$\eta = \frac{\bar{P}_{ac}}{\bar{P}_{dc}} = \frac{V_{CC}^2/8R_L}{V_{CC}(V_{CC}/2R_L)} = 0.25 \qquad\qquad 12.16$$

Actual maximum efficiency, of course, will be less for a real class A amplifier because of $V_{CE(sat)} > 0$ and collector leakage current at cutoff.

The emitter- (or source-) follower is also used as a class A power amplifier. It has an inherently low output resistance, as shown in Chapter 1, and a voltage gain of slightly less than unity. Its analysis generally follows the treatment for the BJT grounded-emitter PA just covered, and although the emitter-follower has considerable power gain, its efficiency is also bounded by 0.25.

When a transformer (here treated as an ideal transformer) is used to couple the load, R_L, to a class A BJT power amplifier, the power conversion efficiency can be made to increase markedly. The reason is that the BJT's collector can be made to see an optimum effective load resistance through correct selection of the transformer turns ratio. Figure 12.10(A) illustrates a single BJT, transformer-coupled, class A power amplifier. The quiescent operating point, Q, for this power BJT is set by selecting I_{BQ} to locate the Q-point in the center of the linear operating region, just below the $P_{D(max)}$ hyperbola. V_{CC} is selected to be less than $V_{sus/2}$. The transformer turns ratio is made so that the BJT's collector "sees" an R'_L value that gives a dynamic load line which intersects the V_{CE} axis at $2V_{CC}$ and the I_C axis at $2I_{CQ}$. Thus $R'_L = 2V_{CC}/2I_{CQ} = V_{CC}/I_{CQ} = R_L(n_1/n_2)^2$ Ω, and $V_{CC}I_{CQ} < P_{D(max)}$.

The efficiency of a transformer-coupled class A power amplifier is double that for a DC class A power amplifier. The average power from the power supply is

$$P_{dc} = V_{CC}I_{CQ} \qquad\qquad 12.17$$

The maximum average ac power delivered to the load, $\bar{P}_{ac}$, is the same maximum power delivered to the ideal transformer's primary. Neglecting saturation and leakage

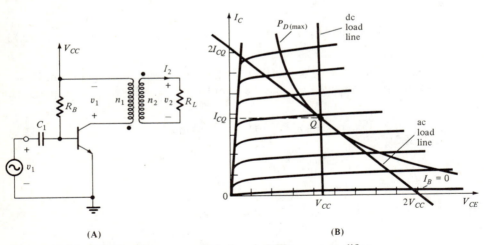

(A)   (B)

Figure 12.10   (A) Transformer-coupled, class A, BJT power amplifier.
(B) Operating characteristics of the transformer-coupled PA. The
Q-point is chosen to lie in the center of the linear region, below the
$P_{D(max)}$ hyperbola. The dc load line is nearly vertical, being determined
by the transformer's primary dc winding resistance.

current, we can write two expressions for $\bar{P}_{ac}$:

$$\bar{P}_{ac} = \frac{1}{2\pi} \int_{-\pi}^{\pi} V_{CC}(1 - \sin \omega t) I_{CQ}(1 + \sin \omega t) \, d\omega t \qquad\qquad 12.18$$

$$\bar{P}_{ac} = \frac{1}{2\pi} V_{CC} I_{CQ} \int_{-\pi}^{\pi} (1 - \sin^2 \omega t) \, d\omega t = \frac{V_{CC} I_{CQ}}{2} \qquad\qquad 12.19$$

Thus the maximum efficiency is

$$\eta = \frac{\bar{P}_{ac}}{\bar{P}_{dc}} = \frac{V_{CC} I_{CQ}/2}{V_{CC} I_{CQ}} = 0.5 \qquad\qquad 12.20$$

This value is double that for a DC, class A power amplifier, seen in Eq. 12.16.

Although class A, transformer-coupled PAs have 50% efficiency and low inherent distortion even without feedback, they are seldom used in new designs.

12.5 Properties of Class AB and Class B Power Amplifiers

Most modern IC power amplifier stages use class AB or B modes of operation for efficiency. They are also generally of DC design; no transformer or capacitor is used to couple the load to the PA's output. Figure 12.11 illustrates several commonly used class AB and class B power amplifier output stage designs. The amount of nonlinearity in class AB and B power amplifier stages is of great interest. Nonlinearity in PA transfer characteristics generates harmonic distortion in the output. Although we have seen that overall negative feedback is effective in reducing total harmonic distortion (THD), good design practice begins by making the PA output stage as linear as possible, before using negative feedback. Figure 12.12 shows the static voltage transfer characteristic of the elementary complementary symmetry amplifier of Fig. 12.11(A). The 1.4 V dead zone is the result of the volt-ampere characteristics of the BJT's base-emitter diodes. It gives rise to crossover distortion in V_o. To eliminate this dead zone, Q_1 and Q_2 must be biased so that they will conduct more than one-half cycle of the sinusoidal input. That is, the PA will be class AB. A common means of obtaining class AB or B biasing is shown in Fig. 12.11(B) through (D). The diodes shown act to offset the base voltage from V_1 by an amount equal to the V_{BE}'s of the BJTs when $I_B > 0$. This effect keeps both upper (npn) and lower (pnp) BJTs slightly conducting when $V_1 = 0$, achieving class AB operation.

The power conversion efficiency of class B BJT power amplifiers is easily calculated. Note that the PA stages in Fig. 12.11 use split supplies. Thus, to calculate P_{dc}, we need to add the average power per cycle contributed by both power supplies. During each sine wave period, each power supply provides current for one half-cycle. The output current is zero over the other half-cycle. The peak value of the power supply load current is

$$I_{DDpk} = \frac{V_{opk}}{R_L} \qquad\qquad 12.21$$

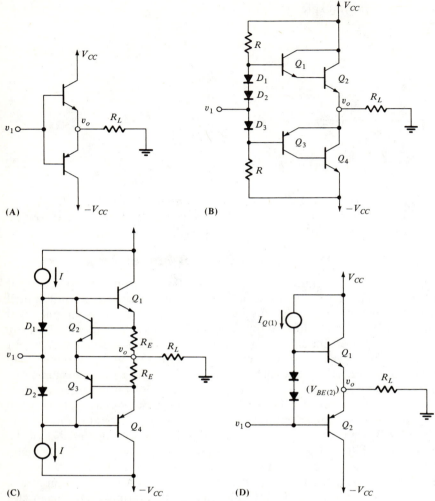

Figure 12.11 **(A)** Simple complementary symmetry, emitter-follower PA output stage, direct-coupled to its load. (See the discussion on its linearities in Sec. 12.5.) **(B)** True class B, quasi-complementary symmetry PA stage. Q_1 and Q_2 comprise an npn Darlington configuration, generally packaged in a common case. Q_3 and Q_4 form a feedback pair that is equivalent to a single power pnp transistor. Q_4 is a power npn transistor, and Q_3 is a low-power pnp device. Diodes D_1 and D_2 bias the Darlington so that it is just conducting when $V_1 = 0$; D_3 acts similarly to bias Q_4 on producing class B or AB operation. **(C)** Complementary power stage with output current limiting by *base current robber* transistors Q_2 and Q_3. **(D)** Offset complementary symmetry PA. Diodes D_1 and D_2 act to eliminate the biasing dead zone seen in the circuit of (A).

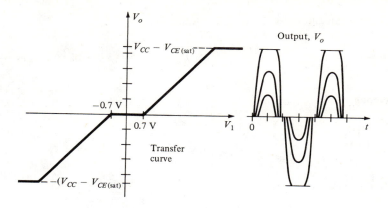

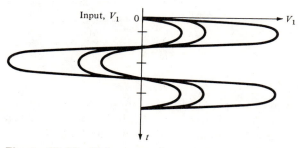

Figure 12.12   Voltage transfer curve for the complementary symmetry BJT amplifier of Fig. 12.11(A). Input and output waveforms are shown. The output is never free from distortion because of the dead zone.

The average power per cycle supplied by the positive V_{CC} supply is thus

$$\bar{P}_{dc+} = \frac{V_{CC}}{2\pi} \int_{-\pi}^{\pi} \frac{V_{opk}}{R_L} (\sin \omega t)\, d\omega t \qquad\qquad 12.22$$

$$\bar{P}_{dc+} = \frac{V_{CC}}{2\pi} \frac{V_{opk}}{R_L} (-\cos \omega t)\Big|_0^{\pi} = \left(\frac{1}{\pi}\right)\left(\frac{V_{CC} V_{opk}}{R_L}\right) \qquad\qquad 12.23$$

and the total power supplied by both supplies is

$$\bar{P}_{dc} = \bar{P}_{dc+} + \bar{P}_{dc-} = \frac{2V_{opk} V_{CC}}{\pi R_L} \qquad\qquad 12.24$$

The average ac power in the load is

$$\bar{P}_{ac} = \frac{V_{opk}^2}{2R_L}$$

12.25

Hence the efficiency is

$$\eta = \frac{\bar{P}_{ac}}{\bar{P}_{dc}} = \frac{\left(\dfrac{V_{opk}^2}{2}\right)\left(\dfrac{1}{R_L}\right)}{\left(\dfrac{2}{\pi}\right)\left(\dfrac{V_{opk}}{R_L}\right)V_{CC}} = \left(\frac{\pi}{4}\right)\left(\frac{V_{opk}}{V_{CC}}\right)$$

12.26

The maximum efficiency occurs when V_{opk} is maximum. V_{opk} cannot exceed $(V_{CC} - V_{CE(sat)})$ for the complementary symmetry BJT power amplifier (see Fig. 12.12). If $V_{CC} \gg V_{CE(sat)}$, which is usually the case, Eq. 12.26 for η reduces to

$$\eta \cong \left(\frac{\pi}{4}\right)\left(\frac{V_{CC}}{V_{CC}}\right) = 0.786$$

12.27

Thus DC class B power amplifier stages are over three times more efficient than DC class A power amplifiers. In addition, they dissipate negligible power when given zero signal input. They are used in practically all new IC power amplifier designs, including power op-amp output stages.

Most audio amplifiers drive loudspeaker loads that have a nominal impedance of 8 Ω; these PAs typically use 40 V supplies ($+V_{CC} = 40$ V, $-V_{CC} = -40$ V). Hence peak load current is 5 A with the 8 Ω load. Of course, the power amplifier BJTs and power supplies must be able to handle this peak current. Suppose a PA designed to provide 5 A peak to an 8 Ω load is connected to a 2 Ω load and then driven to $V_{opk} = V_{CC} = 40$ V. Providing nothing melts, a peak load current of 20 A should flow. If the power supply outputs and the load are not protected by fast-blow fuses, it is probable that one or both of the BJTs will fail, acting as expensive fuses. The PA can be protected from output current overload problems of this sort by using output current–limiting circuitry, such as that illustrated in Fig. 12.11(C). Here external values of R_e are selected so that Q_2 or Q_3 turns on when the emitter (load) current through either Q_1 or Q_4 exceeds a preset limit. Q_2 or Q_3 shunts away the base current drive to the power BJTs, Q_1 or Q_4, limiting the output current through R_L. As an example, if we wish to limit I_L above 7 A, then

$$R_e = \frac{0.65\ \text{V}}{7\ \text{A}} = 0.093\ \Omega$$

12.28

will cause the *base current robbers*, Q_2 and Q_3, to turn on when I_L is greater than 7 A, limiting I_L to slightly above that value.

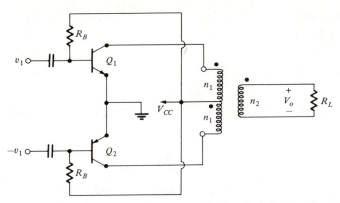

Figure 12.13   npn, transformer-coupled, push-pull PA. The R_b's bias Q_1 and Q_2 bases to 0.65 V for low crossover distortion (class B operation).

When very low impedance loads must be driven, BJTs can be paralleled to handle the increased load currents in the complementary symmetry PA stage shown in Fig. 12.11(C). Or, if DC is not required, and bandwidth requirements can be met, a classical push-pull architecture using a center-tapped power transformer can be used, as illustrated in Fig. 12.13. The push-pull, transformer-coupled PA is usually run class AB to eliminate crossover distortion. As an approximation, we will consider this stage to be running class B to simplify analysis. Each BJT thus conducts over one-half cycle and is cut off over the remainder of the cycle. The collector of each BJT "sees" a load impedance of $R'_L = (n_1/n_2)^2 R_L$ when it conducts. The ac load line for Q_1 is shown in Fig. 12.14. Note that n_1/n_2 is adjusted so that the ac load line lies

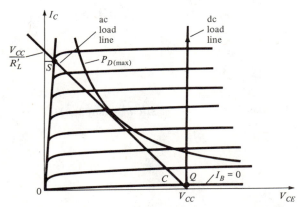

Figure 12.14   Load lines for Q_1 in the transformer-coupled PA of Fig. 12.13.

under the $\bar{P}_{d(max)}$ hyperbola, and the peak collector current (neglecting saturation voltage) is $V_{CC}/R'_L < I_{C(max)}$. The peak current through the load is then

$$I_{Lpk} = \left(\frac{n_1}{n_2}\right)\left(\frac{V_{CC}}{R'_L}\right) = \left(\frac{n_1}{n_2}\right)\left(\frac{V_{CC}}{(n_1/n_2)^2\,R_L}\right) = \sqrt{\frac{V_{CC}I_{C(max)}}{R_L}} \qquad 12.29$$

It is left as an exercise to show that the efficiency of the push-pull, transformer-coupled, class B power amplifier is $\eta = \pi/4$.

12.6 Nonlinearity and Distortion in Power Amplifiers

The static input/output characteristic of a perfectly linear PA may characterized by the equation

$$Y = b + a_1 X \qquad 12.30$$

where

    Y = the output voltage
    X = the input voltage
    a_1 = the voltage gain
    b = a dc offset at the output (assume b can be set to zero).

A perfectly linear amplifier obeys the principle of superposition. That is, if

$$Y_1 = a_1 X_1 \quad \text{and} \quad Y_2 = a_1 X_2 \qquad 12.31$$

then, by superposition,

$$Y = Y_1 + Y_2 = a_1(X_1 + X_2) \qquad 12.32$$

If the I/O characteristic is nonlinear, superposition no longer applies. The I/O characteristic of a nonlinear PA may be described by a power series of the form

$$Y = b + a_1 X + a_2 X^2 + a_3 X^3 + a_4 X^4 + \cdots \qquad 12.33$$

Now, if we let $x(t) = X \sin \omega t$, Eq. 12.33 yields

$$Y = b + a_1 X \sin \omega t + \frac{a_2 X^2}{2}(1 - \cos 2\omega t) + \frac{a_3 X^3}{4}(3 \sin \omega t - \sin 3\omega t) + \cdots$$

$$12.34$$

or

$$y(t) = \left(b + \frac{a_2 X^2}{2}\right) + \left(a_1 X + \frac{3a_3 X^3}{4}\right)\sin \omega t$$
$$- \left(\frac{a_2 X^2}{2}\right)\cos 2\omega t - \left(\frac{a_3 X^3}{4}\right)\sin 3\omega t + \cdots \qquad 12.35$$

Inspection of Eq. 12.35 reveals that harmonics (sinusoidal terms whose frequencies are integral multiples of the fundamental frequency, ω) are generated by the power law I/O nonlinearity. These harmonics are unwanted output components and are the direct result of the distortion from the nonlinearity.

If $x(t)$ has two components,

$$x(t) = X_1 \sin \omega_1 t + X_2 \sin \omega_2 t \qquad 12.36$$

then substitution of Eq. 12.36 into Eq. 12.33, with $a_3, a_4, a_5, \ldots = 0$, yields, after some algebra and the arcane use of trigonometric identities,

$$
\begin{aligned}
y(t) = &\left(b + \frac{a_2 X_1^2}{2} + \frac{a_2 X_2^2}{2} \right) + a_1 X_1 \sin(\omega_1 t) + a_1 X_2 \sin(\omega_2 t) \\
&- \left(\frac{a_2 X_1^2}{2} \right) \cos(2\omega_1 t) - \left(\frac{a_2 X_2^2}{2} \right) \cos(2\omega_2 t) \\
&+ a_2 X_1 X_2 \cos[(\omega_1 - \omega_2)t] - a_2 X_1 X_2 \cos[(\omega_1 + \omega_2)t]
\end{aligned}
\qquad 12.37
$$

We see that the presence of only square-law distortion generates second harmonic terms and sum and difference frequency terms in the output. Clearly, superposition is not working! The sum and difference frequency terms are called intermodulation distortion and are more numerous when the I/O nonlinearity includes cubic and higher-order terms.

Rather than directly measure the $Y = F(X)$ transfer curve for an amplifier and attempt to evaluate a_2, a_3, and so on, it is common practice to test amplifier non-linearity indirectly using a single pure sine wave input and varying X and ω. The harmonics in the output are measured directly using a spectrum analyzer and serve as an indirect measure of amplifier nonlinearity. Modern spectrum analyzers low-pass filter the amplifier's output to prepare it for sampling (digitization). They then digitize a finite number of samples (e.g., 1024) at a known rate (e.g., 10,000 samples/s), and then perform a fast Fourier transform (FFT) operation on the samples. The process is repeated continuously a number of times (e.g., 32), and each time, the FFT obtained is averaged in with those obtained previously. At the end of the averaging process, the magnitude of the averaged FFT is converted to analog form and displayed versus frequency on a CRT, usually using a dB scale (dB of rms voltage). The fundamental and harmonic terms appear as spikes (line spectra) on the CRT. Because the dB scale is logarithmic, direct comparison of the fundamental rms voltage with the harmonics is possible.

The nth harmonic distortion is defined as the ratio of the nth harmonic rms amplitude (A_n) to the rms value of the fundamental component, A_1:

$$D_n = \frac{|A_n|}{|A_1|} \qquad 12.38$$

The total harmonic distortion (THD) is defined as

$$\text{THD} = \sqrt{\sum_{n=2}^{\infty} D_n^2} \qquad\qquad 12.39$$

State-of-the-art DC, class B, high-fidelity audio PAs using large amounts of negative feedback typically have THDs of less than 0.003% at low frequencies and at low output power levels. THD is seen to rise with frequency and output power level, generally to a perceivably unacceptable level. The unacceptable THD level is highly subjective and depends on the listener and what is being listened to; 1% THD is a conservative upper bound in state-of-the-art PAs. Typical plots of THD versus closed-loop voltage gain and THD versus frequency are illustrated in Figs 12.15(A) and (B) for two power op-amps. Note that THD increases with increasing closed-loop voltage gain at constant output power level and frequency. This relationship is to be expected, because, as was seen in Chapter 6, Eq. 6.47, THD is divided by the return difference in a negative feedback amplifier, and so is closed-loop gain. In other words, high loop gain (large βA_v) results in low closed-loop gain and low THD. THD also increases with frequency, largely owing to slew-rate limiting. The rate of change at the amplifier's output is limited to

$$\frac{dV_o}{dt} = \eta \quad \text{V}/\mu\text{s} \qquad\qquad 12.40$$

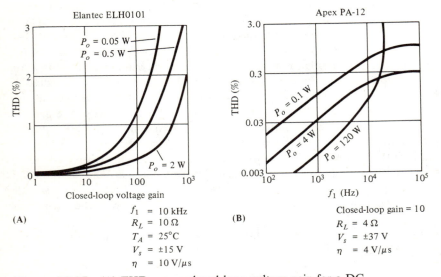

Figure 12.15 (A) THD versus closed-loop voltage gain for a DC power op-amp. (B) THD versus frequency at different output power levels for another DC power op-amp.

At very high frequencies and amplitudes, the output waveform of a slew-rate-limiting amplifier tends to a triangular form. The peak amplitude of this triangular output is no longer proportional to the amplifier's input sine wave, and is also a function of frequency. It can be shown by Fourier series analysis that a pure triangle wave of peak height V_o and radian frequency $\omega = 2\pi/T$ has the harmonics

$$v_o(t) = -V_o \sum_{n=1,3,5\ldots}^{\infty} \left(\frac{2}{\pi n}\right)^2 \cos(n\omega t) \qquad\qquad 12.41$$

Thus we see that the output of a slew-rate-limiting amplifier not only contains severe harmonic distortion, but also has nonlinear rate saturation given by Eq. 12.41.

The plots of THD in Fig. 12.15 appear to offer a contradiction, that is, at a fixed frequency and gain, there is higher THD at lower output power levels. This effect is caused by crossover distortion. Low power levels mean lower output voltages; hence the harmonic distortion terms given by Eq. 12.37 are larger because crossover distortion affects the output signal more at low amplitudes (see Fig. 12.12) than it does at high amplitudes.

The abrupt rise in THD for $P_{\text{out}} = 120$ W near 30 kHz in Fig. 12.15(B) can involve more than slew-rate limiting. It may be due to hard clipping of the output waveform, or failure of the power supplies to maintain their rated output voltages at high output current levels. (Some electronically regulated dc power supplies have sharp output current limiting for their own protection and also can exhibit slew-rate limiting in the regulator loop.)

As a closing observation on distortion in amplifiers, it is easy to see that if the I/O characteristic of a nonlinear PA is perfectly odd in X (i.e., $Y[X] = -Y[-X]$), then all the even-term coefficients in Eq. 12.30 will be zero by definition (i.e., $a_2 = a_4 = a_6 = \cdots = 0$). As a result, it is possible to show that, given a sinusoidal input, the output will contain only odd harmonics (A_1, A_3, A_5, etc.); even harmonics will be zero. The cancellation of even harmonics is another major reason for using push-pull or complementary symmetry PA output stages; they have nearly odd I/O characteristic curves and generally make negligible even harmonics in the output.

12.7 Power Op-Amps

Most op-amp manufacturers now make power op-amps. Power op-amps are generally high-voltage op-amps driving class B, complementary symmetry, BJT, power amplifier stages, all in one case. Op-amp PAs offer all the advantages of conventional op-amps in terms of low sensitivities, stability, ease in shaping a transfer function using active filter design principles, ability to implement nonlinear functions, and so forth. Maximum power op-amp power dissipations range from a few watts to over 125 W.

Applications of power op-amps include compact audio high-fidelity amplifiers, drivers for small to medium dc servomotors, deflection amplifiers for magnetic and electrostatic CRTs, precision temperature controllers, power supplies, and drivers for ultrasound transducers.

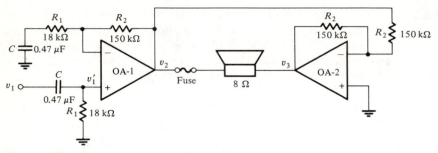

Figure 12.16 Audio PA using two IC power op-amps in a bridge circuit. The load is direct-coupled, although the input signal is reactively coupled to OA-1.

An audio PA using two identical power op-amps is shown in Fig. 12.16. In this circuit, the op-amps should have high output slew rates (over 10 V/μs) to minimize high-frequency, high-amplitude distortion. Also a large amount of feedback should be used; that is, the amplifiers should be run at low closed-loop gain. Assuming the op-amps are ideal, we see that $V_3 = -V_2$. In addition, it is easy to show that

$$V_2 = V_1 \left(\frac{sR_1C[sC(R_1 + R_2) + 1]}{(sR_1C + 1)^2} \right) \qquad 12.42$$

The transfer function of OA-1 is seen to be high-pass with two poles and two zeros, even though the load is direct-coupled. (This transfer function acts to limit low-frequency noise such as turntable rumble at the output.) Note that OA-2 is a slave, driving the right side of the speaker coil to $-V_2$. With a sinusoidal v_2 of peak value V_2, the average power in the load, P_{ac}, will be

$$\overline{P}_{ac} = \frac{\left(\frac{2V_2}{\sqrt{2}} \right)^2}{R_L} = \frac{2V_2^2}{R_L} \qquad 12.43$$

If $+15$ V and -15 V power supplies are used, $V_{2(max)} = 14.5$ V and, by Eq. 12.40, $P_{ac} = 52.56$ W in the 8 Ω load. The peak load current is

$$I_{Lpk} = \frac{14.5 - (-14.5)}{R_L} = 3.63 \text{ A} \qquad 12.44$$

The two 15 V dc power supplies must each be able to supply a peak power of

$$P_{dc(pk)} = (15)(3.63) = 54.38 \text{ W} \qquad 12.45$$

Table 12.1 Characteristics of selected power op-amps

Make/Model	Operating Voltage (V)	Peak Output Current (A)	Slew Rate	Unity GBWP (MHz)	Power Diss. (W)	Power Bandwidth (kHz)	CMRR (dB)
Apex PA84	15–150	0.040	200	75	26	250	130
Apex PA09	10–40	2	400	150	78	2500	100
Apex PA12A	10–50	15	4	4	125	20	100
Apex PA02A	7–19	5	20	4.5	48	350	—
Burr-Brown 3571	15–40	5	3	0.5	33	16	90
Burr-Brown 3573	10–34	5	2.6	1	40	23	110
Burr-Brown OPA501	10–40	10	1.5	1	79	16	110
Elantec ELH0021	15–18	1.2	3	1.2	23	20	90
Elantec ELH0101C	15–22	5	10	5	60	300	100
PMI OP50	15–18	0.095	3	25	—	—	126
Analog Systems MA-329	15–18	3.5	25	450	25	400	90

If a single power op-amp were used, with $+30$ V and -30 V power supplies, the peak power from these supplies would have to be double that drawn from the 15 V supplies (108.8 W). Thus, by using two power op-amps in a bridge circuit, we halve the power requirement for the dc supplies for a given maximum output. Of course, the load is floating with respect to ground, but this is seldom a problem in an audio system.

Table 12.1 summarizes some characteristics of selected state-of-the-art power op-amps. The op-amps vary widely in performance characteristics. Note that a power op-amp used to drive a dc servomotor need not have the heroic (and expensive) characteristics of one used to drive a deflection coil in a CRT terminal. One pays dearly for high slew rate and high power bandwidth; these capabilities are generally not needed in electromechanical servosystem designs.

SUMMARY

The volt-ampere characteristics of power BJTs and MOSFETs have been reviewed, and the linear operating regions of these devices have been described. Problems associated with device power dissipation, heatsink design, and device power derating were covered.

Power amplifier classification and efficiency were treated. Transformers were shown to improve efficiency in reactively coupled amplifiers, but at the cost of limited bandwidth and greater weight. Design of direct-coupled, transformerless PAs was stressed; the complementary symmetry PA stage and its variations were presented.

Sources and measurement of harmonic distortion in PAs were covered; recall that in Chapter 6, negative feedback was shown to reduce total harmonic distortion.

Finally, power op-amps, an IC category that simplifies the design of PA systems, were covered; a table summarizing the characteristics of some available IC power op-amps was given.

PROBLEMS

12.1 An emitter-follower is used as a PA. Let $V_{CE(\text{sat})} = 0$, $R_L = 100 \ \Omega$, $\beta = 19$, $V_{CEQ} = 10$ V, and

$$v_1 = V_1 \sin \omega t$$

(a) Find I_{BQ} required for dc quiescent biasing.
(b) Find the dc power in R_L under quiescent conditions.
(c) Find the maximum average undistorted power in R_L under sinusoidal operating conditions.

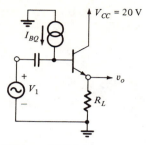

Figure P12.1

12.2 The peak undistorted sinusoidal output voltage of a certain power op-amp is 35 V; the peak undistorted sinusoidal output current is 5 A. The power op-amp is used to drive a 700 Ω load, R_L. Let $v_1 = V_1 \sin \omega t$.

(a) Find the maximum average undistorted ac power in R_L in Fig. P12.2(a).

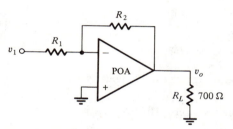

Figure P12.2(a)

(b) In Fig. P12.2(b), the load is now coupled to the power op-amp with an ideal transformer to improve power transfer efficiency. What turns ratio n_1/n_2 is necessary to obtain maximum average undistorted ac power in R_L?

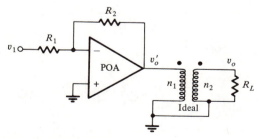

Figure P12.2(b)

12.3 A BJT emitter-follower is coupled to a load with an ideal transformer. The BJT has $\beta_o = 29$, $h_{oe} = h_{re} = 0$, and is biased so that $I_{CQ} = 0.1$ A. Assume $h_{ie} \cong 1/40 I_{BQ}$.

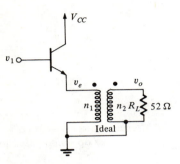

Figure P12.3

(a) Assume small-signal sinusoidal operation. Find the Thevenin source resistance the transformer's primary sees (is driven by) looking up the emitter.
(b) Use the MFSSM to find an expression for v_e/v_1.
(c) Use the MFSSM to find an expression for v_o/v_1.
(d) Find n_1/n_2 for maximum ac power transfer to R_L under small-signal conditions.

12.4 A class B, BJT, complementary symmetry, power output stage drives a 10 Ω load (Fig. P12.4). Assume $V_{CE(\text{sat})}$ for the transistors is zero. The PA stage is to deliver 50 W maximum average power to the 10 Ω load. Let

$$v_1 = V_1 \sin \omega t$$

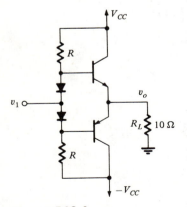

Figure P12.4

(a) Find the smallest V_{CC} that can be used.
(b) Find the peak sinusoidal output voltage for which Q_1 and Q_2 dissipate maximum average collector power, $\bar{P}_c$. $\bar{P}_c = \overline{V_{ce}(t)I_c(t)}$. Find $P_{c(max)}$.
(c) Find the maximum average power that must be supplied by the positive and negative V_{CC} supplies.

12.5 A class B, complementary symmetry, PA stage uses matched transistors with $V_{CE(max)} = 80$ V and $I_{C(max)} = 5$ A. With proper heatsinks, each transistor can dissipate $\bar{P}_{CE} = 50$ W. $V_{CE(sat)} = 0$.

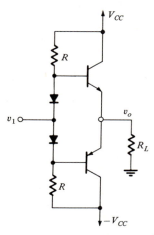

Figure P12.5

(a) What is the maximum average output power that the amplifier can deliver with a sinusoidal signal?
(b) Specify the values of $\pm V_{CC}$ and R_L needed to obtain the output power in part (a). (*Note:* $V_{CE(max)} = 2V_{CC}$ in this circuit.)
(c) Find the maximum $\bar{P}_{CE}$.

12.6 In Fig. P12.6, a BJT, complementary symmetry, class B, power amplifier stage generates 13.10 V_{rms} of total harmonic distortion at its output when driven by a pure sinusoidal input signal so the rms fundamental frequency output voltage is 17.68 V. No feedback is used. This power stage is to be driven by a high-gain, linear differential amplifier; overall negative voltage feedback is to be used.

(a) What loop gain, $-K_v\beta$, is required to reduce the closed-loop system's output THD to 0.03% when the rms fundamental frequency output is again 17.68 V?
(b) Find K_v and β required so that $V_1 = 100$ mV$_{rms}$ produces an average fundamental frequency power of 10 W in R_L.

12.7 The simple PA in Figure P12.7 is made from three identical BJTs. Assume $\beta_o = 50$, $V_{BE(ON)} = 0.7$ V, and $V_{CE(sat)} = 0.3$ V. Sketch and dimension $v_o(t)$ for a sinusoidal input. Assume v_1 is large enough that the output clips show saturation and crossover voltage levels. Also derive an expression for the dead- (crossover-) zone angle.

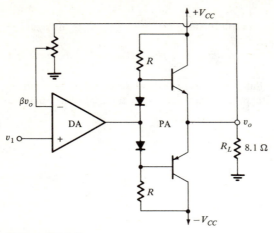

Figure P12.6

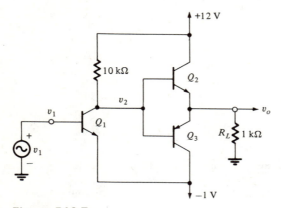

Figure P12.7

12.8 In Fig. P12.8, a class B, complementary symmetry, *bridge* PA is used to drive a 25 Ω resistive load. The transistors Q_1 and Q_3, and Q_2 and Q_4, are driven in phase. Q_2 and Q_4 are driven 180° from Q_1 and Q_3, however.

(a) Assume $V_{CE(sat)} = 0$. Find the peak power dissipated in R_L.

(b) Find the maximum average undistorted ac power in R_L.

(c) Find the maximum average undistorted power dissipated in any one transistor.

(d) What Thevenin source resistance does R_L see when the transistors are in their linear regions? Assume the transistor bases are driven by ideal voltage sources; use MFSSM analysis.

(e) Calculate the amplifier's efficiency, η.

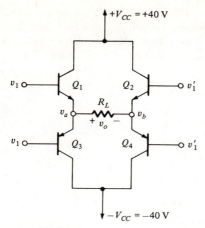

Figure P12.8

12.9 In Fig. P12.9, a class A, push-pull, emitter-follower PA drives a $20\,\Omega$ load. Assume $Q_1 = Q_2$, $V_{CE(sat)} = 1$ V, $h_{fe} \gg 1$, and $I_S = 0.95$ A (dc sinks). For peak positive v_1, Q_1 is just saturated, and Q_2 is cut off.

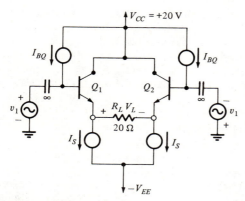

Figure P12.9

(a) Find the maximum peak undistorted power in R_L.

(b) Find the maximum average undistorted ac power in R_L.

(c) Find the maximum average collector power dissipation in Q_1 and Q_2 for the conditions of part (b).

(d) Find the maximum average power the 20 V source must supply for the conditions of part (b). Calculate the amplifier's efficiency.

12.10 The complementary pair of power VMOSFETs in Fig. P12.10 is used to make a class B power amplifier stage, driven by an op-amp. Overall negative feedback is used. Q_1 is an enhancement n-channel device; Q_2 is an enhancement p-channel device. Q_1 and Q_2 are biased to have very small drain currents and equal V_{DS}'s so that $V_L = 0$ under quiescent conditions.

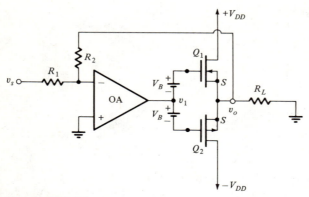

Figure P12.10

(a) Assume that for positive drive ($v_1 > 0$), Q_1 is operating in its linear (saturated drain) region, and Q_2 is cut off. Q_1 is cut off and Q_2 is linear for $v_1 < 0$. Write an expression for the steady-state Thevenin source resistance R_L sees.

(b) Write an expression for the steady-state voltage gain of the PA stage, $K_{pa} = v_o/v_1$.

(c) Assume $f_d = 200\ \Omega$, $g_m = 0.750\ S$, and $R_L = 10\ \Omega$. Also, the op-amp's gain is $v_1/v_i = 10^{-4}$. When $V_o = 25\ V_{rms}$ (at the fundamental frequency), the THD without feedback is 20%. Find the $\beta = R_1/(R_1 + R_2)$ required to reduce the THD to 0.02% when V_o is again 25 V_{rms} at the fundamental frequency.

(d) Find the overall voltage gain of the system, V_o/V_s, for the conditions found in part (c).

12.11 An audio PA is built using two identical power op-amps with the schematic of Fig. 12.16. The op-amps have slew rates of 20 V/μs, open-loop dc gains of 100 dB, a first pole at 45 Hz, a second pole at 1 MHz, a maximum power dissipation of 48 W, and a maximum (saturation) output current of 5 A. Assume a 10 Ω resistive load. Use an ECAP such as Micro-Cap II to describe the amplifier's small-signal frequency response. Plot the frequency response over a range that includes the upper and lower 0 dB points. Use a specific sinusoidal source to observe the frequency-amplitude product(s) that will cause slew-rate limiting and output distortion.

Chapter *13*

Special Circuits and Systems

In this chapter, we will examine the design and application of selected special-purpose analog integrated circuits frequently used in instrumentation, control, and communications systems. They include phase-sensitive rectifiers, true rms converters, and analog multipliers. The analysis and design of an autobalancing impedance-measuring system is presented to illustrate the use of special analog circuits.

13.1 Phase-Sensitive Rectifiers

The phase-sensitive rectifier (PSR), also known as the phase-sensitive detector, synchronous detector, synchronous rectifier, or balanced demodulator, is a versatile subsystem used in many instrumentation, communications, and control systems. For example, the outputs of a number of instrumentation systems, including linear variable differential transformers (LVDTs) and resistive (Wheatstone) bridges given ac excitation, are described as double-sideband/suppressed-carrier (DSBSC) signals. A DSBSC signal, $y(t)$, is formed by multiplying a high-frequency sinusoidal carrier, $x_c(t)$, by a low-frequency modulating signal, $x_m(t)$, as illustrated in

$$y(t) = x_m(t)V_c\sin(\omega_c t) \tag{13.1}$$

If $x_m(t)$ is sinusoidal with frequency $\omega_m < \omega_c$, then we have

$$\begin{aligned} y(t) &= V_m\sin(\omega_m t)V_c\sin(\omega_c t) \\ &= \left(\frac{V_c V_m}{2}\right)\{\cos[(\omega_m + \omega_c)t] - \cos[(\omega_c - \omega_m)t]\} \end{aligned} \tag{13.2}$$

The DSBSC signal, $y(t)$, clearly has two *sideband* components at $(\omega_c + \omega_m)$ and $(\omega_c - \omega_m)$, and no component at the carrier frequency, ω_c, for nonzero ω_m.

One way to recover the modulating signal from $y(t)$ is to use an analog multiplier to form the product of $y(t)$ and the carrier, and then low-pass filter the multiplier's output, $z(t)$. This process is shown in Fig. 13.1. From this figure, we can see that

$$\begin{aligned} z(t) &= \left(\frac{V_c^2}{10}\right)\sin^2(\omega_c t)V_m\sin(\omega_m t) \\ &= \left(\frac{V_c^2 V_m}{20}\right)\sin(\omega_m t) - \left(\frac{V_c^2 V_m}{20}\right)\{\sin[(\omega_m + 2\omega_c)t] - \sin[(2\omega_c - \omega_m)t]\} \end{aligned} \tag{13.3}$$

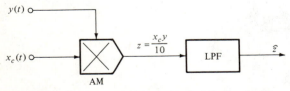

Figure 13.1   Demodulation of DSBSC signal with an analog multiplier and low-pass filter.

The LPF attenuates the high-frequency terms in $z(t)$; its output is thus

$$\hat{z}(t) = \left(\frac{V_c^2 V_m}{20}\right) K_f \sin(\omega_m t) \qquad\qquad 13.4$$

where K_f is the LPF's dc gain. Thus $\hat{z}(t)$ is seen to be proportional to the modulating signal, $V_m\sin(\omega_m t)$.

The same demodulation process can be accomplished using several other PSR circuits. One such demodulator that works well in the audio-frequency range uses a MOS analog switch and two op-amps, as shown in Fig. 13.2(A). The MOS switch

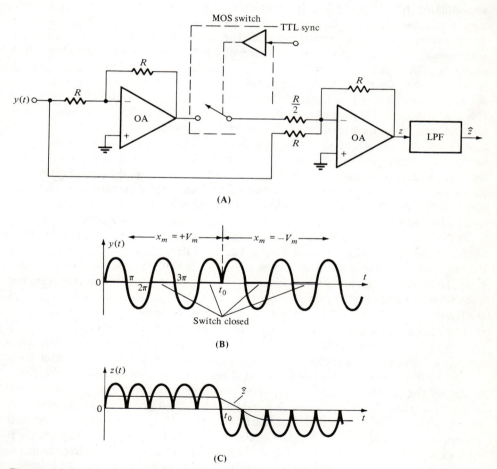

(A)

(B)

(C)

Figure 13.2 (A) Phase-sensitive rectifier using an analog MOS switch (e.g., 4016B) and op-amps. LPF is an active low-pass filter (B) Input waveform showing switch dwell times [controlled by $x_c(t)$]. (C) Output waveforms of the PSR. $\hat{z}(t)$ lags $z(t)$ because of the dynamics of the LPF.

is driven from a TTL square wave in phase with the sinusoidal carrier, $x_c(t)$. The input waveform, $y(t)$, is shown in Fig. 13.2(B). In this case, we have assumed that the modulating signal has a positive dc level, V_m, until $t = t_o$, when it switches to a negative dc level, $-V_m$. The action of the MOS switch causes $z = -y$, $y < 0$, when the switch is open, and $z = -y + 2y = y$, $y > 0$, when the switch is closed, producing the positive full-wave rectified waveform shown on the left side of Fig. 13.2(C). Because the MOS switch action remains phase-locked to $x_c(t)$, the 180° phase shift in $y(t)$ for $t > t_o$ causes the demodulator output, $z(t)$, to become a negative full-wave rectified sine wave. The smooth curve in Fig. 13.2(C) shows the approximate LPF output, $z(t)$, which follows $x_m(t)$ with the time constant of the LPF.

Analog Devices has recently put on the market the AD-630 balanced modulator/demodulator IC. This single-chip electronic system contains a precision voltage comparator, two analog switches (operated by the comparator), two differential input amplifier stages, and an op-amp integrator. The switches connect the output of either input amplifier to the integerator. An external LPF must be used with the AD-630 used as a PSR. Applications of this versatile IC include balanced modulator (to generate a DSBSC signal), balanced demodulator (PSR), precision phase comparator, precision ac rectifier (the ac input is used as its own reference), and lock-in amplifier. The last application allows measurement of a coherent signal buried in noise (both random and coherent). A lock-in amplifier normally provides in-phase and quadrature component outputs of the signal being measured, which in general is phase-locked to the reference signal but has a different phase. The manufacturer illustrates the recovery of an in-phase, 400 Hz signal with a -100 dB rms signal-to-noise ratio with an AD-630 lock-in system. Since commercial lock-in amplifiers generally cost well over \$5000, the AD-630 IC can provide an economical solution to the design of custom dedicated lock-in amplifiers operating in the audio-frequency range.

Phase-sensitive rectification can also be accomplished using electromechanical *chopper* switches, transistor switches, and balanced diode bridges, the latter system finding applications at radio frequencies. Figure 13.3(A) illustrates a bipolar chopper PSR. The electromechanical chopper is usually run at a carrier frequency ranging from 60 Hz to 400 Hz. Modern photoelectric and MOS switch choppers can run in the hundreds of kilohertz. The chopper's waveforms are shown in Fig. 13.3(B). Note that $y(t)$ has a phase shift, ψ, with respect to $x_c(t)$ controlling the switch.

A balanced diode bridge PSR is shown in Fig. 13.3(C). Here the reference voltage, V_r, is in phase with the carrier, $x_c(t)$, and is assumed to be much greater than V_s, the DSBSC signal proportional to $y(t)$. These voltages are coupled to the diode bridge by two center-tapped transformers. The diode bridge PSR works as follows. When V_r is positive, diodes a and b conduct, and c and d are reverse biased. The signal voltage at node B is therefore felt at nodes A and C (along with the forward voltage drops of the diodes). Because of this action, a current, i_r, flows from node B through diodes a and b to nodes A and C and the two halves of the reference transformer secondary to node E, then through the load resistor, R. During the reference voltage negative half-cycle, diodes c and d conduct, and a and b are reverse biased. Now the lower half of the signal transformer secondary delivers signal current to the load, R. The waveform of the voltage across the load resistor thus has the same form

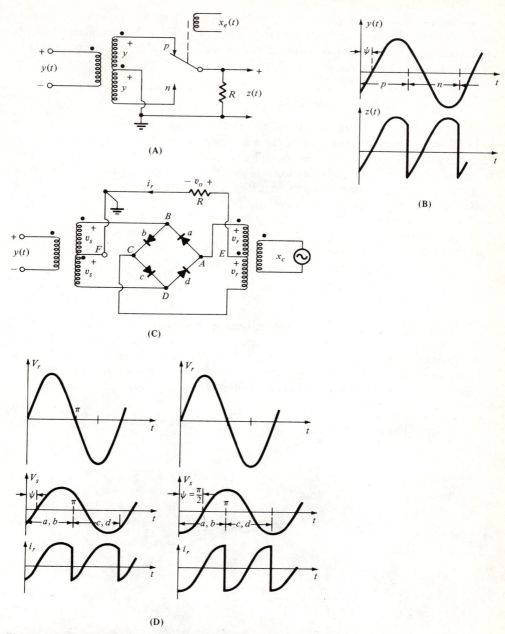

Figure 13.3 **(A)** Chopper PSR. **(B)** Waveforms associated with the chopper PSR. Note that signal $y(t)$ is displaced from the reference voltage, $x_c(t)$, by a phase angle, ψ, reducing the average output, $\hat{z}$, by a factor of $\cos(\psi)$. **(C)** Balanced diode bridge PSR. **(D)** Waveforms seen in the diode PSR. Note that when $\psi = 90°$, $i_r = 0$.

as the output from the chopper PSR described before. The average value of the output waveform in both examples is proportional to $V_s\cos(\psi)$. Proper phasing of the reference signal sets $\cos(\psi) = 1$, and of course from Eq. 13.4, V_s is proportional to $x_m(t)$; hence we have phase-sensitive rectification.

13.2 True rms Conversion

The root mean square (rms) value of a waveform is an important measure of the waveform's power-delivering ability. This is especially true for periodic signals such as sine, square, or triangle waves, as well as for random waveforms. The rms value, determined by the sequential operations of squaring, averaging (usually done by passing the squared waveform through an LPF), and square-rooting, is often called the effective value of a time-varying waveform because the rms waveform value is equal to the value of dc that will deliver the same power to a resistive load.

The following equation illustrates the mathematical steps involved in computing the rms value of a periodic waveform:

$$v_{\text{rms}} = \sqrt{\frac{1}{T} \int_0^T v^2(t)\, dt} \qquad\qquad 13.5$$

Note that the averaging of a periodic waveform need be done over only one period T. If the waveform is random, that is, aperiodic with zero mean, then the mean of its square is approximated by averaging it over a long time or, equivalently, by passing the squared waveform through an LPF with a long time constant. Thus the average of a squared waveform can also be written as

$$\overline{v^2(t)} = \lim_{T \to \infty} \frac{1}{T} \int_0^T v^2(t)\, dt \qquad\qquad 13.6$$

Table 13.1   rms and average values of selected periodic waveforms

Waveform	Average Value of Waveform	rms Value
$v = V_{pk}\sin(\omega t)$	0	$V_{pk}/\sqrt{2}$
$v = V_{pk}\lvert\sin(\omega t)\rvert$	$(2/\pi)V_{pk}$	$V_{pk}/\sqrt{2}$
$v = V_{pk}\sin(\omega t),\ v > 0$		
$\quad v = 0,\ \sin(\cdot) < 0$	$(1/\pi)V_{pk}$	$V_{pk}/2$
Triangle wave, V_{pk},		
$\quad 50\%$ duty cycle	0	$V_{pk}/\sqrt{3}$
Square wave, V_{pk}	0	V_{pk}
$v = V_o + V_{pk}\sin(\omega t)$	V_o	$\sqrt{V_o^2 + V_{pk}^2/2}$

If $v(t) = V_{pk}\sin(\omega t)$, then, by Eq. 13.5, we obtain the well-known result

$$v_{rms} = \sqrt{\frac{1}{T}\int_0^T V_{pk}^2\sin^2(\omega t)\,dt}$$

$$= V_{pk}\sqrt{\frac{1}{T}\int_0^T \frac{1}{2}[1 - \cos(2\omega t)]\,dt}$$

$$= \frac{V_{pk}}{\sqrt{2}} \qquad\qquad 13.7$$

With the basic definition Eq. 13.5, it is easy to calculate the rms value for certain waveforms found in engineering practice. Table 13.1 lists some of these waveforms and their average and rms values. In the case of a random waveform having a Gaussian probability density function with zero mean, the rms value is the waveform's standard deviation, σ.

The earliest and most fundamental means of rms-to-dc conversion was by vacuum thermocouple (Fig. 13.4). An applied voltage causes a current to flow in a small heater resistance element to which is bonded a bimetallic thermocouple junction. An average power, $\overline{P_h}$, is dissipated in the heater due to this current, causing a rise in the temperature of the heater and thermocouple junction above ambient temperature. This process can be expressed mathematically by the equations

$$i_h = \frac{V_1}{R_1 + R_h} \qquad\qquad 13.8$$

$$\overline{P_h} = \overline{i_h^2}R_h \qquad\qquad 13.9$$

$$\Delta T = K_1\overline{P_h} \qquad\qquad 13.10$$

The thermocouple emf is proportional to ΔT:

$$V_o = E_1 - E_2 + E_3 - E_3 = a\,\Delta T + b(\Delta T)^2 + \cdots \qquad\qquad 13.11$$

Hence

$$V_o = \frac{K_1 a R_h \overline{V_1^2}}{(R_1 + R_h)^2} = K_2\overline{V_1^2} \qquad\qquad 13.12$$

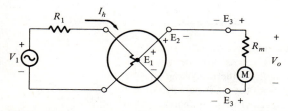

Figure 13.4   Vacuum thermocouple. The dc meter current is proportional to the mean-squared V_1.

Equation 13.12 indicates that the dc thermocouple output voltage is proportional to the mean-squared input voltage. The output voltage, V_o, is dc for a dc V_i input, and also is dc for a sinusoidal V_1 having frequencies ranging from low audio (about 20 Hz) to VHF (about 50 MHz). For V_1 frequencies between dc and 20 Hz, the temperature of the heater and the thermocouple tends to vary with the instantaneous V_1 value, causing an ac component in V_o at twice the input frequency.

To indicate the true rms value, the dc millivoltmeter attached to the thermocouples must have a square-root scale.

Thermocouple voltmeters and ammeters are highly susceptible to destructive overload. That is, it is easy to burn out the heater by exceeding its current rating (the heater becomes an expensive fuse). Some means of protecting the heater from power surges must be employed.

EXAMPLE 13.1

Design of a Feedback, Thermocouple, True rms Voltmeter

The schematic diagram of an electronic, feedback, vacuum thermocouple, true rms voltmeter is shown in Fig. 13.5. In this system, the dc output of a reference thermocouple is subtracted from the dc output of the input thermocouple to form an error signal, V_o, which in turn drives the reference thermocouple. From inspection of Fig.

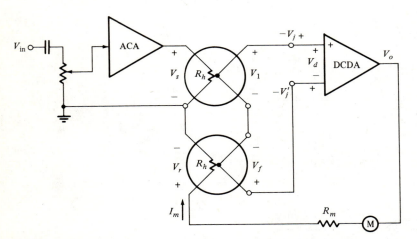

Figure 13.5 Feedback, thermocouple, true rms voltmeter. ACA is a linear, broad-band, ac amplifier. DCDA is a sensitive, chopper-stabilized, DC differential amplifier that sums the thermocouple outputs. M is a dc milliameter calibrated in V_{rms}. Both the ACA and the DCDA have built-in saturation functions to protect the vacuum thermocouple heaters.

13.5, we can write

$$v_s = K_a V_1$$

13.13

The input thermocouple's dc open circuit voltage is proportional to its T, which in turn is proportional to the average power dissipated in R_h. Thus,

$$V_1 = K_t \overline{v_s^2}$$

13.14

and, similarly,

$$V_f = K_t \overline{V_r^2} = K_t R_h^2 \overline{I_m^2}$$

13.15

The output of the DC differential amplifier is simply

$$V_o = K_v V_d = K_v(V_1 - V_f + V_j - V_j)$$

13.16

The meter current is given by Ohm's law:

$$I_m = \frac{K_v(V_1 - V_f)}{R_h + R_m}$$

13.17

Equations 13.14 and 13.15 are substituted into Eq. 13.17, and a quadratic equation in I_m is written:

$$\frac{I_m^2 R_h^2 K_v K_t}{R_h + R_m} + I_m - \frac{K_v K_t \overline{v_s^2}}{R_h + R_m} = 0$$

13.18

Solving Eq. 13.18 for I_m, we find

$$I_m = \left(\frac{1}{2R_h^2}\right)\left[\frac{-(R_h + R_m)}{K_v K_t} + \sqrt{\left(\frac{R_h + R_m}{K_v K_t}\right)^2 + 4R_h^2 \overline{v_s^2}}\right]$$

13.19

In the rms meter design, $K_v K_t$ is made much greater than $(R_h + R_m)$, so the expression for I_m reduces to

$$I_m \cong \left(\frac{1}{2R_h^2}\right)\left(+2R_h\sqrt{\overline{v_s^2}}\right) = \frac{v_{s(rms)}}{R_h}$$

13.20

Equation 13.20 tells us that the dc d'Arsonval milliameter in the circuit of Fig. 13.5 deflects proportionally to the true rms value of the input thermocouple heater voltage, v_s, which of course is proportional to v_{in}. This system does true rms-to-dc conversion with less than 0.1% error and is the basis for a well-known commercial true rms voltmeter.

●

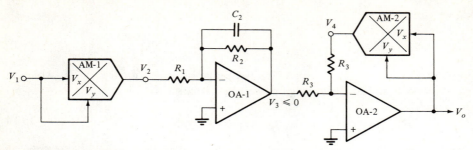

Figure 13.6 True rms converter using two op-amps and two analog multipliers.

Recently, analog signal-processing technology has made possible direct rms-to-dc conversion using IC modules as shown in Fig. 13.6. In this circuit,

$$V_2 = \frac{V_1^2}{10}$$

13.21

and

$$V_3 \cong -\frac{\overline{V_1^2}R_2}{10R_1}$$

13.22

The square-rooting operation follows the equations

$$V_4 = \frac{V_o^2}{10}$$

13.23

$$V_3 + V_4 = 0$$

13.24

Substitution of Eq. 13.23 into Eq. 13.24 yields

$$V_o = \sqrt{-10V_3}$$

13.25

Thus V_3 must be negative for square-rooter stability. This is the case because the LPF (averager) formed by op-amp 1, R_1, R_2, and C_2 inverts the positive signal V_2. The rms converter output can finally be written as

$$V_o = \sqrt{\frac{\overline{V_1^2}R_2}{R_1}}$$

13.26

Analog Devices has recently introduced a true analog rms-to-dc converter in a single, dual in-line package IC. A block diagram of their AD-536/636 IC rms-to-dc converter is shown in Fig. 13.7. The input voltage is first full-wave rectified by an op-amp absolute value circuit (this rectification enables squaring to be done in the first

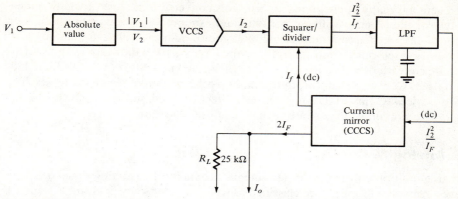

Figure 13.7 Block diagram of the AD-536/636 IC rms-to-dc converter.

quadrant). The absolute value output, V_2, drives an op-amp VCCS. Now a current I_2 proportional to $|V_1|$ drives the squarer/divider, which performs both the squaring and square-rooting functions in one stage by using feedback. The feedback current, I_f, is divided into the squared input current, I_2^2, using log/antilog circuit functions. The output from the divider stage, I_2^2/I_f, is averaged by an RC low-pass filter. This filtered signal, I_2^2/I_F, drives a current mirror, whose outputs are I_F and $I_o = 2I_F$. The I_o of the AD-536A is scaled to be 400 $\mu A/V_{1(rms)}$. If the R_L pin is grounded, $V_o = 1$ V dc/$V_{1(rms)}$ at the "I_o" pin. Operation of this versatile circuit can be described mathematically:

$$I_2 = K_1|V_1|$$ 13.27

After squaring and division, we have

$$\frac{I_2^2}{I_f} = \frac{K_1^2 V_1^2}{I_f}$$ 13.28

The dc current mirror output is

$$I_f = \frac{I_2^2}{I_f} = \frac{\overline{V_1^2} K_1^2}{I_f}$$ 13.29

Hence I_f is proportional to the rms v_1:

$$I_f = K_1 \sqrt{\overline{V_1^2}}$$ 13.30

and

$$V_o = R_L I_f = R_L K_1 \sqrt{\overline{V_1^2}}$$ 13.31

when the R_L pin is grounded.

There are many applications for true rms conversion modules, including measurement of the effective value of periodic and random waveforms, automatic gain control circuits in telecommunications and radio, adaptive filters for audio noise reduction, and self-optimizing control systems: The -3 dB bandwidth for present IC rms-to-dc converters ranges from 1.3 MHz (AD-636J) to 8 MHz (AD-637 J) for full-scale output. Converter bandwidth decreases as the rms input signal decreases. The Analog Devices IC rms-to-dc converters are direct-coupled, so they can be calibrated using a precision dc voltage source, which is a useful feature.

13.3 Analog Multipliers

An analog multiplier forms the instantaneous product of two analog signals, usually voltages. Analog multipliers are classified as one-, two-, or four-quadrant systems. A one-quadrant analog multiplier multiplies two positive signals. A two-quadrant multiplier forms the product of a positive signal and a signal that can have either polarity. Both signals can have either polarity in a four-quadrant multiplier.

The output of an analog multiplier is given in general by

$$V_o = K_m V_x V_y + (a_o + a_x V_x + a_y V_y) + g(V_x, V_y) \qquad\qquad 13.32$$

The $(a_o + a_x V_x + A_y V_y)$ term represents linear offset terms, which generally can be set to zero. The $g(V_x, V_y)$ term represents nonproduct nonlinearity and may be thought of as an infinite series of terms of the form $b_1 V_x V_y + b_2 V_x V_y + b_3 V_x V_y + \cdots$. The output of an ideal analog multiplier is simply

$$V_o = K_m V_x V_y \qquad\qquad 13.33$$

In practice, K_m generally equals 0.1, and V_o, V_x, and V_y range from -10 V to $+10$ V.

Analog multipliers have many applications, as we have seen. These uses include, but certainly are not limited to, phase detection in PLLs, modulation, demodulation, mixing (also known as frequency conversion or heterodyning), variable gain elements for use in tracking and adaptive active filters, voltage-controlled attenuators, analog simulation of nonlinear systems dynamics, automatic gain control circuits, and audio noise reduction systems. Used along with op-amps, analog multipliers can form analog division circuits and square-rooters. Their use in true rms-to-dc conversion circuits was covered in the last section. Figure 13.8 summarizes some uses of analog multipliers.

Analog multipliers used in early analog computers were generally servomultipliers, time division multipliers, or quarter-square multipliers. Servomultipliers used electromechanical components (motors, gears, synchros, etc.) in a feedback system. Although precise, they had low bandwidth because of mechanical component inertia; they also were expensive.

Time division multipliers were all-electronic, and were generally two-quadrant systems in which a positive V_y altered the duty cycle (on-time) of a chopper switch operating on the bipolar signal, -10 V $\le V_x \le +10$ V. The variable duty cycle,

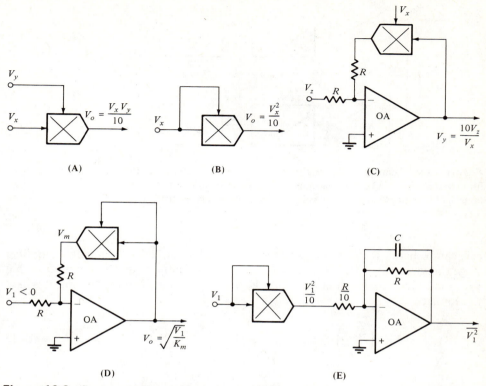

Figure 13.8   Some operations with analog multipliers. **(A)** Multiplication of two signals. **(B)** Squaring a signal. **(C)** Division of V_z by V_x. Note that as $V_x \to 0$, the system saturates. **(D)** Square-rooting. Note that $V_1 < 0$ for stability. **(E)** Mean-squared output V_1. $\overline{V_1^2}$ can be square-rooted by the circuit of (D) to give $V_{1(\text{rms})}$.

chopped waveform was then demodulated by low-pass filtering, giving a V_o that follows Eq. 13.33.

Quarter-square analog multipliers are best described by the block diagram in Fig. 13.9. It is easy to show that the output of the quarter-square multiplier is given by

$$V_o = K_1(4V_x V_y) \qquad\qquad 13.34$$

Quarter-square multiplier bandwidth ranged typically from dc to high audio frequencies.

Modern four-quadrant, IC, analog multiplier designs are based on a variable-transconductance, BJT, differential amplifier circuit known as the Gilbert multiplier "cell." The basic operation of a modern analog multiplier circuit, and the Gilbert

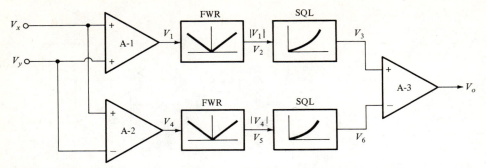

Figure 13.9   Quarter-square multiplier. FWR is a precision, op-amp, full-wave rectifier. SQL is a one-quadrant, precision, square-law non-linearity. The amplifiers sum or subtract their input signals as shown.

cell, can be better understood by first examining the simple BJT differential amplifier stage shown in Fig. 13.10. We assume that this differential amplifier is biased so that $V_{BE(1)}$ and $V_{BE(2)}$ are much greater than V_T. From the Ebers-Moll equations, it is easy to see that

$$V_{BE(1)} = V_T\ln\left(\frac{I_{C(1)}}{I_{s(1)}}\right) \quad \text{and} \quad V_{BE(2)} = V_T\ln\left(\frac{I_{C(2)}}{I_{s(2)}}\right) \qquad 13.35$$

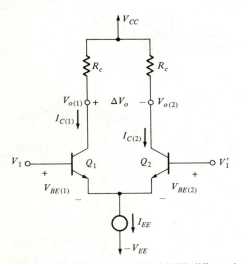

Figure 13.10   Symmetrical BJT differential amplifier with application in analog multiplication.

The differential amplifier is designed to be symmetrical, so $Q_1 = Q_2$, and $I_{s(1)} = I_{s(2)} = I_s$. By Kirchhoff's voltage law,

$$V_1 - V_{BE(1)} + V_{BE(2)} - V'_1 = 0 \qquad\qquad 13.36$$

Substituting the relations for $V_{BE(1)}$ and $V_{BE(2)}$ into Eq. 13.36 and defining the differential input voltage as

$$V_{1d} = \frac{V_1 - V'_1}{2} \qquad\qquad 13.37$$

we find that

$$\frac{I_{C(1)}}{I_{C(2)}} = e^{2V_{1d}/V_T} \qquad\qquad 13.38$$

Now, by Kirchhoff's current law and the definition of BJT alpha, we observe that

$$-(I_{E(1)} + I_{E(2)}) = I_{EE} = \frac{1}{\alpha_F}(I_{C(1)} + I_{C(2)}) \qquad\qquad 13.39$$

Equation 13.39 can be used with Eq. 13.38 to obtain the relations

$$I_{C(1)} = \frac{\alpha_F I_{EE}}{1 + e^{-2V_{1d}/V_T}} \quad \text{and} \quad I_{C(2)} = \frac{\alpha_F I_{EE}}{1 + e^{2V_{1d}/V_T}} \qquad\qquad 13.40$$

The differential output voltage of the amplifier is seen to be

$$\begin{aligned} \Delta V_o &= V_{o(1)} - V_{o(2)} = (V_{CC} - I_{C(1)}R_c) - (V_{CC} - I_{C(2)}R_c) \\ &= (I_{C(2)} - I_{C(1)})R_c \end{aligned} \qquad\qquad 13.41$$

Substituting relations 13.40 into Eq. 13.41 and using some algebra yields

$$\Delta V_o = -\alpha_F R_c I_{EE}\left[\frac{1}{\dfrac{1}{\sinh(2V_{1d}/V_T)} + \dfrac{1}{\tanh(2V_{1d}/V_T)}}\right] \qquad\qquad 13.42$$

When $|2V_{1d}/V_T| \gg 1$, Eq. 13.42 reduces to the well-known relation

$$\Delta V_o \cong -(\alpha_F R_c I_{EE})\tanh\left(\frac{2V_{1d}}{V_T}\right) \qquad\qquad 13.43$$

If $|2V_{1d}/V_T| \ll 1$, then Eq. 13.42 becomes

$$\Delta V_o \cong -\left(\frac{\alpha_F R_c I_{EE}}{V_T}\right) V_{1d} \qquad\qquad 13.44$$

Now, if the current source I_{EE} is made variable around some average value $\overline{I_{EE}}$, described by

$$\overline{I_{EE}} = G_m V_2 + \overline{I_{EE}} \qquad\qquad 13.45$$

then Eq. 13.44 can be rewritten in the form of Eq. 13.46:

$$\Delta V_o = -\left[\left(\frac{\alpha_F R_c G_m}{V_T}\right) V_{1d} V_2 + \left(\frac{\alpha_F R_c \overline{I_{EE}}}{V_T}\right) V_{1d}\right] \qquad\qquad 13.46$$

The first product term on the right side of Eq. 13.46 is the desired output; however, it is based on the operating conditions that $V_{BE(1)}$ and $V_{BE(2)}$ are much greater than V_T and $|2V_{1d}/V_T| \ll 1$. Also, there is an unwanted first-order term in the differential output voltage of the differential amplifier.

Modern IC analog multipliers have been designed to remove the restriction on small V_{1d} so that V_{1d} can be ≤ 10 V, and effectively to cancel the first-order terms. These designs are based on the Gilbert multiplier cell, shown in Fig. 13.11. The basic

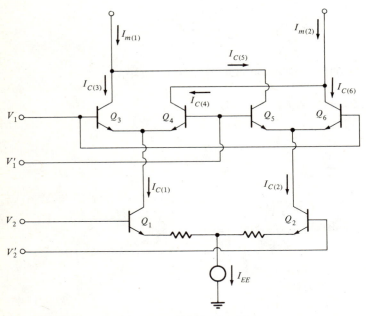

Figure 13.11   Basic Gilbert multiplier "cell."

Gilbert cell also suffers from the fact that $\Delta I = (I_{m(1)} - I_{m(2)})$ is proportional to the products of nonlinear (exponential) terms.

Using the approach we applied to the analysis of the simple DA multiplier of Fig. 13.10, and assuming that $|2V_{1d}/V_T|$ and $|2V_{2d}/V_T|$ are much less than one, we can show, with some algebra, that the current difference for the basic Gilbert multiplier cell is given by

$$\Delta I = I_{m(1)} - I_{m(2)} \cong \left(\frac{\alpha_F^2 \overline{I_{EE}}}{V_T^2}\right) V_{1d} V_{2d} \qquad 13.47$$

Note that there are no first-order terms in this relation for ΔI; these terms have canceled out, and the output is proportional to the true product of V_{1d} and V_{2d}.

One strategy for expanding the effective range of the input signals to the Gilbert cell is to use a pair of matched transistors connected as diodes, as shown in Fig. 13.12. The matched transistors in this figure are Q_7 and Q_8. Transistors Q_9 and Q_{10}

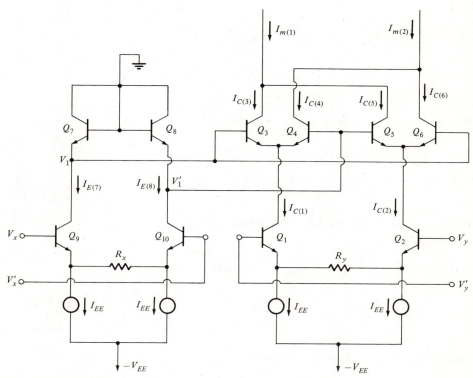

Figure 13.12 Basic IC analog multiplier schematic. The design utilizes the Gilbert cell (Q_3, Q_4, Q_5, and Q_6). Q_7 and Q_8 act as diodes to linearize and extend the V_x input voltage range. (Q_1, Q_2) and (Q_9, Q_{10}) serve as differential input voltage transconductors. The circuit is a simplified version of the Motorola MC-1595 analog multiplier.

serve as differential voltage-to-current converters. To examine how Q_7 and Q_8 expand the input voltage range of the multiplier, we assume that $I_{E(7)}$ and $I_{E(8)}$ are much larger than the base currents into the Gilbert cell; that is, $I_{E(7)} = I_{C(9)}$ and $I_{E(8)} = I_{C(10)}$. Now, from the basic diode equations,

$$V_1 = -V_T \ln\left(\frac{I_{E(7)}}{I_s}\right) \quad \text{and} \quad V_1' = V_T \ln\left(\frac{I_{E(8)}}{I_s}\right) \qquad 13.48$$

V_{1d} is easily seen to be

$$V_{1d} \sim \left(\frac{V_T}{2}\right) \ln\left(\frac{I_{E(8)}}{I_{E(7)}}\right) \qquad 13.49$$

Equation 13.49 for $V_{1(d)}$ is substituted into the following equations for the Gilbert cell:

$$I_{C(3)} = \frac{\alpha_F I_{C(1)}}{1 + e^{-2V_{1d}/V_T}} \qquad I_{C(4)} = \frac{\alpha_F I_{C(1)}}{1 + e^{2V_{1d}/V_T}}$$
$$\hspace{8cm} 13.50$$
$$I_{C(5)} = \frac{\alpha_F I_{C(2)}}{1 + e^{2V_{1d}/V_T}} \qquad I_{C(6)} = \frac{\alpha_F I_{C(2)}}{1 + e^{-2V_{1d}/V_T}}$$

The Gilbert cell difference current, ΔI_m, is found to be

$$\Delta I_m = I_{m(1)} - I_{m(2)} = (I_{C(3)} + I_{C(5)}) - (I_{C(4)} + I_{C(6)})$$
$$\cong \frac{(I_{C(1)} - I_{C(2)})(I_{E(8)} - I_{E(7)})}{I_{E(8)} + I_{E(7)}} \qquad 13.51$$

which is valid for large-signal current differences so long as the BJT V_{BE}'s are much greater than V_T.

Now, if (Q_9, Q_{10}) and (Q_1, Q_2) act like differential current sources, it can be shown (Grebene, 1984) that the differential Gilbert cell current becomes

$$\Delta I_m \cong \left(\frac{2}{\sqrt{I_{EE}R_xR_y}}\right)(V_x - V_x')(V_y - V_y') \qquad 13.52$$

If load resistors are used in the Gilbert cell $I_{m(1)}$ and $I_{m(2)}$ lines, we have

$$\Delta V_o \cong -\left(\frac{2R_c}{I_{EE}R_xR_y}\right)(V_x - V_x')(V_y - V_y') \qquad 13.53$$

in which the gain constant, $2R_c/I_{EE}R_xR_y$, is made to be 0.1.

Modern IC analog multipliers based on the Gilbert cell are offered by several manufacturers. Most of these ICs are laser-trimmed to achieve low four-quadrant error ($\pm 0.25\%$ for the AD-534). There are many sources of errors in the operation of IC analog multipliers; at high frequencies, relative phase differences between the X- and Y-channels will cause errors. Application notes for the Motorola MC-1595/1495 multiplier state that a $3°$ phase shift between V_x and V_y will result in a 5% vector error in the output. Other sources of error in multipliers come from offset voltages, finite source impedances for V_x and V_y, and large-signal nonlinearities. Because of phase shift (vector) errors, analog multiplier frequency response can be specified for different error conditions; the MC-1595/1495 multipliers are specified for a -3 dB bandwidth at 3 MHz, a $3°$ relative phase shift between V_x and V_y at 750 kHz, and 1% absolute error due to phase shift at 30 kHz. A more modern multiplier design, such as the Analog Devices AD-534, has complex specifications for error versus frequency, but is down -3 dB at 1 MHz. The Burr-Brown MPY634 multiplier is down -3 dB at 10 MHz and has a $\pm 0.5\%$ maximum four-quadrant error. The AD-539 is a two-channel, two-quadrant multiplier. When used with wide bandwidth op-amps, its -3 dB frequency is typically about 25 MHz, although it can be used up to 60 MHz.

In closing, we point out that modern IC analog multipliers work well from dc to video frequencies and are used as building blocks for hybrid and IC complex circuits where they serve as balanced modulators (generating DSBSC signals) or as phase detectors. IC analog multipliers are versatile nonlinear analog building blocks for a variety of applications, and their use should be part of every analog circuit designer's repertoire.

13.4 Autobalancing Impedance Measurement

In this section, we describe the design of an electronic system that senses relatively fast changes in the conductance of a medium or object, while maintaining zero output (null) for slow changes in G through a self-balancing mechanism. This autobalance conductance measurement (ACM) system has been used successfully to count fish passing through a bypass pipe, and to measure human heart and respiratory function (as an autobalancing plethysmograph bridge).

The organization of the ACM system is illustrated in Fig. 13.13. A sine wave oscillator generates the measurement signal, V_1. This signal was 180 Hz in the fish counter version of the ACM circuit just mentioned, and 75 kHz in the autobalancing plethysmograph. A small ac current flows through the test medium, which has an admittance $Y_x = \bar{Y}_x + \Delta Y(t)$. Op-amp 1 converts this current to a voltage, V_x, which is one input to a differential amplifier. V_x is given by the well-known relation

$$V_x = -\frac{Y_r}{Y_x} V_1 \qquad\qquad 13.54$$

At the same time, an ac reference signal, V_r, with approximately zero phase difference with respect to V_x, is generated and subtracted from V_x by the differential amplifier.

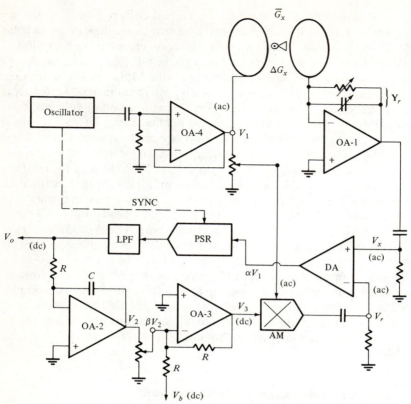

Figure 13.13 Schematic diagram of an autobalancing conductance measurement system.

The amplitude of V_r is set by a potentiometer and an analog multiplier with dc control input V_3. Thus,

$$V_r = V_1 \alpha \left(\frac{V_3}{10} \right)$$

$\hspace{12cm}$ 13.55

The self-balancing behavior of this circuit is determined from the ac error signal, V_e, at the DA output. V_e is converted to dc by a PSR, low-pass filter subsystem. The dc output of the LPF, V_o, can have either polarity, depending on the relative amplitudes of V_r and V_x. In the steady state, when the ACM system is balanced, $V_o = 0$.

V_o is integrated by the circuit of op-amp 2, whose output is attenuated and added to a dc offset voltage by op-amp 3 to generate the dc control voltage, V_3. The integrator is required in the feedback loop to force the system to come to static (dc) balance.

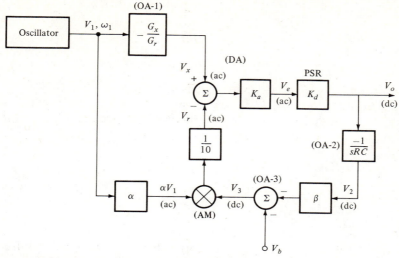

Figure 13.14   Systems block diagram for the ACM system.

Figure 13.14 shows an equivalent systems block diagram of the ACM system, derived from Fig. 13.13. From the block diagram, we can see that the output voltage can be written

$$V_o = K_a K_d (V_x - V_r)$$
$$= K_a K_d \left(-V_1 \frac{G_x}{G_r} - \frac{\alpha V_1 V_3}{10} \right) \tag{13.56}$$

(Here we assume for simplicity that Y_x and Y_r are pure conductances, and that the phase-sensitive rectifier's low-pass filter dynamics are negligible.)

The dc control voltage, V_3, can be written

$$V_3 = -\left(V_b - \frac{\beta V_o}{sRC} \right) \tag{13.57}$$

When Eq. 13.57 is substituted into Eq. 13.56, we find, after some algebra,

$$V_o = \frac{sK_a K_d V_1 \left(\dfrac{\alpha V_b}{10} - \dfrac{\overline{G_x} + \Delta G_x}{G_r} \right)}{s + \dfrac{K_a K_d V_1 \alpha \beta}{10RC}} \tag{13.58}$$

In this transfer function, both V_b and $\overline{G_x}$ are constants. Application of the Laplace final-value theorem shows that the steady-state value of V_o is zero.

If V_b is set such that

$$V_b = \frac{10G_x}{\alpha G_r} \qquad\qquad 13.59$$

then the integrator output, V_2, has a symmetrical swing around 0 V for $G_x = \overline{G_x} \pm \Delta G_x$. Substituting Eq. 13.59 into Eq. 13.58, and noting that the parametric conductance change, $\Delta G_x(t)$, is the system input, we finally arrive at the transfer function for the ACM system:

$$\frac{V_o}{\Delta G_x}(s) = -\frac{\dfrac{sK_aK_dV_1}{G_r}}{s + \dfrac{K_aK_dV_1\alpha\beta}{10RC}} \qquad\qquad 13.60$$

Note that this transfer function is of high-pass form; that is, above the break frequency given by

$$\omega_o = \frac{K_aK_dV_1\alpha\beta}{10RC} \qquad\qquad 13.61$$

it has a flat frequency response with a gain of $20\log(V_1K_aK_d/G_r)$ dB. Below ω_o, the frequency response, $V_o/G_x(j\omega)$, rises at $+6$ dB/octave.

A step increase in G_x, ΔG_x, at $t = 0$ produces a negative exponential decay voltage pulse, given by

$$V_o(t) = -\Delta G_x\left(\frac{V_1K_aK_d}{G_r}\right)e^{-(V_1K_aK_d\alpha\beta/10RC)t} \qquad\qquad 13.62$$

The ACM system is very sensitive; inspection of Eq. 13.62 shows that if $G_r = 10^{-4}$, $K_a = 10^2$, $K_d = 10$, and $V_1 = 3$ V, then a conductance step of $G_x = 3.33 \times 10^{-8}$ S will produce a 1 V peak $V_o(t)$.

SUMMARY

In this chapter, we have illustrated the designs and applications of some special analog circuits and systems. Several designs for the phase-sensitive rectifier were described, and applications of it to communications, control, and instrumentation were discussed.

Systems for true rms-to-dc conversion were introduced, including vacuum thermocouples, op-amps and analog multipliers, and dedicated rms-to-dc conversion ICs. Applications for rms-to-dc converters cited included noise measurements and control of adaptive active filters.

The evolution of the design of modern analog multipliers was presented. The Gilbert cell was shown to be the core of contemporary IC analog multipliers. Means of extending the dynamic range of IC multiplier inputs and canceling error terms in the outputs were described. We reviewed applications of analog multipliers in non-linear signal-processing systems and in adaptive active filters.

Finally, we presented the design of a self-balancing conductance-measuring system. This system shows how various analog circuit building blocks can be integrated to make a novel special-purpose system. Use is made of several op-amp functions including low-pass filtering, buffering, and integration. Other special-purpose analog circuits such as a differential instrumentation amplifier, a phase-sensitive rectifier, an analog multiplier, and an IC sinusoidal oscillator were also used. The self-balancing conductance-measuring system has been successfully used in such varied applications as counting fish and measuring human respiration and heart rate.

Note: No problem section appears with this chapter because of the descriptive nature of the chapter contents.

Chapter 14

Digital Interfaces

Microcomputers are now well established as systems used for controlling analog data acquisition, for processing acquired data, and as controllers in closed-loop feedback systems. Accordingly, interfaces exist that allow digital-to-analog conversion, and analog-to-digital conversion. In discussing conversion interfaces, it is expedient first to consider digital-to-analog converters (DACs), because these systems are used in several designs for analog-to-digital converters (ADCs). Data conversion from digital to analog form, or from analog to digital form, is generally done periodically, which has great significance in terms of the signal-processing dynamics, as will be shown in the next section. A noise-free analog signal sample theoretically has infinite resolution. Once the analog signal sample has been converted to digital form, it is represented in the digital system by a finite number of binary bits (e.g., 10), which places a limit on the resolution of the sample (one part in 1024 for 10 bits). This rounding-off of the converted signal is called quantization, and the resulting errors, quantization noise. Quantization noise was considered in Chapter 8, Sec. 8.11.

Integrated circuit ADCs and DACs exist in a profusion of specifications. They are available in quantization levels ranging from 6 to 20 bits, and conversion speeds suitable for slow dc signal acquisition to video-frequency conversion rates. Most DACs and ADCs accept or produce parallel digital data for speed and efficiency, although a few are designed to accept or produce serial digital data to minimize wiring. Serial converters are not common, however, and will not be considered in this chapter.

14.1 The Sampling Theorem

Because data conversion is generally a periodic process, we will first examine what happens in the frequency domain when an analog signal, $x(t)$, is periodically (and ideally) sampled. An ideal sampling process acquires a data sequence from $x(t)$ only and exactly at the sampling instants when $t = nT_s$. n is an integer ranging from $-\infty$ to $+\infty$, and T_s is the sampling period. An ideal sampling process is mathematically equivalent to impulse modulation; it is illustrated in Fig. 14.1. Periodic conversion of an analog signal to infinite-resolution digital form can be represented by a multiplication of the continuous analog signal, $x(t)$, by a periodic train of ideal, unit impulses. This multiplication produces a periodic number sequence, $x^*(t)$, at the sampler output. In the frequency domain, $\mathbf{X}^*(j\omega)$ is given by the convolution of $\mathbf{X}(j\omega)$ and $\mathbf{P}_T(j\omega)$. An easy way to characterize $\mathbf{X}^*(j\omega)$ is to represent the periodic function $\mathbf{P}_T(t)$ by its complex Fourier series form in

$$P_T(t) = \sum_{n=-\infty}^{\infty} \delta(t - nT_s) = \sum_{n=-\infty}^{\infty} C_n e^{-jn\omega_s t} \qquad 14.1$$

where

$$\omega_s = \frac{2\pi}{T_s} \qquad 14.2$$

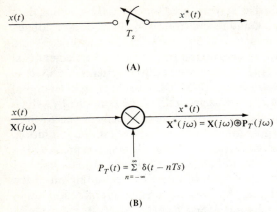

(A)

(B)

Figure 14.1   **(A)** Symbol for an ideal sampling process (analog-to-digital conversion). T_s = the sampling period. **(B)** Impulse modulation equivalent to ideal sampling. $X^*(j\omega)$ = the Fourier transform of the sampled signal. $X^*(j\omega)$ can be expressed as the complex convolution of $X(j\omega)$ and $P_T(j\omega)$.

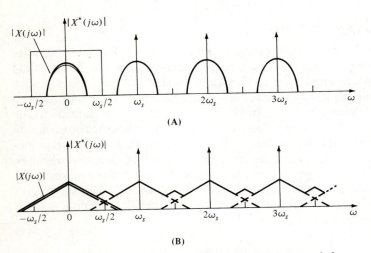

(A)

(B)

Figure 14.2   **(A)** Fourier transform magnitude of a sampled, bandwidth-limited analog signal. Note that the baseband spectrum, $X(j\omega)$, is repeated at integer multiples of the sampling frequency as a result of sampling. The original $x(t)$ can be recovered from $x^*(t)$ by passing $x^*(t)$ through an ideal LPF (rectangle). **(B)** Fourier transform magnitude of a sampled broadband signal. Note that $x(t)$ cannot be recovered from $x^*(t)$ even with an ideal LPF. $X^*(j\omega)$ is said to be aliased in this case.

and the complex-form Fourier series coefficients are given by

$$C_n \triangleq \frac{1}{T_s} \int_{-T_s/2}^{T_s/2} P_T(t)e^{jn\omega_s t} \, dt = \frac{1}{T_s} \qquad \text{14.3}$$

Thus the Fourier series for the pulse train is

$$P_T(t) = \frac{1}{T_s} \sum_{n=-\infty}^{\infty} e^{-jn\omega_s t} \qquad \text{14.4}$$

The sampler output is the product of Eq. 14.4 and $x(t)$:

$$x^*(t) = x(t)\left(\frac{1}{T_s} \sum_{n=-\infty}^{\infty} e^{-jn\omega_s t}\right)$$

$$= \frac{1}{T_s} \sum_{n=-\infty}^{\infty} x(t)e^{-jn\omega_s t} \qquad \text{14.5}$$

The Fourier theorem for complex exponentiation is

$$F\{y(t)e^{-jat}\} \triangleq Y(j\omega - ja) \qquad \text{14.6}$$

Using this theorem, we can Fourier transform Eq. 14.5 for $x^*(t)$:

$$X^*(j\omega) = \frac{1}{T_s} \sum_{n=-\infty}^{\infty} X(j\omega - jn\omega_s) \qquad \text{14.7}$$

Equation 14.7 for $X^*(j\omega)$ is called the Poisson sum form. It is highly significant because it allows us to visualize the effects of ideal sampling in the frequency domain. Figure 14.2 illustrates two cases of sampling in the frequency domain. In Fig. 14.2(A), the baseband spectrum magnitude $|X(j\omega)|$, goes to zero sharply before one-half the sampling frequency. In Fig. 14.2(B), $|X(j\omega)|$ has spectral components beyond $\omega_s/2$. Notice how the original signal, $X(j\omega)$ can be reconstructed from the sampled spectrum in Fig. 14.2(A) by use of an ideal LPF. Perfect reconstruction is not possible in the case illustrated in Fig. 14.2(B), where $X(j\omega)$ has spectral components exceeding $\omega_s/2$. The overlap and summation of the sampled spectra in this figure is called *aliasing*. Aliasing is a significant source of error in data conversion systems.

All analog-to-digital systems must operate on input signals that obey the *Nyquist criterion*; that is, $|X(j\omega)|$ must have no significant spectral power above one-half the sampling frequency. An anti-aliasing, sharp cutoff LPF is generally used to condition $x(t)$ before it is sampled and digitized to ensure that no aliasing takes place. All modern digital signal-processing equipment, such as FFT spectrum analyzers and digital oscilloscopes, use analog anti-aliasing filters to condition their inputs before sampling and data conversion.

Anti-aliasing filters are generally high-order, linear phase LPFs that attenuate the input signal at least 40 dB at the Nyquist frequency of the digital system. Many designs are possible for high-order anti-aliasing filters. For example, Chebychev filters maximize the attenuation cutoff rate at the cost of some passband ripple. Chebychev filters can achieve a given attenuation cutoff slope with a lower order (fewer poles) than other filter designs. It should be noted that in the limit as passband ripple approaches zero, the Chebychev design approaches the Butterworth form of the same order, which has no ripple in the passband. Chebychev filters are designed in terms of their order, n, their cutoff frequency, and the maximum allowable peak-to-peak ripple (in decibels) in their passband.

If ripples in the frequency response stopband of an anti-aliasing filter are permissible, then elliptic or Cauer filter designs may be considered. With stopband ripple allowed, even sharper attenuation in the transition band than obtainable with Chebychev filters of a given order can be obtained. Elliptic LPFs are specified in terms of their order, their cutoff frequency, the maximum peak-to-peak passband ripple, and their minimum stopband attenuation.

Bessel or Thomson filters are designed to have linear phase in the passband. They generally have a "cleaner" transient response, that is, less ringing and overshoot at their outputs, given transient inputs.

As an example of anti-aliasing filter design, Franco (1988) shows a sixth-order Chebychev anti-aliasing filter made from three Sallen and Key (quadratic) low-pass modules. This filter was designed to have an attenuation of 40 dB at the system's 20 kHz Nyquist frequency, a corner frequency of 12.8 kHz, a -3 dB frequency of 13.2 kHz, and ± 1 dB passband ripple.

Following analog-to-digital conversion and digital processing, an output reconstruction filter (also an LPF is used to convert the periodic number sequence output from the digital processor to a continuous analog signal. The most basic form of reconstruction filter, implemented with a latched-input DAC, is the zero-order hold (ZOH). Frequency domain effects of the ZOH and other output reconstruction filters are discussed in Sec. 14.3.

14.2 Digital-to-Analog Converters

The DAC is an electronic system that converts a digitally coded signal to an analog voltage according to a conversion law. For simplicity, we will assume here that all DAC inputs have pure base-2 binary coding.

Several types of DAC circuits are used. However, we will consider only one in detail, the R-$2R$, current-scaling DAC. This circuit has low switching transients and is relatively fast. A simplified schematic of a voltage-driven, N-bit, R-$2R$, current-scaling DAC is shown in Fig. 14.3. V_R is the DAC's reference voltage, generally $+5$ V or $+10$ V, although in some DACs, V_R can be a time-varying bipolar signal, allowing two-quadrant multiplication of V_R times the digital input. The DAC is called a multiplying DAC (MDAC) when operated in this mode.

Note that in the DAC architecture of Fig. 14.3, the currents $I_{o(1)}$ and $I_{o(2)}$ are allowed to flow to ground at all times; hence there is constant current through the $2R$ resistors, and the voltages $V_1 \ldots V_N$ remain constant. Note that one or both outputs can be connected to the virtual ground at the summing junction of an op-

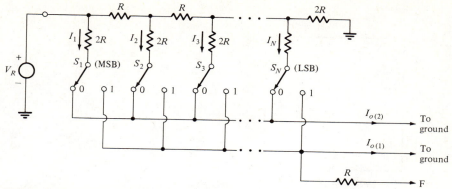

Figure 14.3 R-$2R$, current-scaling DAC ladder. $S_1 \cdots S_N$ are SPOT MOS switches actuated by the binary input signal.

amp current-to-voltage converter. The reference source must supply a current $I_R = V_R/R$. (Examination of the R-$2R$ ladder shows that V_R "sees" $R\ \Omega$ looking into the V_1 node.) By inspection, we see that $I_1 = I_R/2$, $I_2 = I_R/4$, and so on. Hence the maximum $I_{o(1)}$ occurs when the binary input is all 1's, and is given by

$$I_{o(1)\text{max}} = \left(\frac{V_R}{R}\right)\left(\frac{1}{2^1} + \frac{1}{2^2} + \frac{1}{2^3} + \cdots + \frac{1}{2^N}\right)$$

$$= \left(\frac{V_R}{R}\right)\left(\frac{2^N - 1}{2^N}\right) \tag{14.8}$$

The current output of this DAC is, in general,

$$I_{o(1)} = \left(\frac{V_R}{R}\right) \sum_{k=1}^{N} \frac{D_k}{2^k} \tag{14.9}$$

It is easy to show that

$$I_{o(2)} = \overline{I_{o(1)}} = \left(\frac{V_R}{R}\right)\left(\frac{2^N - 1}{2^N} - \sum_{k=1}^{N} \frac{D_k}{2^k}\right) \tag{14.10}$$

where $D_k = 0$ or 1, and D_1 is the MSB, and D_N the LSB, of the input word.

Current-scaling DAC ladders are generally used with op-amps to convert $I_{o(1)}$ or $I_{o(2)}$ to an output voltage. Figure 14.4(A) illustrates unipolar binary operation of the R-$2R$, current-scaling DAC. In this DAC connection, it is easy to see that

$$V_o = -R\left[\left(\frac{V_R}{R}\right) \sum_{k=1}^{N} \frac{D_k}{2^k}\right] = -\left(\frac{V_R}{2}\right) \sum_{k=1}^{N} \frac{D_k 2^{k-1}}{2^N} \tag{14.11}$$

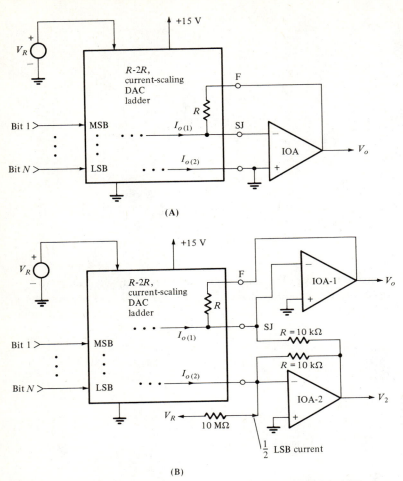

Figure 14.4 **(A)** Unipolar binary DAC operation. **(B)** Offset binary DAC operation. $R = 10$ kΩ. SJ = the summing junction at virtual ground.

Table 14.1 Unipolar binary DAC output with $N = 8$

D	V_o
0 0 0 0 0 0 0 0	0
0 0 0 0 0 0 0 1	$-V_R/256$
1 0 0 0 0 0 0 0	$-V_R(128/256)$
1 1 1 1 1 1 1 1	$-V_R(255/256)$

Table 14.1 shows V_o versus D values for the unipolar binary-connected DAC. Note that the maximum output is $-V_R(2^N - 1)/2^N$ rather than $-V_R$. Each LSB step is

$$\Delta V_o = \frac{-V_R}{2^N} \qquad\qquad 14.12$$

In *offset binary* operation of the R-$2R$, current-scaling DAC ladder, two op-amps are used to realize an output voltage that ranges approximately over $\pm V_R$. The output of the offset binary DAC in the general case can be written

$$V_o = -RI_{o(1)} - V_2 = -RI_{o(1)} - (-RI_{o(2)}) + \frac{\Delta V_o}{2}$$
$$= -V_R \sum_{k=1}^{N} \frac{D_k}{2^k} + V_R\left(\frac{2^{N-1}}{2^N} - \sum_{k=1}^{N} \frac{D_k}{2^k}\right) + \frac{\Delta V_o}{2}$$
$$= -2V_R \sum_{k=1}^{N} \frac{D_k}{2^k} + V_R\left(\frac{2^{N-1}}{2^N}\right) + \frac{\Delta V_o}{2} \qquad\qquad 14.13$$

The $\Delta V_o/2$ term is a $\frac{1}{2}$ LSB step used to make the transfer characteristic of this DAC an odd function. Table 14.2 gives typical V_o versus D values for this offset binary DAC.

In practical IC DAC designs, current sources are often used to drive the R-$2R$ ladder. Figure 14.5 illustrates a DAC in which matched transistors and emitter resistors are used to make N equal-value current sources which are connected either to the R-$2R$ ladder or to ground, depending on the state of the bit controlling that switch. The R-$2R$ ladder acts as a binary attenuator to scale the output current. The emitter resistors, R_e, are made sufficiently large to minimize the effects of differences in the V_{BE} of individual transistors. It is easy to show that the R-$2R$ ladder acts on

Table 14.2 Coding and output for a 10-bit, offset binary DAC

D	V_o
1 1 1 1 1 1 1 1 1 1	$-V_R(511/512)$
1 0 0 0 0 0 0 0 0 1	$-V_R(1/512)$
1 0 0 0 0 0 0 0 0 0	0
0 1 1 1 1 1 1 1 1 1	$+V_R(1/512)$
0 0 0 0 0 0 0 0 0 0	$+V_R(511/512)$

Note: The 1/2 LSB trim voltage in Eq. 14.13 was neglected in finding V_o for simplicity

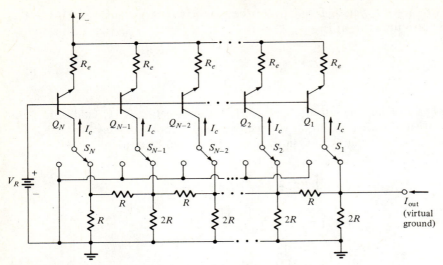

Figure 14.5 *N*-bit, *R-2R* DAC that uses switched, equal-value current sources. S_1 is operated by the MSB, S_N by the LSB. V_R sets the currents, I_c. It is easy to show that $I_c = (V_R - V_{BE} - V_-)/\alpha_F R_e$.

each current source I_c, to give

$$I_{out} = 2I_c \left(\frac{D_1}{2^1} + \frac{D_2}{2^2} + \frac{D_3}{2^3} + \cdots + \frac{D_N}{2^N} \right)$$ 14.14

where

  D_1 = the most significant input bit

  D_N = the least significant input bit

Note that the use of low R values in the ladder increases switching speed.

   Often DACs make use of differential BJT current switches in their designs to obtain higher switching speeds by avoiding large voltage swings at the switching nodes. The architecture of a differential switch is shown in Fig. 14.6. The currents I_C are directed either to ground or to the *R-2R* ladder elements, depending on the states of the digital input word. Note that the switch's base current, I_B, can affect I'_C. Various schemes to compensate for this error have been proposed; one of the cleverest involves using a feedback bias circuit, details of which may be found in Chapter 14 of Grebene (1984).

   In discussing measures of DAC performance, we will first consider the dynamic response of the DAC output to an abrupt change in the state of the digital input word. The worst-case output transient usually occurs around $V_o = V_R/2$, when D goes from 1000 0000 to 0111 1111, and so forth. The output transients caused by bit switching are called *glitches*. DAC output glitches can produce fuzzy CRT displays

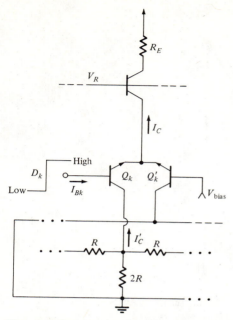

Figure 14.6 Differential BJT current switch commonly used in
R-$2R$, current-scaling DACs. D_k = the kth digital input. The actual kth
ladder current, $I'_c = I_c - I_{Bk}$.

on data terminals and digital TVs when glitchy DACs are used to drive the CRT
deflection systems. Glitches also represent a source of high-frequency noise in measurement, communications, and control systems using DACs. To minimize glitches, care
must be taken in selecting the output op-amp(s). These op-amps must have appropriate
slew-rate and step response characteristics. Glitches can be made worse by using an
output op-amp with a step response that "rings." Careful design of the DAC's analog
switches and their timing can go far toward eliminating glitches. Since all the IC
resistors in the R-$2R$ ladder have capacitances to ground, switching strategies that
keep these parasitic capacitors charged to constant voltages also help in reducing
output glitches. In the case of CRT displays, it is possible to blank the electron beam
for the few hundred nanoseconds it takes the glitch to die out and V_o to reach its
new steady-state value. Another strategy is to follow the DAC output with an analog
sample-and-hold circuit, and sample V_o after it reaches its steady state. (In this procedure we assume that the sample-and-hold circuit itself has a rapid and glitch-free
response.)

   Settling time is generally used as a measure of DAC response speed. It is generally
in the hundreds of nanoseconds for current output DACs (without op-amps) to reach
within $\frac{1}{2}$ LSB of steady-state output. Development of video-frequency op-amps has
made fast, voltage output DACs common. For example, the Analog Devices AD-7240

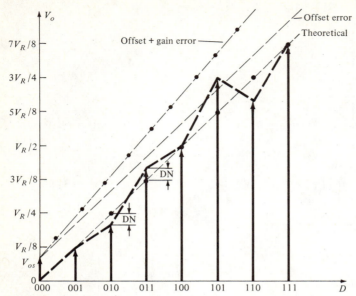

Figure 14.7   Transfer characteristic of a 3-bit, unipolar binary DAC, showing types of error. (*Note:* Nonmonotonic behavior at $D = 101$.)

12-bit, voltage output MDAC is reported to settle to 0.01% of steady state in only 550 ns in response to a digital input change.

The transfer characteristic of a 3-bit ($N = 3$), unipolar binary DAC is shown in Fig. 14.7. The dashed line shows the output voltage levels when there is a dc offset voltage error (V_{os} is added to each output equally). V_{os} is generally trimmable to zero at a desired operating temperature (e.g., 25°C) but has a tempco in spite of clever IC design (e.g., 5 ppm/°C). The dot-dash line illustrates the effect of gain error plus offset voltage. Gain error affects the output voltage scale relative to V_R [gain error can be due to $V_R(T)$]. Gain, too, is trimmable to a desired value but will drift with temperature (e.g., gain tempco = 20 ppm/°C).

Any non-uniformity in the R-$2R$ ladder can result in nonlinearity in the conversion process. *Differential nonlinearity* (DN) is the measure of the deviation of each output step size from the theoretical value of that step, given that offset and gain errors have been nulled. DN should be as small as possible, certainly less than $\pm\frac{1}{2}$ LSB. If $|\text{DN}| > 1$ LSB, then the DAC output is said to be nonmonotonic, a serious nonlinear flaw in which an output for a 1 LSB increment in D is actually less than the unincremented output. In other words, the DAC transfer curve is double-valued. *Integral nonlinearity* (IN) is another specification, defined as the worst-case deviation of the DAC output curve (solid vertical lines in Fig. 14.7) from a straight line connecting $V_o = 0$ with $V_o = V_{o(\text{max})}$. IN, too, should be less than $\pm\frac{1}{2}$ LSB.

Although we have focused our attention on R-$2R$ types of DACs, it is important to know that another class of DAC exists which uses MOS, switched-capacitor tech-

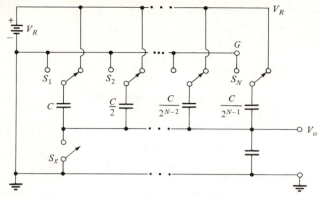

Figure 14.8 Switched-capacitor or charge-scaling DAC. The entire circuit is realized with MOS integrated circuit technology.

nology. These DACs, known as *switched-capacitor* or *charge-scaling* DACs, generate an analog output voltage based on the distribution of steady-state charge in a capacitive voltage divider, and the relation, $V = Q/C$.

Figure 14.8 is a simplified schematic of a switched-capacitor DAC. In the reset mode, all the MOS switches are connected to ground, discharging all the capacitors. During the period in which the input word is read, switch S_g is open, and selected binary-weighted scaling capacitors are connected to the reference voltage source, V_R. For a pure binary input, the scaling capacitors corresponding to zeros in the input word are left connected to ground. The capacitors so connected form a capacitive voltage divider in which V_o is given by

$$V_o = \left(\frac{V_R}{C_T}\right) \sum_{k=1}^{N} D_k\left(\frac{C}{2^{k-1}}\right) \qquad 14.15$$

where C_T is the total capacitance of the switched array, given by

$$C_T = \sum_{k=1}^{N} \frac{C}{2^{k-1}} + \frac{C}{2^{N-1}} = 2C \qquad 14.16$$

and D_k are the bits (0, 1) in the input binary word. D_1 is the MSB, and D_N is the LSB.

A fundamental limitation to the bit size of the digital word that can be converted is created by the ratio of areas between the largest (MSB) capacitor and the smallest (LSB) capacitor on the switched-capacitor DAC chip. This ratio is easily shown to be 2^{N-1}.

If the smallest capacitor is 2 pF, then the total DAC capacitance, C_T, is 512 pF for an 8-bit DAC. It is apparently not practical to have N greater than 8 for this class of ADC because of the hefty glitch when the capacitors are reset to zero charge,

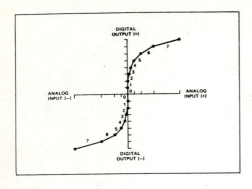

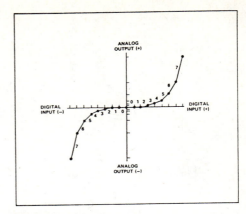

Figure 14.9 **(A)** Bell μ-255 compander encoding curve for ADCs.
(B) Complementary μ-255 decoding curve for companding DACs.

the switch-capacitor DAC is mostly used as a subsystem in the design of MOS analog-to-digital converters.

Another specialized type of DAC with a nonlinear conversion characteristic is the *companding* DAC. Companding DACs are used in modern digital telecommunications systems to enhance the dynamic range of a digital channel. They are used as feedback elements in telecommunications ADCs to compress the analog signal into a relatively efficient digital code. At the receiving end of the digital channel, a companding DAC restores (expands) the digital code into a wide dynamic range analog signal. The term *companding* is derived from the role this class of DAC has in compressing analog signals in digitization, and expanding them at the receiver to restore signal linearity. Figure 14.9(A) illustrates the sigmoid shape of the compander encoding transfer characteristic, and Fig. 14.9(B) shows the complementary, decoding I/O curve.

One example of the encoding law is the Bell μ-255 law transfer characteristic, given by

$$V_e = \frac{\ln(1 + 255|V_{\text{in}}|)}{\ln 256} \, \text{sgn}(V_{\text{in}}) \qquad\qquad 14.17$$

where $-1 \text{ V} \le V_{\text{in}} \le +1 \text{ V}$ and V_e is the continuous (analog) equivalent of the encoding (compressing) ADC output. Note that the 8-bit digital code generated by a compressing ADC gives a 72 dB dynamic range for both V_{in} polarities. The actual nonlinearity relating compander ADC digital output to V_{in} has an eight-segment (chord), piecewise-linear approximation with 16 linear steps in each segment, for both polarities. The digital code generated by the encoding ADC (or used by the decoding DAC) has 8 bits. The first bit is determined by (determines) polarity; the next 3 bits are selected by (select) the specific chord segment, and the last 4 bits are selected by (select) the particular step within a given chord.

The companding DAC's input/output curve is described by an exponential function, given by the inverse of the encoding law:

$$V_o = \frac{256^{|V_a|} - 1}{255} \, \text{sgn}(V_a)$$

14.18

In this function; V_a is the analog equivalent input voltage, with a range $-1 \le V_a \le 1$, so $-1 \le V_o \le 1$. The actual input to this DAC is, of course, the 8-bit digital code just described. For more details on the design and use of companding DACs in telecommunications, see the *PMI Telecommunications Applications Handbook* (1981) and Chapter 14 in Grebene (1984).

Applications of DACs are manifold. As will be seen in Sec.14.4, they are an integral component in the designs of certain ADCs. They are also used to interface a number of useful functions done by computers to the analog world. These applications include analog waveform generation, controlling CRT displays, programming power supplies, music synthesis, voice synthesis, interfacing digital controllers with proportional transducers such as dc servomotors, and music reproduction from digital disks. MDACs were seen in Chapter 10 to be useful in the design of tracking and adaptive active filters as variable gain elements.

14.3 Hold Dynamics

The digital input to a DAC is generally periodic; in most cases, the DAC's analog output from the nth digital input is held constant until the $(n + 1)$th digital input updates it. and so on. This process generates a stepwise output waveform if D is changing. The process is called holding, and this one, in particular, is called a *zero-order hold* (ZOH). The frequency domain dynamics of a ZOH can be obtained from consideration of its impulse response (shown in Fig. 14.10). The impulse response of the ZOH, $h_o(t)$, is its output given a 1 input at $t = 0$. The transfer function of the ZOH can be written by inspection of Fig. 14.10. $h_o(t)$ can be decomposed into a positive unit step at $t = 0$, and a negative unit step at $t = T_s$. These two steps are Laplace transformed to yield

$$\mathbf{H}_o(s) = \mathscr{L}\{h_o(t)\} = \frac{1}{s} - \left(\frac{1}{s}\right)e^{-sT_s} = \frac{1 - e^{-sT_s}}{s}$$

14.19

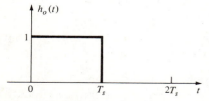

Figure 14.10 Impulse response of a zero-order (boxcar) hold.

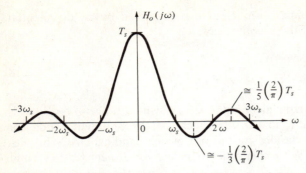

Figure 14.11   Frequency response magnitude of the zero-order hold.

The frequency response of the ZOH is found by letting $s = j\omega$ in Eq. 14.19, and then multiplying the numerator and denominator by $T_s e^{+j\omega T_s/2}$. This operation gives

$$\mathbf{H}_o(j\omega) = T_s \frac{e^{j\omega T_s/2} - e^{-j\omega T_s/2}}{(2j)(\omega T_s/2)} e^{-j\omega T_s/2} \qquad 14.20$$

which, by the Euler relation for $\sin(x)$, reduces to

$$\mathbf{H}_o(j\omega) = T_s \frac{\sin(\omega T_s/2)}{\omega T_s/2} e^{-j\omega T_s/2} \qquad 14.21$$

A plot of $\mathbf{H}_o(j\omega)$ versus ω is shown in Fig. 14.11. The zeros of $\mathbf{H}_o(j\omega)$ occur for

$$\frac{\omega T_s}{2} = n\pi \qquad 14.22$$

where $n = 1, 2, 3, \ldots$, or at

$$\omega_o = \frac{n2\pi}{T_s} \qquad 14.23$$

Thus the overall process of sampling (periodically digitizing) a signal and reconverting it to analog form with a DAC and ZOH gives rise to an overall transfer function given by the product of Eq. 14.21 and Eq. 14.7. If the ZOH output is $x(t)$, then, neglecting quantization, we have

$$\mathbf{X}(j\omega) = \sum_{n=0}^{\infty} \frac{\sin(\omega T_s/2)}{\omega T_s/2} X(j\omega - jn\omega_s)e^{-j\omega T_s/2} \qquad 14.24$$

We have seen that an ideal LPF can reconstruct sampled data that obey the Nyquist criterion. The ZOH is a realizable LPF, far from being ideal. The frequency response function Eq. 14.24 can be visualized by multiplying Fig. 14.11 times Fig. 14.2(A). In

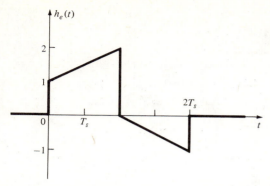

Figure 14.12 Impulse response of a first-difference extrapolator hold.

addition to the baseband spectrum, remnants of $X(j\omega - jn\omega_s)$ leak through because of the nature of the $\sin(x)/x$ function.

Other types of holds have been devised that more closely approximate the ideal analog signal by generating linear extrapolations between the sampling instants. Such extrapolator hold systems possess *hybrid* dynamics; that is, their behavior is describable in the time domain by a combination of linear differential and difference equations. Therefore, in the frequency domain, their transfer functions are rational in both s- and e^{-sT} terms, implying both digital and analog filtering. An extrapolator hold requires a digital memory to store past digital samples used to generate the linear ramp component of its output.

A simple extrapolator hold uses the first difference of input samples to generate a linear ramp component of output voltage which is added to the ZOH output, generating a closer approximation (in some cases) to the analog signal, $x(t)$. The weighting function (unit impulse response) of the first-difference extrapolator hold, $h_e(t)$, is shown in Fig. 14.12. Decomposition of the hold weighting function of Fig. 14.12 into its component ramps and steps leads to the transfer function:

$$H_e(s) = \frac{1 - e^{-sT}}{s} + \frac{1 - 2e^{-sT} + e^{-2sT}}{s^2} - \frac{(1 - e^{-sT})(e^{-sT})}{s} \qquad 14.25$$

which is seen to have several components, one being the basic ZOH, the others giving rise to the ramp portions of $h_e(t)$.

Proceeding as we did in the treatment of Eqs. 14.20 and 14.21, we obtain the frequency response function for the first-difference, extrapolator hold transfer function:

$$H_e(j\omega) = T_s^2 \left[\frac{\sin^2(\omega T_s/2)}{(\omega T_s 2)^2} \right] (1 + j\omega)(e^{-j\omega T}) \qquad 14.26$$

Notice that the frequency response magnitude of this hold drops off sharply to zero for ω near integer multiples of the sampling frequency because of the $(\sin x/x)^2$ function.

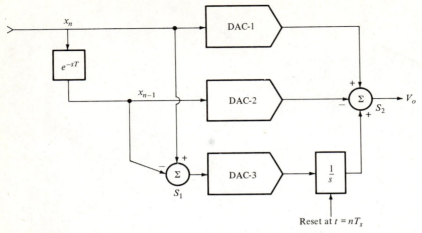

Figure 14.13 Implementation of a first-difference extrapolator hold transfer function. Summer S_1 performs the numerical summation of terms to estimate the interpolating ramp slope. Summer S_2 adds the analog signals to compose the extrapolator hold output. The input data, x_n, is periodic, and the three DACs operate in the latching mode, holding their outputs constant until they are updated at the next sampling period.

   Implementation of this hold's transfer function can be done with three latching DACs, the outputs of which are summed according to Eq. 14.25. The implementation is illustrated in Fig. 14.13. The integrator generates the ramp component in Eq. 14.25; its output is reset to zero at the end of each sampling period by a transistor switch.
   The output of the first-difference extrapolator hold is generally smoother than that of a ZOH, but its complexity generally does not warrant its use if one can sample faster, thereby reducing the conversion errors from the ZOH.

14.4 Analog-to-Digital Converters

Various types of ADCs have evolved to meet various data acquisition requirements, ranging from digital multimeters to video converters used in real-time digital picture processing. The major types of ADCs include: (1) successive approximation converters, (2) tracking (servo) types, (3) dual-slope integrating converters, (4) flash (parallel) converters, and (5) dynamic range, floating point ADCs. We will briefly describe the organization, features, and applications of these five classes of ADCs.

14.4.1 Successive Approximation (SA) ADCs

A block diagram of an SA ADC is shown in Fig. 14.14. Note that this ADC uses a DAC and an analog comparator in its design. SA ADCs are fast enough for audio-frequency applications, with conversion times typically ranging from 5 μs to 15 μs.

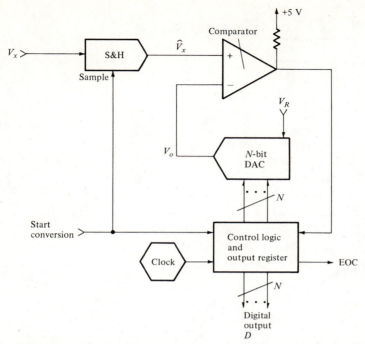

Figure 14.14 Block diagram of a successive approximation ADC.
S&H is an analog sample-and-hold module to "freeze" V_x in order to
prevent it from changing during the conversion process. EOC is a
digital status line that signals when conversion is complete or when
data are valid on the D_{out} lines.

The conversion cycle of an 8-bit SA ADC begins with the analog input signal,
V_x, being sampled and held at $t = 0$. Then the output register is cleared (all $D_k = 0$);
next D_1 is set to 1 (MSB), with all other $D_k = 0$, making the DAC analog output
$V_o = V_R(128/256) = V_R/2$. The comparator performs the operation $\text{sgn}(V_x - V_R/2)$. If
$\text{sgn}(V_x - V_R/2) = 1$, then D_1 is kept 1; if $\text{sgn}(V_x - VR/2) = -1$, then $D_1 = 0$. This
completes the first (MSB) cycle in the conversion process. Next, D_2 is set to 1 (D_1
retains the value found in the first cycle). The comparator tests to see if $V_x > V_o$
($V_o = V_R/2 \cdot D_1 + V_R/4$). If yes, $D_2 = 1$; if no, $D_2 = 0$, completing the second bit's
conversion cycle. This process continues until all 8 bits are converted. It is summarized
by the flowchart in Fig. 14.15. Conversion time is fixed, being proportional to the
number of bits converted.

14.4.2 Tracking or Servo ADCs

A block diagram of a tracking ADC is shown in Fig. 14.16. The tracking ADC is a
relatively slow system best suited for conversion of dc and low-frequency ac signals

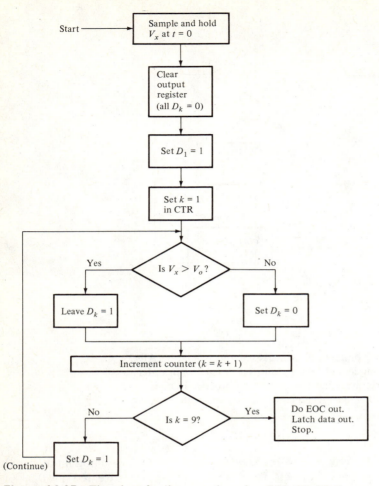

Figure 14.15 Flowchart for the operation of an 8-bit SA ADC.
$D_1 = $ MSB, and $D_8 = $ LSB.

to digital form. The waveforms encountered in its operation are shown in Fig. 14.17. Note that in the steady state, V_o has a ± 1 LSB dither around V_x, because the counter counts one count up, then one count down. Because of this steady-state output dither, the LSB line is not used in the digital output.

The maximum slew rate of the servo ADC is easily found from a knowledge of the clock period, T_c, the number of DAC bits, N, and the maximum V_o, $V_{o(max)}$. The slew rate in volts per second gives the maximum $|dV_o/dt|$ that this ADC can follow without gross error. It is found from

$$\eta = \frac{V_{o(max)}}{2^N T_c} \text{ V/s}$$ 14.27

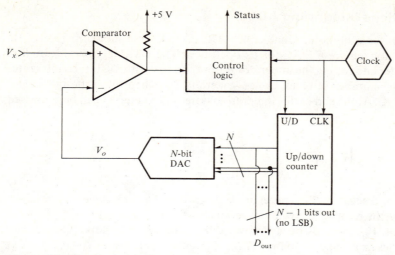

Figure 14.16   Block diagram of a tracking (servo) ADC.

If $N = 10$ bits, $T_c = 5 \times 10^{-7}$ s, and $V_{o(max)} = 10$ V, then the slew rate is found to be $\eta = 0.01953$ V/μs. This rate is relatively slow. If the tracking ADC were reset after each conversion, it could convert a 10 V signal at a maximum rate of 1953 conversions/s. Another way of looking at the conversion rate of this type of ADC is to assume that $V_x(t) = 5 + 5 \sin(\omega t)$. Thus the maximum $|dV_x/dt| = 5\omega$ must be less than the slew rate; that is, $5\omega < 1.953 \times 10^4$ V/s, or $\omega < 3.906 \times 10^3$ rad/s, or $f_{max} = \omega/2\pi = 622$ Hz for conversions to proceed without gross error on all parts of the $V_x(t)$ waveform.

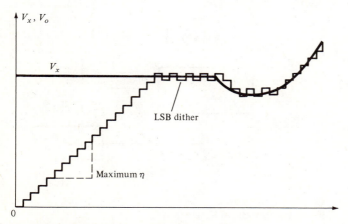

Figure 14.17   Tracking ADC waveforms.

14.4.3 Dual-Slope Integrating ADCs

A block diagram for a unipolar, dual-slope ADC (DSADC) is shown in Fig. 14.18. This ADC does not require a DAC for its operation. The DSADC operates by first clearing its counter and setting the integrator output, V_2, to 0 with S_2. It then integrates $V_x \geq 0$ for a fixed number of clock cycles, generally 2^N, where N is the bit resolution of the DSADC. At the end of this integration time, T_1, voltage V_2 can be expressed as

$$V_2(T_1) = -\frac{1}{RC}\left\langle \int_0^{T_1} \frac{V_x(t)\,dt}{T_1} \right\rangle T_1 = -\frac{T_1}{RC}\langle \overline{V_x(T_1)} \rangle \qquad 14.28$$

where $T_1 = 2^N T_c$.

(For reasons to be discussed, T_1 is often made $\frac{1}{60}$ s.) After 2^N clock cycles are reached, the counter is again reset to 0, switch S_1 connects the integrator to $-V_R$, and the integrator output, V_2, proceeds linearly toward 0 V with positive slope. The counter counts clock cycles from time T_1 to time $(T_1 + T_2)$ when V_2 reaches 0 V. At $(T_1 + T_2)$, the counter count, M, is latched into an output register. At T_2, we can write

$$-\frac{T_1}{RC}\langle \overline{V_x(T_1)} \rangle + \frac{T_2 V_R}{RC} = 0 \qquad 14.29$$

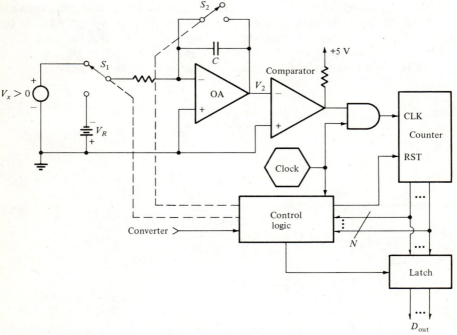

Figure 14.18   Unipolar, dual-slope, integrating ADC.

However,

$$T_2 = MT_c$$

14.30

so we can write

$$\frac{MT_cV_R}{RC} = \frac{2^NT_c\langle \overline{V_x(T_1)}\rangle}{RC}$$

14.31

which leads to

$$M = \frac{\langle \overline{V_x(T_1)}\rangle}{V_R} 2^N$$

14.32

Thus the down count time is proportional to the average of the input signal over T_1, $\overline{V_x(T_1)}$.

This conversion process is seen to be independent of slow changes in the parameters R, C, and T_c. Its accuracy does depend on a highly stable V_R, however. If T_1 is made an integer number of power line cycles, there will be no net contribution to V_2 from 60 Hz hum contaminating V_x. Power line noise can be rejected by as much as 70 dB through this mechanism. DSADCs are widely used in dc digital multimeters; they are available as LSI ICs, and most have autozeroing modes in which input offset voltage measured with $V_x = 0$ is converted and used for offset correction.

Figure 14.19 illustrates the organization of a DSADC having an offset binary output code; this ADC accepts bipolar dc input signals. Its operation is similar to that of the unipolar DSADC. First the integrator integrates the output of op-amp 1 for T_1 s (2^N clock cycles). Thus we can write

$$V_2(T_1) = \frac{1}{2RC} \int_0^{T_1} (V_R + V_x(t))\, dt = \frac{T_1V_R}{2RC} + \frac{T_1}{2RC} \int_0^{T_1} \frac{V_x(t)\, dt}{T_1}$$
$$= \frac{T_1V_R}{2RC} + \frac{T_1\langle \overline{V_x(T_1)}\rangle}{2RC}$$

14.33

This positive voltage is then integrated down to zero, and the number of clock cycles required to do this is stored in the output latch. Hence we can write, noting that $T_1 = 2^NT_c$,

$$\frac{2^NT_cV_R}{2RC} + \frac{2^NT_c\overline{V_x}}{2RC} - \frac{MT_cV_R}{RC} = 0$$

14.34

Solving for M, we find

$$M = 2^N\left(\frac{1 + \overline{V_x}/V_R}{2}\right)$$

14.35

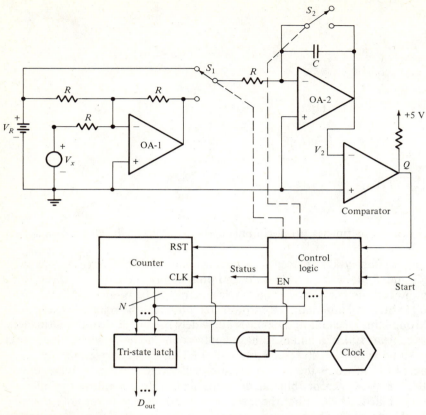

Figure 14.19   Block diagram of a dual-slope integrating ADC for bipolar signals. The DSADC has offset binary output.

Note that when $\overline{V}_x = -V_R$, $M = 0$; when $\overline{V}_x = 0$, $M = 2^N/2$; and when $V_x = V_R$, $M = 2^N$, giving a true offset binary output code.

    Charge-balancing ADCs (CBADCs) are similar in architecture to DSADCs but operate quite differently on dc input signals. A block diagram of a unipolar CBADC is shown in Fig. 14.20. The operating cycle proceeds as follows. Prior to the start of the conversion cycle, S_2 is closed, $V_2 = 0$, and the output counter is reset to 0. When conversion begins, V_2 ramps negatively; when V_2 reaches the comparator reference voltage, V_o, the comparator output, Q, changes state. This transition causes switch S_1 to connect current source I_R to the summing junction for exactly one-half clock cycle, $T_c/2$. During this time, an amount of charge, Q_r, is extracted from the summing junction. Q_r is simply

$$Q_r = I_R \frac{T_c}{2}$$

14.36

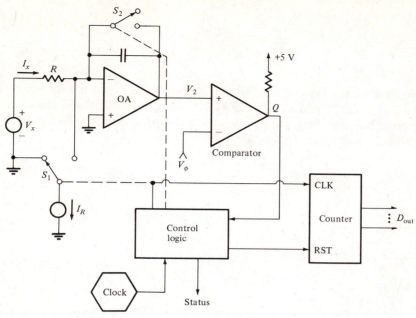

Figure 14.20   Block diagram of a charge-balancing ADC.

Since $I_R > (I_{x(max)} = V_{x(max)}/R)$, V_2 ramps up sharply toward 0 V until S_1 opens after $T_c/2$ s; then V_2 goes negative again because $V_x > 0$. When V_2 again reaches $V_o < 0$, S_1 again closes for $T_c/2$ s and V_2 ramps positive, and so on. The process continues for a total of 2^N clock cycles, where N is the CBADC bit resolution. The count M on the output counter is proportional to V_x, as is easily seen if we use Kirchhoff's current law and note that the net charge accumulated at the summing junction is zero. A charge balance equation can be written:

$$Q_x = \frac{V_x}{R} 2^N T_c = I_R \frac{M T_c}{2}$$
14.37

Solving Eq. 14.37 for M, we find

$$M = \frac{(\overline{V}_x/R)2^N}{I_R/2}$$
14.38

Now I_R is scaled so that $M = 2^N$ when $\overline{V}_x = V_{x(max)}$. This means that

$$I_R = 2\left(\frac{V_{x(max)}}{R}\right)$$
14.39

Hence Eq. 14.38 can be rewritten as

$$M = \left(\frac{\overline{V_x}}{V_{x(max)}}\right) 2^N \qquad\qquad 14.40$$

As is the case for the DSADC, calibration of the CBADC does not depend on absolute values of R, C, T_c, and V_o. The CBADC does require stability in the clock signal's duty cycle, however, for accurate operation. This ADC is inherently slow and is well suited for dc multimeter applications.

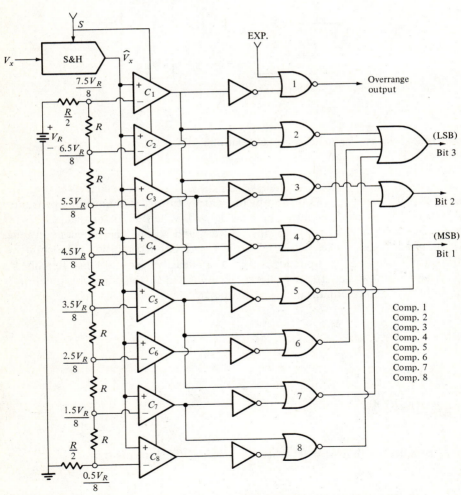

Figure 14.21   Schematic of a 3-bit flash ADC with binary decoding.

14.4.4 Flash (Parallel) ADCs

Flash ADCs (FADCs) are the fastest and most expensive ADC type. Some operate at conversion rates in excess of 100 megasamples/s (one conversion every 10 ns). FADCs are used in modern broadband digital oscilloscopes and digital signal-processing systems. They generally use high-speed bipolar logic (ECL) gates and are fabricated as LSI or hybrid LSI ICs. With the present state-of-the art in IC fabrication, it is difficult to build an LSI FADC with more than 8-bit resolution. Techniques exist, however, for combining two FADCs of lower resolution to make a higher-resolution converter. Note that 2^N analog comparators are used in the design of an N-bit FADC, along with suitable high-speed decoding logic to give a binary output. Figure 14.21 illustrates a 3-bit FADC. Here, $(2^3 - 1)$ comparators are used to code the output; the eighth comparator is used to sense an overrange input $(V_x > 7.5V_R/8)$ and provide a signal to enable stacking two or more 3-bit FADCs to obtain more binary resolution. The input/output characteristic of the 3-bit FADC are shown in Fig. 14.22.

It appears at this time not to be practical to implement FADC designs with more than 8 bits on a single chip. Problems exist with signal propagation delays, the input capacitance of 256 analog comparators, dc offset trims for this large number of comparators, and so forth. Comparator inputs can be buffered with follower circuits

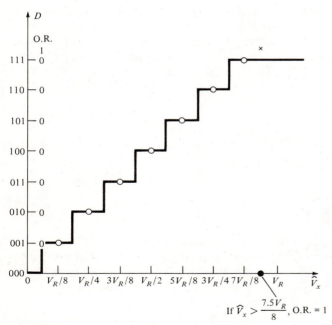

Figure 14.22 I/O characteristic of the 3-bit FADC of Fig. 14.21. Note the overrange output (NOR #1).

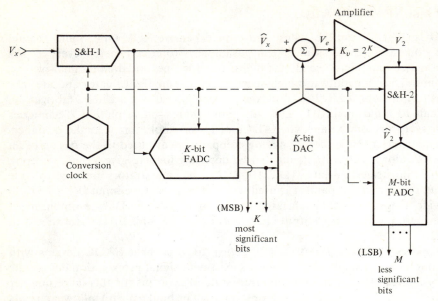

Figure 14.23   Two-step FADC. $K + M = N$ bits.

to reduce their input admittances, and autozero offset cancellation circuitry can be implemented with each comparator, but at the expense of circuit complexity and cost.

One way to increase flash converter resolution is to use the two-step architecture shown in Fig. 14.23. Let $K = M = 6$ bits, so 12-bit resolution will be obtained for the system in which $0 \leq V_x \leq V_{FS}$. The K most significant bits output of the K-bit DAC will be

$$V_1 = V_{FS} \sum_{i=1}^{K} \frac{D_i}{2^i} \qquad\qquad 14.41$$

The input V_2 to the lesser bits FADC is

$$V_2 = 2^K \left(\bar{V}_x - V_{x(\max)} \sum_{i=1}^{K} \frac{D_i}{2^i} \right) \qquad\qquad 14.42$$

The voltage difference in parentheses is V_e and can be as large as $V_x/2^K$; hence the second converter generates a binary code on the quantized remnant, V_e, to give a net $N = (K + M)$-bit output. Because the K- and M-bit FADCs use $(2^K - 1)$ and $(2^M - 1)$ comparators, respectively, the system is, in general, less expensive than a single N-bit FADC. The total coded output of this system is

$$V_x = V_{FS} \left(\sum_{i=1}^{K} \frac{D_i}{2^i} + \sum_{j=k+1}^{M} \frac{D_j}{2^j} \right) = V_{FS} \sum_{k=1}^{N} \frac{D_k}{2^k} \qquad\qquad 14.43$$

Table 14.3 MN-5420 DRFP ADC exponent switching points and coding for $V_x > 0$

PGA Switching Voltage, V_x	PGA Gain	Exponent	LSB Voltage
$+5 \text{ V} \leq V_x \leq 2.5 \text{ V}$	1	1000	2.44 mV
$2.5 \text{ V} < V_x \leq 1.25 \text{ V}$	2	0111	1.22 mV
$1.25 \text{ V} < V_x \leq 0.625 \text{ V}$	4	0110	610 μV
$0.625 \text{ V} < V_x \leq 0.3125 \text{ V}$	8	0101	305 μV
$0.3125 \text{ V} < V_x \leq 0.15625 \text{ V}$	16	0100	153 μV
$0.15625 \text{ V} < V_x \leq 0.078125 \text{ V}$	32	0011	76 μV
$0.078125 \text{ V} < V_x \leq 0.039063 \text{ V}$	64	0010	38 μV
$0.039063 \text{ V} < V_x \leq 0.019531 \text{ V}$	128	0001	19 μV
$0.019531 \text{ V} < V_x \leq 0 \text{ V}$	256	0000	9.5 μV

14.4.5 Dynamic Range, Floating Point ADCs

A 20-bit, dynamic range, floating point ADC (DRFP ADC) having better than 1 ppm resolution has been developed by Micro Networks. This ADC (MN-5420) has a 16-bit output consisting of a 12-bit mantissa and a 4-bit exponent. The MN-5420 can do 3.2×10^5 conversions/s. It consists of a 4-bit FADC that codes 4 bits of exponent and sets the gain of a nine-range, programmable gain amplifier (with gains 1, 2, 4, ..., 128, 256), followed by a 1 μs, 12-bit ADC. The output of the 12-bit ADC is the ADC's mantissa and is presented in two's complement coding. When the analog input signal is small, the gain of the programmable gain amplifier is high; for example, for $V_x < 19.5$ mV, the gain is 256. The value of a mantissa LSB at this gain is 9.5 μV. The maximum input voltage of the MN-5420 DRFP ADC is ±5 V. Table 14.3 gives this ADC's gain switching points and the mantissa LSB values.

When V_x goes negative, exponent coding is the same as for positive V_x in the table; however, the MSB of the 12-bit mantissa goes from 0 to 1. Figure 14.24 illustrates the organization of the MN-5420 DRFP ADC. The principal use of this novel ADC design is in digital signal processing.

SUMMARY

It should be pointed out that the choice of ADC type is generally governed by the maximum conversion speed required and the accuracy needed for a given application. Maximum conversion rate, in turn, is determined by the highest frequency to be converted in the analog input signal, f_{max}, and the Nyquist criterion. Conservative design requires that the conversion rate be four to eight times f_{max}, although conversion as slow as two and one-half times the Nyquist frequency can be accomplished by

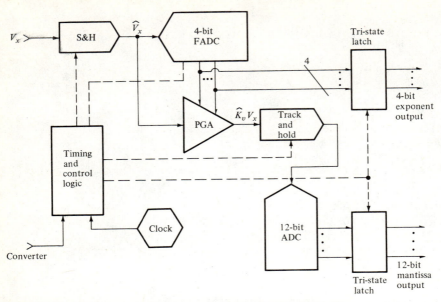

Figure 14.24   Block diagram of the MN-5420 DRFP ADC. The programmed gain amplifier (PGA) has an autozeroing feature (not shown).

putting a zero in the anti-aliasing filter's transfer function at the sampling frequency. A quantitative determination of the best sampling rate in a design involves such factors as error criteria, and is beyond the scope of this chapter, the cutoff characteristics of the anti-aliasing filter, and the high-frequency behavior of the analog signal's power density spectrum.

Cost is another important criterion in the selection of ADCs. Although a successive approximation ADC works well at audio-frequency rates, it might be cost-competitive with a dual-slope integrating ADC when considered for use in a dc data acquisition system. However, most dual-slope integrating ADCs have the autozero feature, which successive approximation ADCs do not. Thus choices are complex when slow data conversion systems are being designed; many factors must be considered. The designer's choice narrows considerably once conversion rates of over 10^5 samples/s are reached. The reader who wishes to explore the circuit design of data conversion systems in greater detail should see Chapters 14 and 15 of the text by Grebene (1984), Chapter 11 in Franco (1988), and also read some of the DAC and ADC manufacturers' handbooks and application notes. See, for example, the *Data Acquisition and Conversion Handbook* published by Datel-Intersil.

Appendix A

Signal Flow Graph Reduction by Mason's Rule

In this appendix, we review the use of Mason's signal flow graphs and the application of Mason's general gain formula in finding the transfer functions of complex, multiloop feedback systems. S. J. Mason devised signal flow graphs in 1953, and he developed his general gain formula in 1956. Signal flow graphs have found wide use in the analysis of electronic circuits and systems, and of course, in control systems. Use of signal flow graphs is restricted to linear systems.

A signal flow graph has two components: (1) unidirectional branches between pairs of nodes, each with an assigned branch gain or transmittance factor; and (2) nodes, which are summing points for branch outputs. A branch output is the product of the branch's transmittance with the input node variable. These properties are illustrated using Fig. A.1. From Fig. A.1(A), we can write

$$y_1 = T_1 x_1 \qquad y_2 = T_2 x_1 \qquad y_3 = T_3 x_1 \qquad \text{A.1}$$

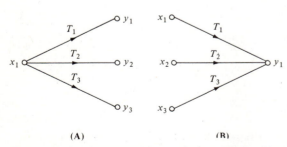

(A)   (B)

Figure A.1 **(A)** Signal flow graph with the x_1 node as the signal source point. **(B)** Signal flow graph with the y_1 node as the output.

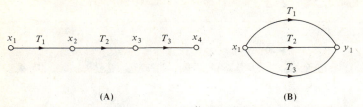

(A)   (B)

Figure A.2   **(A)** Signal flow graph cascaded or series path. **(B)** Signal flow graph parallel path.

Fig. A.1(B) gives us

$$y_1 = x_1 T_1 + x_2 T_2 + x_3 T_3 \qquad\qquad\qquad\qquad\qquad \text{A.2}$$

Cascaded branches and nodes are shown in Fig. A.2(A). Figure A.2(B) illustrates parallel branches between two nodes. The transmission between x_1 and x_4 in the cascaded signal flow graph of Fig. A.2(A) is simply

$$\frac{x_4}{x_1} = T_1 T_2 T_3 \qquad\qquad\qquad\qquad\qquad\qquad \text{A.3}$$

The transmission of the parallel graph is just

$$\frac{y_1}{x_1} = T_1 + T_2 + T_3 \qquad\qquad\qquad\qquad\qquad \text{A.4}$$

As an example of formulating a signal flow graph to analyze an electronic circuit, consider the simple JFET amplifier of Fig. A.3(A) at mid-frequencies. From inspection of the MFSSM of this amplifier, shown in Fig. A.3(B), and using Ohm's law and Kirchhoff's voltage law, we can write the following equations:

$$v_s = i_d R_S \qquad\qquad\qquad\qquad\qquad\qquad\qquad \text{A.5}$$

$$v_{gs} = v_1 - v_s \qquad\qquad\qquad\qquad\qquad\qquad\qquad \text{A.6}$$

$$i_d = \frac{v_{gs}}{R_D + r_d + R_S} \qquad\qquad\qquad\qquad\qquad \text{A.7}$$

$$v_o = -i_d R_D \qquad\qquad\qquad\qquad\qquad\qquad\qquad \text{A.8}$$

A signal flow graph can be constructed based on these equations, and then solved to obtain the amplifier's gain, v_o/v_1; this graph is shown in Fig. A.4. The graph has one loop, owing to the negative voltage feedback introduced by the unbypassed source resistor, R_S, as illustrated by Eq. A.6. Algebraic solution of Eqs. A.5 through

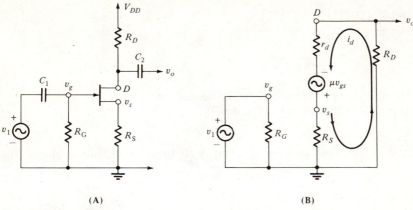

(A)   (B)

Figure A.3 (A) JFET amplifier with an unbypassed source resistor, R_s. (B) MFSSM of the JFET amplifier.

A.8 for the amplifier's mid-frequency, small-signal gain yields

$$\frac{v_o}{v_1} = \frac{-\mu R_D}{R_D + r_d + R_S(\mu + 1)} \qquad\qquad \text{A.9}$$

Fortunately, a systematic procedure for finding signal flow graph gains was developed by Mason in 1956. Mason's gain formula is

$$\frac{v_k}{v_j} = H_{jk} = \frac{\displaystyle\sum_{n=1}^{N} F_n \Delta_n}{\Delta_g} \qquad\qquad \text{A.10}$$

where

H_{jk} = the *net transmission* from the jth (input) node to the kth (output) node

F_n = the gain of the nth *forward path;* a forward path is a cascade of branches beginning on the jth (input) node and ending on the kth (output) node along which no node is passed through more than once (Note that a signal flow graph can have multiple forward paths which can share common nodes.)

Δ_g = signal flow graph *determinant*

= 1 − (sum of all individual loop gains) + (sum of products of pairs of non-touching loop gains) − (sum of products of nontouching loop gains taken three at a time) + $\cdots$

Δ_n = *cofactor* for the nth forward path

= Δ_g evaluated for nodes that do not touch the nth forward path

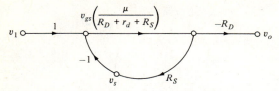

Figure A.4 Signal flow graph describing the volt-ampere relations in the JFET amplifier MFSSM shown in Fig. A.3(B).

Application of Mason's gain formula (Eq. A.10) to the signal flow graph of Fig. A.4 gives us

$$n = 1 \tag{A.11}$$

$$F_1 = 1\left(\frac{\mu}{R_D + r_d + R_S}\right)(-R_D) \tag{A.12}$$

$$\Delta_g = 1 - (-1)\left(\frac{\mu}{R_D + r_d + R_S}\right)(R_S + 0) \tag{A.13}$$

$$\Delta_1 = 1 \tag{A.14}$$

Substituting these relations into Eq. A.10, we get

$$\frac{v_o}{v_1} = \frac{\dfrac{-\mu R_D}{R_D + r_d + R_S}}{1 + \dfrac{\mu R_S}{R_D + r_d + R_S}} = \frac{-\mu R_D}{R_D + r_d + R_S(\mu + 1)} \tag{A.15}$$

which is the same gain seen in Eq. A.9, which was derived algebraically.

We further illustrate the use of Mason's gain formula with three additional examples. The signal flow graph for the first example is shown in Fig. A.5(A). Following the procedure just outlined, we can write

$$n = 1 \tag{A.16}$$

$$F_1 = ac \tag{A.17}$$

$$\Delta_g = 1 - (-e - ab - cd) + 0 \tag{A.18}$$

$$\Delta_1 = 1 \tag{A.19}$$

Therefore,

$$\frac{v_o}{v_1} = \frac{ac}{1 + e + ab + cd} \tag{A.20}$$

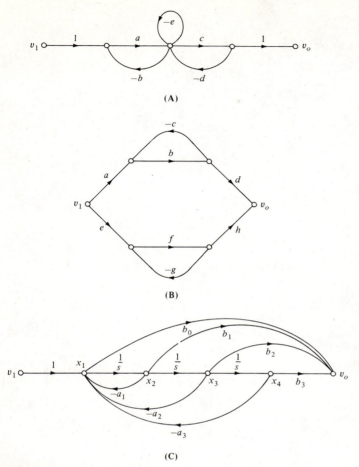

Figure A.5   Signal flow graphs.

The second graph, illustrated in Fig. A.5(B), is seen to have

$$n = 2 \qquad\qquad\qquad\qquad\qquad\qquad\qquad \text{A.21}$$

$$F_1 = abd \qquad F_2 = efh \qquad\qquad\qquad\qquad \text{A.22}$$

$$\begin{aligned} \Delta_g &= 1 - (-bc - gf) + [(-bc)(-gf)] \\ &= 1 + bc + gf + bcgf \end{aligned} \qquad\qquad \text{A.23}$$

$$\Delta_1 = 1 - (-fg) \qquad \Delta_2 = 1 - (-bc) \qquad \text{A.24}$$

Thus, by Mason's rule, we have

$$\frac{v_o}{v_1} = \frac{abd(1 + fg) + efh(1 + bc)}{1 + bc + fg + bcgf} \qquad \text{A.25}$$

The third graph, shown in Fig. A.5(C), has the components

$$n = 4 \qquad \text{A.26}$$

$$F_1 = b_0 \qquad F_2 = \frac{b_1}{s} \qquad F_3 = \frac{b_2}{s^2} \qquad F_4 = \frac{b_3}{s^3} \qquad \text{A.27}$$

$$\Delta_g = 1 - \left(\frac{-a_1}{s} - \frac{a_2}{s^2} - \frac{a_3}{s^3} \right) + 0$$

$$= \frac{s^3 + a_1 s^2 + a_3 s + a_3}{s^3} \qquad \text{A.28}$$

$$\Delta_1 = \Delta_2 = \Delta_3 = \Delta_4 = 1 \qquad \text{(all loops touch all forward paths)} \qquad \text{A.29}$$

Thus, by Eq. A.10

$$\frac{v_o}{v_1} = \frac{b_0 s^3 + b_1 s^2 + b_2 s + b_3}{s^3 + a_1 s^2 + a_2 s + a_3} \qquad \text{A.30}$$

Appendix B

Electronic Circuit Analysis by Computer

With the advent of a new generation of personal computers having such refinements as high-density, double-sided, double-density floppy disk storage, large amounts (≥ 256K) of RAM, arithmetic coprocessors, hard disk storage (≥ 20MB), and 16- and 32-bit CPU chips, we are witnessing the rapid development of engineering analysis and design software to run on these PCs. Formerly, one had to go to a large mainframe computer to run programs dealing with electronic circuit analysis, such as Berkeley SPICE or IBM's ECAP, originally developed to run on the System 360.

Modern PC-based electronic circuit analysis programs (ECAPs) are generally organized on a nodal basis. This means that the unknowns solved for by the PC using an admittance matrix are the node voltages in the circuit under analysis. In one modern program, circuit elements are entered by an interactive graphics process so that the computer stores a schematic diagram of the circuit under analysis as well as solves for the desired node voltages. Circuit elements handled by modern ECAPs include linear passive circuit elements (resistors, inductors, and capacitors) as well as nonlinear devices including pn junction diodes, switches, and large-signal models for BJTs and FETs. Transformers, transmission lines, and op-amps are often included in a program's modeling options and in its *parts library*. Independent sources are generally steady-state sinusoidal voltage or current generators (used for frequency response calculations), dc generators used for biasing and operating point studies, and transient sources. Most programs allow VCCS and VCVS dependent sources to be used, as well as CCCSs and CCVSs. In some programs, voltage sources always have a nonzero series. (Thevenin) resistance which must be specified; an ideal voltage source cannot be used. Similarly, a current source may require a nonzero Norton conductance to be specified.

ECAP analysis modes typically include dc (to examine quiescent biasing conditions), steady-state ac (for Bode plots), and transient (to study pulse rise times and propagation delays). Most ECAPs written for PCs have graphical output capabilities that allow graphical representation of output data to be made, including time-domain, transient responses and Bode magnitude and phase plots for the circuit under analysis.

B.1 When to Use ECAPs

When the time and effort required to analyze an electronic circuit with pencil, paper, and programmable pocket calculator exceeds the time and effort required to enter the same circuit's parameters into an ECAP and select the analysis mode and output format, then it is clearly time seriously to consider the ECAP as a steady electronic engineering partner. This crossover point in effort generally occurs when the circuit under analysis has three or more nodes that cannot be uncoupled by any of the circuit theorems or approximations described in this text.

Another consideration in choosing to do circuit analysis by computer is the ability of certain programs to provide high-resolution, report-ready graphics of the programs's calculations (e.g., Bode plots or transient response), with the PC's graphics printer.

The cost of ECAPs that run on PCs varies from the sublime (free) to the ridiculous (over $1000). At this time, we see a pleasant trend of ECAP prices converging on values comparable to those of electronics textbooks.

An ECAP is the only practical means of evaluating analog IC designs having many resistors, capacitors, transistors, diodes, and nodes. As an example, consider the NE/SE5539 video-frequency op-amp design. This op-amp has a gain-bandwidth product of 1.2 GHz, a slew rate of 600 V/μs, an $A_{v(ol)}$ of 50 dB, and a full-power bandwidth of 48 MHz. It is a direct-coupled IC with 11 npn BJTs, 20 resistors, and 22 nodes. Its analysis would provide an awesome task for pencil, paper, and calculator. Alternatively, one could build the circuit using discrete elements and bench-test it, a time-consuming and expensive approach. Clearly, an ECAP simulation of this circuit is the only valid approach if the IC is being evaluated as a new design.

Another use of ECAPs is in the verification and sensitivity analysis of analog active filter designs. Op-amps can be modeled quite precisely, and temperature effects can be included in most ECAPs by specifying component tempcos and circuit operating temperature.

B.2 Circuit Simulation by ECAPs

Resistors, inductors, and capacitors are typically modeled as ideal linear elements obeying Ohm's law. That is,

$$V = IR \tag{B.1}$$

$$V = L\frac{dI}{dt} \tag{B.2}$$

$$V = \frac{1}{C}\int I\,dt \tag{B.3}$$

Simulation of nonlinear elements such as diodes, BJTs, and FETs varies widely between programs, however, as does the representation of FETs, BJTs, and op-amps. Some of the simpler programs have no specific models for transistors and op-amps; the simulator must create them as small-signal models using resistors, inductors, capacitors, and appropriate controlled sources. The larger and more complex ECAPs include parts libraries in which both specific small-signal models and nonlinear models for transistors, op-amps, and sources are stored, and can be called up by number or name.

Table B.1 summarizes six currently available ECAPs written to run on PCs. These programs range widely in cost and capability. SPICE, which originated as a batch process language for large mainframe computers, was developed and released in 1975 by the University of California at Berkeley. The Berkeley SPICE source code is in the public domain, and several software companies have adapted it to run on PCs. SPICE has many nice features, but in its original form, it is unwieldy to edit.

Table B.1 Partial listing and evaluation of contemporary electronic circuit analysis programs (ECAPs) for PCs

Note: Evaluations are for several IBM PC models using MS-DOS 3.1 or 3.2. Abbreviations used in the evaluations are as follows:

C: Coprocessor (8087, 80287, or 80387) required. (C): Coprocessor recommended but not required.

CR $= x$: Cost range (Fall 1988), with $x = 0$ = free (public domain), $x = 1$ = <\$100, $x = 2$ = \$100 to \$250, $x = 3$ = \$250 to \$500, $x = 4$ = \$500 to \$1000, $x = 5$ = >\$1000.

D $= x$: Documentation effectiveness, with $x = 1$ = poorly written, hard to understand, $\ldots, x = 5$ = well written, easy to understand, good examples.

ED $= x$: Editing ease, with $x = 1$ = very difficult, batch-mode program, $\ldots, x = 5$ = easy editing, lots of paths from menu to menu, screen prompts, etc.

EU $= x$: Overall ease of use, including time required to become "moderately proficient," with $x = 1$ = hard, $\ldots, x = 5$ = easy.

FDD: Floppy disk drives.

FFDE: Free-format data entry allowed.

G $= x$: Graphics quality (screen and printer), with $x = 0$ = none, $x = 1$ = poor, $\ldots, x = 5$ = excellent.

GP: Graphics printer required.

HD: Hard disk required.

MC: Maximum number of single components allowed (approximate).

MD: Menu-driven.

MN: Maximum number of nodes allowed (approximate).

MOA: Maximum number of op-amp macros allowed.

MT: Maximum number of transistor macros allowed.

OS: Operating system.

(*continued*)

Table B.1 Continued

Micro-Cap II, v.4.00 (Professional Version), 1988
Spectrum Software, 1021 South Wolfe Road, Sunnyvale, CA 94086, (408) 738-4387

System Requirements: ≥ 512K RAM; 2 FDD or HD; DOS 2.0 or higher; HERC, CGA, or EGA graphics; GP.

Features: CR = 4; D = 4; ED = 5; EU = 5; FFDE; G = 5 (EGA and HERC); MD; MN = 100; MOA = 30; MT = 50. Professional version of Micro-Cap II in which circuit is entered in graphic form on up to 4 pages of CRT screens. Library uses sophisticated large-signal models of diodes, BJTs (Ebers-Moll model #2 with Early effect), MOSFETs (Schichmann-Hodges model used in SPICE), and op-amps. Also has polynomial dependent sources, transmission lines, transformers, programmable sources, independent sources, and controlled switches. Does dc biasing, transient response, Fourier analysis, frequency response (Bode plots show magnitude, phase, and group delay). No noise analysis. Runs in color.

Micro-Cap II, v.5.10 (Student Version), 1988
Spectrum Software, 1021 South Wolfe Road, Sunnyvale, CA 94086, (408) 738-4387

System Requirements: ≥ 512K RAM; ≥ 1 FDD or HD; MS-DOS 3.0 or higher; HERC, CGA, or EGA graphics; GP.

Features: CR = 1; D = 5; ED = 5; EU = 5; G = 5 (HERC); MD; MN = 25; MT = 15; MOA = 5. Student version has limited library. *Manual for Student Micro-Cap II,* by M. S. Roden (Addison-Wesley and Benjamin/Cummings, Reading, MA, 1989) aids beginning students. Program runs in color.

IS-SPICE, v.1.0, 1987
Intusoft, P.O. Box 6607, San Pedro, CA 90734-6607, (213) 833-0710

System Requirements: 640K RAM; C; HD, DOS 2.0 or higher. Program requires about 512K RAM.

Features: Berkeley SPICE program adapted to run on PCs. CR = 1; D = 3; ED = 4; EU = 2; FFDE; G = 5; MD; MN = 200 (50 given as a practical limit based on run time). Uses Gummel-Poon BJT model with 40 parameters, a 42-parameter nonlinear MOSFET model, a 12-parameter large-signal JFET model, and a 14-parameter diode model. Intusoft offers accessory programs for use with IS-SPICE. SPICE-NET, v.1.0, is a schematic entry program that makes netlists for SPICE. INTU-SCOPE is the graphics output and waveform-processing module. PRE-SPICE is a program that manages the IS-SPICE and INTU-SCOPE programs; it contains a full-screen editor and facilities to run Monte-Carlo analysis and library files for devices. Not a program for beginners.

P-SPICE, v.2.03 (Classroom Version), 1985

MicroSim Corp., 23175 La Cadena Drive, Laguna Hills, CA 92653, (714) 770-3022

System Requirements: $\geq$512K RAM; C; 2 FDD or HD; DOS 2.0 or higher; EGA graphics; GP.

Features: CR = 0; D = 4; ED = 1; EU = 1; G = 5; MT = 10; MN = ?; MOA = 2? Classroom version of P-SPICE. Has an abbreviated library of components: 2N2222 and 2N2907 BJTs; IN752, IN754, and IN759 zeners; IN914 and IN916 diodes; IRF150 and IRF9130 MOSFETs; LM111 comparator; LM741 op-amp. Includes a graphics output program, PROBE. Does ac frequency response, transient response, dc operating point, dc sensitivity analysis, noise bandwidth calculations, and group delay. No distortion analysis with classroom version. Runs batch mode. Not for beginners.

ECA v.2.30, 1987

Tatum Labs, 1478 Mark Twain Court, Ann Arbor, MI 48103-9709, (313) 663-8810

System Requirements: $\geq$512K RAM, (C); DOS 2.0 or higher; HERC, CGA, or EGA graphics; GP.

Features: CR = 4; D = 4; ED = 4; FFDE; G = 3, MN = 500. Fully interactive ECAP with editing. Does ac frequency response (only with linear small-signal models for transistors); dc biasing, transient response, and Fourier analysis done with nonlinear large-signal transistor models. Uses Weldner and Davis large-signal model for JFETs and MOSFETs. Ebers-Moll based BJT large-signal models used as well as *h*-parameter or hybrid-pi small-signal models. Requires netlist for data entry. Approaches P-SPICE in its capabilities. A student version, ECA-2S, is available in CR = 1.

Some suppliers of SPICE programs offer interface or preprocessor program modules to facilitate editing. The current cost of P-SPICE and its excellent graphics postprocessor, PROBE, is about $1400. This system is clearly overpriced when compared to other, comparable ECAPs now on the market that have excellent editing capabilities. Fortunately for students (and professors) studying analog electronic circuits, in the fall of 1986, MicroSim Corp. released an abridged, public domain (free) version of P-SPICE and PROBE. This version is the P-SPICE ECAP described in Table B.1.

Other software companies offer versions of SPICE that have interactive editing capabilities. I-G SPICE from AB Associates, Tampa, FL, is a full-featured ECAP that also has a schematic capture package; at present, it runs only on mainframe machines. Intusoft, San Pedro, CA, offers Intu-Scope, an interactive postprocessor for SPICE; it also offers a version of SPICE and an editing preprocessor called PRE-SPICE, all of which run on PCs.

The problem with presenting an up-to-date summary of PC ECAPs is that there is intense development in this area at present. Prices are dropping, and software companies are competing with one another for the student and industrial markets. Although Micro-Cap II was the only PC ECAP with schematic circuit entry, we expect to see this feature on other ECAPs in the near future. Thus the material in Table B.1 probably has a short practical half-life. Nevertheless, the reader should be aware of what is offered in the PC ECAP marketplace.

B.3 Example of Electronic Circuit Analysis Using Micro-Cap II (Student Version) on a PC

We will examine a circuit that has been analyzed in the time and frequency domains using the student version (v.5.10) of Spectrum Software's Micro-Cap II. The circuit is a low-noise, JFET/op-amp, video-frequency amplifier, shown in Fig. B.1. A 2N4416 low-noise, radio-frequency JFET is used as a headstage for a high-frequency, low-noise op-amp (AD-380). The op-amp is operated as a current-to-voltage converter. Its summing junction node is very nearly at small-signal ground potential, giving no significant Miller effect for the JFET, and hence wide bandwidth.

As a first step in the analysis, we must create an AD-380 op-amp model in the op-amp section of the program's parts library. Reference to the op-amp's data sheet gives us parameter values relevant to the system's pulse and frequency responses:

DC GAIN	40,000
FIRST POLE	1000 Hz
SECOND POLE	100 MHz
R_{IN}	$10^{11}\ \Omega$
R_{OUT}	$100\ \Omega$
SLEW RATE	3.3×10^8 V/s

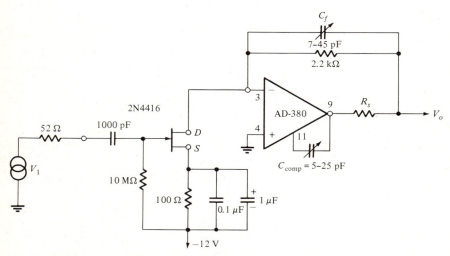

Figure B.1 JFET/AD-380 op-amp low-noise amplifier schematic.

There is no library model for JFETs in this program. Micro-Cap II does support large-signal, high-frequency models for MOSFETs and BJTs, however. Here we have chosen to represent the JFET by a high-frequency small-signal model by using a polynomial source from the program's library. Polynomial sources are versatile controlled sources in which the output can be either a voltage or a current. The input to the polynomial source, X, can be a voltage difference, a voltage sum, a product of voltages, a quotient of voltages, or a current. The output, Y, is given by

$$Y = A + BX^C + DX^E + FX^G \qquad\qquad \text{B.4}$$

where

$Y =$ either voltage or current

$X =$ the input (e.g., voltage difference)

$A \cdots G =$ constants specified by the user

Thus the polynomial source, G_1, models the VCCS in the JFET's small-signal model. In G_1, $B = -0.003$ S is the FET's transconductance at its operating point (the minus sign is used because the $g_m v_{gs}$ current source is directed toward the JFET's source terminal). The exponent $C = 1$ and the other constants are zero.

The schematic for the amplifier is drawn directly on the terminal's CRT using simple commands and the cursor keys. Figure B.2 illustrates a screen dump of the schematic to a graphics printer. The nodes are numbered automatically by the program, and no netlist is required. Dotted lines do not connect to solid or dotted lines that they cross.

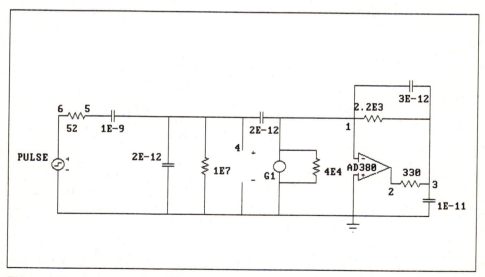

Figure B.2   Amplifier schematic as entered into Micro-Cap II. The effect of C_{comp} is included in the library model of the op-amp.

Table B.2   Analysis limits

Lowest frequency	1E6
Highest frequency	1E8
Lowest gain (db)	−10
Highest gain (db)	30
Lowest phase shift	−360
Highest phase shift	0
Lowest group delay	1E-9
Highest group delay	1E-6
Input node number	5
Output node number	3
Minimum accuracy (%)	5
Auto or Fixed frequency step (A, F)	A
Temperature (Low/High/Step)	27
Number of cases	1
Output: Disk, Printer, None (D, P, N)	N
Save, Retrieve, Normal run (S, R, N)	N

Micro-Cap II has four analysis modes: (1) transient (time domain), (2) ac analysis (giving Bode plots of amplitude, phase, and group delay vs. frequency), (3) dc analysis (for biasing nonlinear circuits), and (4) Fourier analysis. Table B.2 shows the ac analysis limits used, and Fig. B.3 illustrates a screen dump from the ac analysis; the solid line is the circuit's magnitude response, the open squares represent the phase,

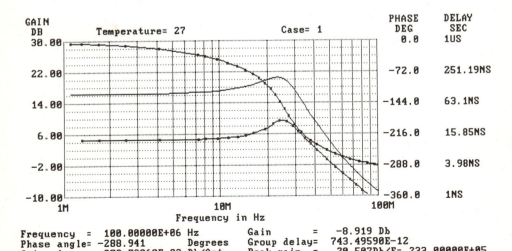

```
Frequency   =   100.00000E+06 Hz        Gain      =    −8.919 Db
Phase angle = −288.941        Degrees    Group delay=  743.49590E-12
Gain slope  = −973.72060E-02 Db/Oct      Peak gain =    20.507Db/F= 233.00000E+05
```

Figure B.3   Frequency response analysis data dump.

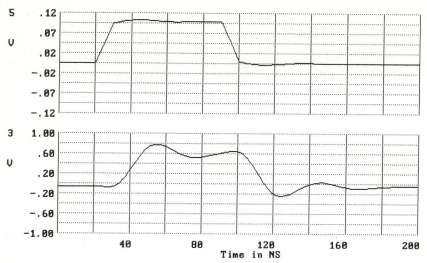

Figure B.4   Pulse response of the amplifier from the graphics printer.

and the solid squares give the group delay. The -3 dB frequency is about 35 MHz, and f_T is about 60 MHz.

The time domain response of the circuit is illustrated in Fig. B.4. The trapezoidal input pulse is measured at node 5 and is shown above the node 3 output voltage. Some over- and undershoot are observed. Pulse and frequency response performance of this amplifier can be fine-tuned by manipulation of the op-amp's poles (with C_{comp}), and by adjustment of the series resistor, R_s, and the feedback capacitor, C_f. The schematic and analysis data can be saved on the program disk.

In summary, the student version of Micro-Cap II is versatile, inexpensive, easy to use, and has excellent graphics outputs. It will run on an IBM PC or equivalent system with one disk drive and 512K RAM. Schematics large enough to fill four linked screens can be analyzed, and up to 25 nodes, 15 BJTs, 15 MOSFETs, and five op-amps can be handled. BJT large-signal models are based on the Ebers-Moll model; MOSFETs are simulated using the Schichmann-Hodges level-1 model used in SPICE. There are no specific JFET models in either the student or professional versions of Micro-Cap II. To simulate an n-channel JFET, for example, one must use an n-channel depletion MOSFET with the appropriate constants and suitable small-signal diodes connnected from the gate to the drain and source nodes.

Appendix C

Two-Port Models for Linear Active Electronic Systems

Most amplifiers can be characterized by circuits having two terminal pairs (input and output) or ports. An exception is the differential amplifier (DA), which has three ports (two input and one output). However, the two input signals can be resolved into common-mode and difference-mode input components which appear at a single, equivalent input port (see Chapter 5). In this appendix, we review various means of describing two-port circuits. The analyses we use assume circuit linearity and hence can be performed on circuits containing small-signal models of diodes and transistors.

The general two-port system is shown in Fig. C.1(A). Figure C.1(B) illustrates the four parameters used to describe the y-parameter model. In this model, the input, v_1, will affect the output current, i_2, and the output voltage, v_2, will affect the input current, i_1. Thus we can write the y-parameter node equations:

$$i_1 = y_{11}v_1 + y_{12}v_2 \qquad\qquad\qquad\qquad\qquad \text{C.1}$$

$$i_2 = y_{21}v_1 + y_{22}v_2 \qquad\qquad\qquad\qquad\qquad \text{C.2}$$

Equations C.1 and C.2 can be put into standard matrix form:

$$\begin{bmatrix} i_1 \\ i_2 \end{bmatrix} = \begin{bmatrix} y_{11} & y_{12} \\ y_{21} & y_{22} \end{bmatrix} \begin{bmatrix} v_1 \\ v_2 \end{bmatrix} \qquad\qquad\qquad \text{C.3}$$

Here v_1 and v_2 are the independent variables, i_1 and i_2 are the dependent variables, and y_{ik} is called the short-circuit admittance matrix. From consideration of Eqs. C.1 and C.2, we can write

$$y_{11} = \left.\frac{i_1}{v_1}\right|_{v_2=0} \qquad \text{(short-circuit input admittance)} \qquad \text{C.4}$$

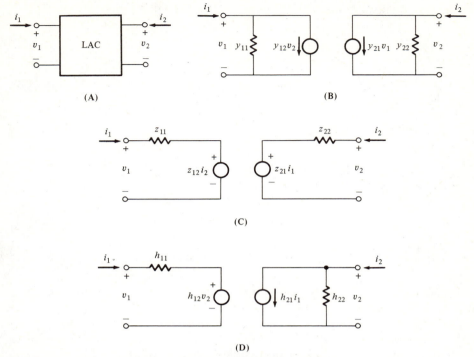

Figure C.1 **(A)** Two-port linear active circuit (LAC). **(B)** Short-circuit, y-parameter characterization of a two-port LAC. **(C)** Open-circuit, z-parameter characterization of a two-port LAC. **(D)** Thevenin input, Norton output, h-parameter characterization of a two-port LAC.

$$y_{12} = \left.\frac{i_1}{v_2}\right|_{v_1=0} \qquad \text{(short-circuit reverse transfer admittance)} \qquad \text{C.5}$$

$$y_{21} = \left.\frac{i_2}{v_1}\right|_{v_2=0} \qquad \text{(short-circuit forward transfer admittance)} \qquad \text{C.6}$$

$$y_{22} = \left.\frac{i_2}{v_2}\right|_{v_1=0} \qquad \text{(short-circuit output admittance)} \qquad \text{C.7}$$

The short-circuit, y-parameter characterization of a linear two-port circuit, shown in Fig. C.1(B), is seen to consist of two Norton equivalent circuits.

If i_1 and i_2 are now treated as the independent variables, we can express the behavior of the linear two-port circuit in terms of open-circuit z-parameters:

$$v_1 = z_{11}i_1 + z_{12}i_2 \qquad \text{C.8}$$

$$v_2 = z_{21}i_1 + z_{22}i_2 \qquad \text{C.9}$$

These open-circuit, z-parameter equations can be put in matrix form:

$$\begin{bmatrix} v_1 \\ v_2 \end{bmatrix} = \begin{bmatrix} z_{11} & z_{12} \\ z_{21} & z_{22} \end{bmatrix} \begin{bmatrix} i_1 \\ i_2 \end{bmatrix} \qquad \text{C.10}$$

From consideration of Eqs. C.8 and C.9, we can define the open-circuit impedance matrix elements, z_{ik}:

$$z_{11} = \frac{v_1}{i_1}\bigg|_{i_2 = 0} \qquad \text{(open-circuit input impedance)} \qquad \text{C.11}$$

$$z_{12} = \frac{v_1}{i_2}\bigg|_{i_1 = 0} \qquad \text{(open-circuit reverse transfer impedance)} \qquad \text{C.12}$$

$$z_{21} = \frac{v_2}{i_1}\bigg|_{i_2 = 0} \qquad \text{(open-circuit forward transfer impedance)} \qquad \text{C.13}$$

$$z_{22} = \frac{v_2}{i_2}\bigg|_{i_1 = 0} \qquad \text{(open-circuit output impedance)} \qquad \text{C.14}$$

The open-circuit impedance characterization of the linear two-port circuit leads to a Thevenin representation, shown in Fig. C.1(C).

A third important model for linear two-port circuit behavior is given by the hybrid- (or h-) parameter characterization. h-parameters are used extensively to provide mid-frequency small-signal model descriptions for BJTs. The Thevenin input, Norton output h-model is illustrated in Fig. C.1(D). (A Norton input, Thevenin output h-model is also possible but will not be treated here.) The h-parameter equations are

$$v_1 = h_{11}i_1 + h_{12}v_2 \qquad \text{C.15}$$
$$i_2 = h_{21}i_1 + h_{22}v_2 \qquad \text{C.16}$$

The h-matrix coefficients are thus

$$h_{11} = \frac{v_1}{i_1}\bigg|_{v_2 = 0} \qquad \text{(short-circuit input resistance)} \qquad \text{C.17}$$

$$h_{12} = \frac{v_1}{v_2}\bigg|_{i_1 = 0} \qquad \text{(open-circuit reverse transfer voltage gain)} \qquad \text{C.18}$$

$$h_{21} = \frac{i_2}{i_1}\bigg|_{v_2 = 0} \qquad \text{(short-circuit forward current gain)} \qquad \text{C.19}$$

$$h_{22} = \frac{i_2}{v_2}\bigg|_{i_1 = 0} \qquad \text{(open-circuit output admittance)} \qquad \text{C.20}$$

A fourth linear two-port circuit which is often used to model BJT behavior is the hybrid-pi model, shown in Fig. C.2(A). It is composed of three passive conductances and a VCCS. The values of these four parameters can easily be expressed

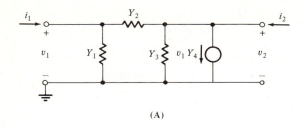

(A)

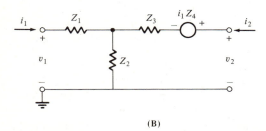

(B)

Figure C.2   **(A)** Hybrid-pi two-port circuit. **(B)** T-equivalent two-port circuit. Note that Z_3 and $i_1 Z_4$ can be made into a Norton equivalent circuit.

in terms of the y-parameter two-port model characterized by Eqs. C.1 and C.2 and Fig. C.1(B). Inspection of the hybrid-pi circuit shows that the conductance of the left arm of the "pi" appears in parallel to ground with the conductance at the top of the pi when the output is shorted to ground ($v_2 = 0$). Thus the equivalent input conductance must be

$$y_{11} = Y_1 + Y_2 = (y_{11} + y_{12}) + (-y_{12}) \qquad\qquad \text{C.21}$$

Similarly, when $v_1 = 0$,

$$y_{22} = Y_3 + Y_2 = (y_{22} + y_{12}) + (-y_{12}) \qquad\qquad \text{C.22}$$

is seen looking into the output port. Now when a voltage source v_2 is applied to the output port, a short-circuit input current, i_1, flows; i_1 is given by

$$i_1 = -v_2(-y_{12}) \qquad\qquad \text{C.23}$$

Hence the reverse transfer admittance of the hybrid-pi two-port is just y_{12}. In the same manner, the forward transfer admittance of the hybrid-pi circuit is easily seen to be

$$y_{12} = \frac{i_2}{v_1} = -(-y_{12}) + (y_{21} - y_{12}) \qquad\qquad \text{C.24}$$

Figure C.2(B) illustrates a T-equivalent two-port model. It is left as an exercise to demonstrate that the T circuit's impedances can be written in terms of the impedances of the z-parameter, two-port model of Fig. C.1(C).

Still another format for the characterization of two-port circuit behavior is the ABCD, transmission, or chain parameter model, defined by the equations

$$v_1 = Av_2 + B(-i_2) \tag{C.25}$$

$$i_1 = Cv_2 + D(-i_2) \tag{C.26}$$

A second chain parameter format is given by

$$v_2 = \mathscr{A}v_1 + \mathscr{B}(-i_1) \tag{C.27}$$

$$i_2 = \mathscr{C}v_1 + \mathscr{D}(-i_1) \tag{C.28}$$

The two chain parameter formats are useful in describing cascaded two-ports, such as those found in modular IC amplification systems, or in systems with feedback. Figure C.3 illustrates two cascaded two-ports. To find the forward transmission properties of these two-ports, we use the script chain parameter equations C.24 and C.25. For the two-port on the left in Fig. C.3, we have

$$\begin{bmatrix} v_2 \\ i_2 \end{bmatrix} = \begin{bmatrix} \mathscr{A} & \mathscr{B} \\ \mathscr{C} & \mathscr{D} \end{bmatrix} \begin{bmatrix} v_1 \\ -i_1 \end{bmatrix} \tag{C.29}$$

For the two-port on the right, we have

$$\begin{bmatrix} v_2' \\ i_2' \end{bmatrix} = \begin{bmatrix} \mathscr{A}' & \mathscr{B}' \\ \mathscr{C}' & \mathscr{D}' \end{bmatrix} \begin{bmatrix} v_1' \\ -i_1' \end{bmatrix} \tag{C.30}$$

Now it is easy to see that $v_1' = v_2$ and $i_1' = -i_2$. Hence Eqs. C.26 and C.27 can be cascaded:

$$\begin{bmatrix} v_2' \\ i_2' \end{bmatrix} = \begin{bmatrix} \mathscr{A}' & \mathscr{B}' \\ \mathscr{C}' & \mathscr{D}' \end{bmatrix} \begin{bmatrix} \mathscr{A} & \mathscr{B} \\ \mathscr{C} & \mathscr{D} \end{bmatrix} \begin{bmatrix} v_1 \\ -i_1 \end{bmatrix} \tag{C.31}$$

or

$$\begin{bmatrix} v_2' \\ i_2' \end{bmatrix} = \begin{bmatrix} \mathscr{A}'' & \mathscr{B}'' \\ \mathscr{C}'' & \mathscr{D}'' \end{bmatrix} \begin{bmatrix} v_1 \\ -i_1 \end{bmatrix} \tag{C.32}$$

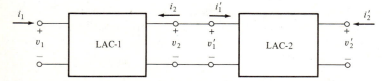

Figure C.3   Two cascaded two-port LACs, describable in terms of ABCD parameters.

where

$$\mathscr{A}'' = \mathscr{A}\mathscr{A}' + \mathscr{B}'\mathscr{C} \qquad\qquad\qquad \text{C.33}$$

$$\mathscr{B}'' = \mathscr{A}'\mathscr{B} + \mathscr{B}'\mathscr{D} \qquad\qquad\qquad \text{C.34}$$

$$\mathscr{C}'' = \mathscr{C}'\mathscr{A} + \mathscr{D}'\mathscr{C} \qquad\qquad\qquad \text{C.35}$$

$$\mathscr{D}'' = \mathscr{C}'\mathscr{B} + \mathscr{D}'\mathscr{D} \qquad\qquad\qquad \text{C.36}$$

Thus the output voltage and current are given in terms of the input voltage and current:

$$v_2' = \mathscr{A}''v_1 + \mathscr{B}''(-i_1) \qquad\qquad\qquad \text{C.37}$$

$$i_2' = \mathscr{C}''v_1 + \mathscr{D}''(-i_1) \qquad\qquad\qquad \text{C.38}$$

If the cascaded two-ports are driven by a Thevenin or Norton circuit at (v_1, i_1), then a relation between v_1, v_s, and i_1 is established. It is then possible to develop an expression for v_2'/v_s in terms of R_s, $\mathscr{A}''$, $\mathscr{B}''$, $\mathscr{C}''$, and $\mathscr{D}''$.

The reverse characteristics of the two cascaded two-ports of Fig. C.3 can be described using the ABCD chain parameter equations.

$$\begin{bmatrix} v_1 \\ i_1 \end{bmatrix} = \begin{bmatrix} A & B \\ C & D \end{bmatrix} \begin{bmatrix} v_2 \\ -i_2 \end{bmatrix} \qquad\qquad \text{C.39}$$

Also,

$$\begin{bmatrix} v_1' \\ i_1' \end{bmatrix} = \begin{bmatrix} A' & B' \\ C' & D' \end{bmatrix} \begin{bmatrix} v_2' \\ -i_2' \end{bmatrix} \qquad\qquad \text{C.40}$$

However, we note that $v_1' = v_2$ and $i_1' = -i_2$, which leads to

$$\begin{bmatrix} v_1 \\ i_1 \end{bmatrix} = \begin{bmatrix} A & B \\ C & D \end{bmatrix} \begin{bmatrix} A' & B' \\ C' & D' \end{bmatrix} \begin{bmatrix} v_2' \\ -i_2' \end{bmatrix} \qquad\qquad \text{C.41}$$

This matrix can be reduced to

$$v_1 = A''v_2' + B''(-i_2') \qquad\qquad\qquad \text{C.42}$$

$$i_1 = C''v_2' + D''(-i_2') \qquad\qquad\qquad \text{C.43}$$

where

$$A'' = AA' + BC' \qquad\qquad\qquad \text{C.44}$$

$$B'' = AB' + BD' \qquad\qquad\qquad \text{C.45}$$

$$C'' = CA' + DC' \qquad\qquad\qquad \text{C.46}$$

$$D'' = CB' + DD' \qquad\qquad\qquad \text{C.47}$$

If the right-hand two-port is terminated in a load resistor, R_L, then $v_2' = -i_2 R_L$, and analysis can proceed using Eqs. C.42 through C.47.

Obviously, the many model formats used in the analysis of two-ports can lead to confusion and chaos unless one uses rules to govern the choice of format. We have seen that there may be merit in using chain parameters when considering cascaded two-ports. h-parameters are widely used in the analysis of low- and mid-frequency BJT circuits; hybrid-pi, BJT small-signal models are used at high frequencies; and the z- and y-parameter models may be useful in other cases. For ease in using the various two-port models, a listing of matrix element equivalences is given in Table C.1. Note that in this table, corresponding matrix elements in a given row of matrices are equal. For example,

$$y_{11} = \frac{z_{22}}{\Delta_z} = \frac{1}{h_{11}} = \frac{D}{B} = \frac{\mathscr{A}}{\mathscr{B}} \qquad\qquad \text{C.48}$$

As an illustration of the use of Table C.1, it is useful to express the y-parameters in the hybrid-pi two-port model of Fig. C.2(A) in terms of the h-parameters given by Eqs. C.17 through C.20. From these equivalences, we see that the left-hand admittance of the hybrid-pi model can be written as

$$Y_1 = y_{11} + y_{12} = \frac{1 + h_{12}}{h_{11}} \qquad\qquad \text{C.49}$$

The top admittance is

$$Y_2 = -y_{12} = \frac{h_{12}}{h_{11}} \qquad\qquad \text{C.50}$$

The right-hand admittance can be written as

$$Y_3 = y_{22} + y_{12} = h_{22} - \frac{h_{12}(1 + h_{21})}{h_{11}} \qquad\qquad \text{C.51}$$

The VCCS of the hybrid-pi model is found to be

$$Y_4 = g_m = y_{21} - y_{12} = \frac{h_{21} + h_{12}}{h_{11}} \qquad\qquad \text{C.52}$$

See Chapter 1 for a more detailed representation of hybrid-pi parameters in terms of BJT common-emitter h-parameters.

When a single linear two-port network is driven by a Norton or Thevenin source and is terminated in a load resistance, R_L, manipulation of the defining equations leads to expressions for the driving-point impedance, Z_{in}, looking into the (v_1, i_1) port; the output impedance, Z_{out}, seen by R_L; the voltage ratio, v_2/v_1; and the current ratio, i_2/i_1. These parameters are summarized for the various model formats in Table C.2.

Table C.1   Two-port matrix conversions

	$[y_{jk}]$	$[z_{jk}]$	$[h_{jk}]$	$\begin{bmatrix} A & B \\ C & D \end{bmatrix}$	$\begin{bmatrix} \mathscr{A} & \mathscr{B} \\ \mathscr{C} & \mathscr{D} \end{bmatrix}$
$[y_{jk}]$	$\begin{matrix} y_{11} & y_{12} \\ y_{21} & y_{22} \end{matrix}$	$\begin{matrix} \dfrac{z_{22}}{\Delta_z} & \dfrac{-z_{12}}{\Delta_z} \\[6pt] \dfrac{-z_{21}}{\Delta_z} & \dfrac{-z_{11}}{\Delta_z} \end{matrix}$	$\begin{matrix} \dfrac{1}{h_{11}} & \dfrac{-h_{12}}{h_{11}} \\[6pt] \dfrac{h_{21}}{h_{11}} & \dfrac{\Delta_h}{h_{11}} \end{matrix}$	$\begin{matrix} \dfrac{D}{B} & -\dfrac{\Delta_A}{B} \\[6pt] \dfrac{-1}{B} & \dfrac{A}{B} \end{matrix}$	$\begin{matrix} \dfrac{\mathscr{A}}{\mathscr{B}} & -\dfrac{1}{\mathscr{B}} \\[6pt] \dfrac{-\Delta_{\mathscr{A}}}{\mathscr{B}} & \dfrac{\mathscr{D}}{\mathscr{B}} \end{matrix}$
$[z_{jk}]$	$\begin{matrix} \dfrac{y_{22}}{\Delta_y} & \dfrac{-y_{12}}{\Delta_y} \\[6pt] \dfrac{-y_{21}}{\Delta_y} & \dfrac{y_{11}}{\Delta_y} \end{matrix}$	$\begin{matrix} z_{11} & z_{12} \\ z_{21} & z_{22} \end{matrix}$	$\begin{matrix} \dfrac{\Delta h}{h_{22}} & \dfrac{h_{12}}{h_{22}} \\[6pt] \dfrac{-h_{21}}{h_{22}} & \dfrac{1}{h_{22}} \end{matrix}$	$\begin{matrix} \dfrac{A}{C} & \dfrac{\Delta_A}{C} \\[6pt] \dfrac{1}{C} & \dfrac{D}{C} \end{matrix}$	$\begin{matrix} \dfrac{\mathscr{D}}{\mathscr{C}} & \dfrac{1}{\mathscr{C}} \\[6pt] \dfrac{\Delta_{\mathscr{A}}}{\mathscr{C}} & \dfrac{\mathscr{A}}{\mathscr{C}} \end{matrix}$
$[h_{jk}]$	$\begin{matrix} \dfrac{1}{y_{11}} & \dfrac{-y_{12}}{y_{11}} \\[6pt] \dfrac{y_{21}}{y_{11}} & \dfrac{\Delta_y}{y_{11}} \end{matrix}$	$\begin{matrix} \dfrac{\Delta_z}{z_{22}} & \dfrac{z_{12}}{z_{22}} \\[6pt] \dfrac{-z_{21}}{z_{22}} & \dfrac{1}{z_{22}} \end{matrix}$	$\begin{matrix} h_{11} & h_{12} \\ h_{21} & h_{22} \end{matrix}$	$\begin{matrix} \dfrac{B}{D} & \dfrac{\Delta_A}{D} \\[6pt] \dfrac{-1}{D} & \dfrac{C}{D} \end{matrix}$	$\begin{matrix} \dfrac{\mathscr{B}}{\mathscr{A}} & \dfrac{1}{\mathscr{A}} \\[6pt] \dfrac{-\Delta_{\mathscr{A}}}{\mathscr{A}} & \dfrac{\mathscr{C}}{\mathscr{A}} \end{matrix}$
$\begin{bmatrix} A & B \\ C & D \end{bmatrix}$	$\begin{matrix} \dfrac{-y_{22}}{y_2} & \dfrac{-1}{y_{21}} \\[6pt] \dfrac{-\Delta_y}{y_{21}} & \dfrac{-y_{11}}{y_{21}} \end{matrix}$	$\begin{matrix} \dfrac{z_{11}}{z_{21}} & \dfrac{\Delta_z}{z_{21}} \\[6pt] \dfrac{1}{z_{21}} & \dfrac{z_{22}}{z_{21}} \end{matrix}$	$\begin{matrix} \dfrac{-\Delta_h}{h_{21}} & \dfrac{-h_{11}}{h_{21}} \\[6pt] \dfrac{-h_{22}}{h_{21}} & \dfrac{-1}{h_{21}} \end{matrix}$	$\begin{matrix} A & B \\ C & D \end{matrix}$	$\begin{matrix} \dfrac{\mathscr{D}}{\Delta_{\mathscr{A}}} & \dfrac{\mathscr{B}}{\Delta_{\mathscr{A}}} \\[6pt] \dfrac{\mathscr{C}}{\Delta_{\mathscr{A}}} & \dfrac{\mathscr{A}}{\Delta_{\mathscr{A}}} \end{matrix}$
$\begin{bmatrix} \mathscr{A} & \mathscr{B} \\ \mathscr{C} & \mathscr{D} \end{bmatrix}$	$\begin{matrix} \dfrac{-y_{11}}{y_{12}} & \dfrac{-1}{y_{12}} \\[6pt] \dfrac{-\Delta_y}{y_{12}} & \dfrac{-y_{22}}{y_{12}} \end{matrix}$	$\begin{matrix} \dfrac{z_{22}}{z_{12}} & \dfrac{\Delta_z}{z_{12}} \\[6pt] \dfrac{1}{z_{12}} & \dfrac{z_{11}}{z_{12}} \end{matrix}$	$\begin{matrix} \dfrac{1}{h_{12}} & \dfrac{h_{11}}{h_{12}} \\[6pt] \dfrac{h_{22}}{h_{12}} & \dfrac{\Delta_h}{h_{12}} \end{matrix}$	$\begin{matrix} \dfrac{D}{\Delta_A} & \dfrac{B}{\Delta_A} \\[6pt] \dfrac{C}{\Delta_A} & \dfrac{A}{\Delta_A} \end{matrix}$	$\begin{matrix} \mathscr{A} & \mathscr{B} \\ \mathscr{C} & \mathscr{D} \end{matrix}$

Table C.2   Two-port network input/output relationships

	$[y_{ik}]$	$[z_{ik}]$	$[h_{ik}]$	$\begin{bmatrix} A & B \\ C & D \end{bmatrix}$	$\begin{bmatrix} \mathscr{A} & \mathscr{B} \\ \mathscr{C} & \mathscr{D} \end{bmatrix}$
Z_{in}	$\dfrac{y_{22} + Y_L}{\Delta_y + y_{11}Y_L}$	$\dfrac{\Delta_z + z_{11}Z_L}{Z_L + z_{22}}$	$\dfrac{\Delta_h + h_{11}Y_L}{h_{22} + Y_L}$	$\dfrac{B + AZ_L}{D + CZ_L}$	$\dfrac{\mathscr{B} + \mathscr{D}Z_L}{\mathscr{A} + \mathscr{C}Z_L}$
Z_{out}	$\dfrac{y_{11} + Y_S}{\Delta_y + y_{22}Y_S}$	$\dfrac{\Delta z + z_{22}Z_S}{z_{11} + Z_S}$	$\dfrac{h_{11} + Z_S}{\Delta_h + h_{22}Z_S}$	$\dfrac{B + DZ_S}{A + CZ_S}$	$\dfrac{\mathscr{B} + \mathscr{A}Z_S}{\mathscr{D} + \mathscr{C}Z_S}$
$\dfrac{v_2}{v_1}$	$\dfrac{-y_{21}}{y_{22} + Y_L}$	$\dfrac{-z_{21}Z_L}{\Delta_z + z_{11}Z_L}$	$\dfrac{-h_{21}}{\Delta_h + h_{11}Y_L}$	$\dfrac{Z_L}{B + AZ_L}$	$\dfrac{\Delta_{\mathscr{A}}}{\mathscr{D} + \mathscr{B}Y_L}$
$\dfrac{i_2}{i_1}$	$\dfrac{y_{21}Y_L}{\Delta_y + y_{11}Y_L}$	$\dfrac{-z_{21}}{z_{22} + Z_L}$	$\dfrac{h_{21}Y_L}{h_{22} + Y_L}$	$\dfrac{-1}{D + CZ_L}$	$\dfrac{-\Delta_{\mathscr{A}}}{\mathscr{A} + \mathscr{C}Z_L}$

Note that the overall two-port voltage gain can be expressed in terms of the driving-point impedance, Z_{in}, and the voltage ratio, v_2/v_1:

$$K_v = \frac{v_2}{v_S} = \left(\frac{Z_{in}}{Z_1 + Z_S} \right) \left(\frac{v_2}{v_1} \right)$$ C.53

For example, for designated y-parameters,

$$K_v = \frac{v_2}{v_s} = \left(\frac{\dfrac{y_{22} + Y_L}{\Delta_y + y_{11} Y_L}}{\dfrac{y_{22} + Y_L}{\Delta_y + y_{11} Y_L} + Z_s} \right) \left(\frac{-y_{21}}{y_{22} + Y_L} \right)$$ C.54

This expression can be reduced to

$$K_v = \frac{v_2}{v_s} = \frac{-y_{21}}{y_{22} + Y_L + Z_s(\Delta_y + y_{11} Y_L)}$$ C.55

for the terminated y-parameter model of a two-port driven by a Thevenin source with open-circuit voltage v_s and $R_{eq} = Z_s$. If $Y_L = 0$, we have the open-circuit (output) voltage gain.

Overall current gain can be written as

$$K_i = \frac{i_L}{i_s} = \frac{i_L}{v_s/Z_s} = \left(\frac{Z_s}{Z_s + Z_{in}} \right) \left(\frac{-i_2}{i_1} \right)$$ C.56

where

Z_{in} = the input impedance

$-i_2/i_1$ = the negative current ratio for the two-port

The two-port parameter power gain, K_p, can be calculated from the definition

$$K_p = \frac{i_L v_L}{i_1 v_1}$$ C.57

for a terminated two-port. It is easy to see that K_p can be expressed as the product of the voltage ratio (v_2/v_1) and the negative current ratio ($-i_2/i_1$).

To summarize, we have seen that various two-port models can be used to describe the behavior of linear circuits. Mid-frequency small-signal models of BJTs and FETs are two-port circuits, and thus a variety of two-port circuit models can be used to characterize transistor small-signal behavior. The most commonly used two-port models for BJTs are the common-emitter and common-base h-parameter system, and the hybrid-pi model. The latter model is widely used at high frequencies.

Appendix D

Answers to Selected Problems

Chapter 1: **(1.1)** $R_B = 1.643$ MΩ. $R_C = 5.660 \times 10^3$ Ω. **(1.3)** $I_{CQ} = 7 \times 10^{-4}$ A, $I_{BQ} = 1.010 \times 10^{-5}$ A, $V_{EQ} = 0.7071$ V, $V_{CEQ} = 7.293$ V, $R_1 = 95.55$ kΩ. **(1.5)** $R_D = 2.871$ kΩ. **(1.7)** $R_S = 436.7$, $I_{DQ} = 1.852$ mA, $V_{GSQ} = 0.639$ V. **(1.10)** $h_{21} = h_{ie}/(h_{ie} + R_B)$, $h_{22} = G_C + (1 + h_{fe})/(h_{ie} + R_B)$, $h_{11} = R_B h_{1e}$, $h_{12} = (h_{fe}R_B - h_{ie})/(h_{ie} + R_B)$. **(1.11)** $K_v = 2R_L(1 + h_{fe})/[h_{ie} + 2R_L(1 + h_{fe})]$. **(1.12)** (a) $h_{11} = h_{ie1}$, $h_{21} = 1$, $h_{22} = h_{oe2}$, $h_{12} = -(1 + h_{fe1} + h_{fe1}h_{fe2})$; (b) $Kv = 0.9821$, $R_{\text{out}} = 1.784$ (low). **(1.13)** (a) $h_{11} = h_{ie1}$, $h_{12} = +h_{fe1}(1 + h_{fe2})$, $h_{22} = h_{oe2}$, $h_{21} = 0$; (b) $K_v = h_{fe1}(1 + h_{fe2})/h_{ie1}(h_{22} + G_E)$. **(1.15)** $K_v = g_m(1 + h_{fe})R_L/[1 + g_mh_{ie} + g_mR_E(1 + h_{fe})]$. **(1.17)** $R_o = R_C$, $K_v = h_{fe}R_Ch_{ie}(1 + h_{fe})R_E/\{h_{ie}[h_{ie} + 2R_E(1 + h_{fe})]\}$, $R_{in} = h_{ie}[h_{ie} + 2R_E(1 + h_{fe})]/[h_{ie} + R_E(1 + h_{fe})]$. **(1.19)** (b) $K_v = -g_{m1}R_Dr_{d1}(g_{m2}r_{d2} + 1)/[(\mu + 1)r_{d1} + r_{d2} + R_D]$. **(1.20)** (b) $K_v = -4.063$. **(1.21)** $K_v = g_mR_F/[1 + R_S(g_m + g_d)]$. **(1.23)** $v_o = -R_D(v_1 + v_1')/(r_d + 2R_D)$, $R_o = 1/(G_D + 2g_d)$. **(1.26)** $R_{in} = r_d + R_S(\mu + 1)$. **(1.28)** $h_{11} = 909.1$ Ω, $h_{12} = 90.82$, $h_{21} = 0.09091$, $h_{22} = 9.182 \times 10^{-3}$ S. **(1.31)** $y_{11} = G$, $y_{21} = -(B + 1)G$, $y_{12} = -G$, $y_{22} = (B + 1)G$. **(1.35)** $V_1/I_1 = R_L/n^2$. **(1.37)** $V_1/I_1 = sCG^2$.

Chapter 2: **(2.1)** $R_{in} = R_S + 1/g_m$, $R_o = R_D$, $K_v = g_mR_D/(1 + g_mR_S)$. **(2.4)** (c) $G_M = g_m/2$; motor sees $G_N = 2(g_m + g_d + G_S)$. **(2.6)** (b) $v_s = gm(v_1 + v_1')/(G_S + 2g_d + 2g_m)$; (c) $A_D = -g_mR_F$, $A_C = g_mR_DR_F/[r_d + 2R_S(g_mR_D + 1)]$. **(2.8)** $V_o/V_1 = -R_3G_2/(sCR_1R_3G_2 + 1)$.

Chapter 3: **(3.5)** GBWP = 25 MHz, $\tau_a = 3.820 \times 10^{-3}$ sec. **(3.6)** $\omega_b = \omega_0\sqrt{2^{1/N} - 1}$ **(3.7)** $\omega_b = \omega_0/\sqrt{2^{1/N} - 1}$.

Chapter 4: **(4.1)** (a) $A_o = -12.007$; (b) $A_o = -g_mr_dR_D/(R_D + r_d)$, $V_o/V_1 = -12.007(1 - s\,5.893 \times 10^{-10})/(s^2\,1.0613 \times 10^{-17} + s\,2.8087 \times 10^{-8} + 1)$; (c) $V_o/V_1 = (\mu + 1)R_D/[R_1(\mu + 1) + r_d + R_D]$; (d) $V_o/V_1 = \mu R_S/[R_s(\mu + 1) + r_d]$. **(4.4)** (b) $V_o/V_g = (1 + sC_{gs}/g_m)/[1 + s(C_{gs} + C_o)/g_m]$; (c) $V_o/V_1 = (1 + sC_{gs}/g_m)/[s^2R_1C_o(C_{gd} + C_{gs})/g_m + s(C_{gs} + C_o)/g_m + 1]$. **(4.8)** $V_o/V_1 = g_mR_F/(1 + g_mR_s)$ mid-band gain.

513

Chapter 5: (5.1) $A_C = 0.01$, $A_D = 40.01$, CMRR = 4001. **(5.3)** (a) $R_E = 3.821 \times 10^5 \ \Omega$;
(b) $A_{dd} = -955.2$, $A_{cc} = -2.083 \times 10^{-2}$, CMRR = 93.2 dB, **(5.4)** $R_E = 7.600 \times 10^7 \ \Omega$,
$A_{dd} = -6.9034$, $A_{cd} = A_{dc} = 0$, $A_{cc} = -2.298 \times 10^{-5}$, CMRR = 109.6 dB.
(5.7) $A_{cc} = +\beta R_c/[h_{ie} + 2R_B + (\beta + 1)R_1]$, $A_{dd} = +\beta R_c/[h_{ie} + (\beta + 1)R_1]$.
(5.9) (b) $A_{dd} = -\beta R_D/h_{ie}$, $A_{cc} = -\beta R_D/[h_{ie} + (\beta + 1)2R_E]$, CMRR $= 1 + (\beta + 1)2R_E/h_{ie}$,
(5.12) (c) CMRR $= A_{dd}(1 + 0.5K \ \beta A_{cc})/A_{cc}(1 + 0.5A_{dd}K\beta)$, $K = -2/A_{dd}\beta$ for infinite CMRR.
(5.13) $A_{dd} = 0$, $A_{cc} = 2g_mg'_mR'_D/(2g_d + g'_m + G_D)$. **(5.16)** (a) DM source sees $R_{in} = h_{ie}$;
(b) CM source sees $R_{in} = h_{ie} + 2R_E(h_{fe} + 1)$; (c) v_1 sees $R_{in} = h_{ie}[h_{ie} + 2R_E(h_{fe} + 1)]/$
$[h_{ie} + R_E(h_{fe} + 1)]$. **(5.20)** (b) $R_{eq} = 21.55$.

Chapter 6: (6.1) $K_1 = -\mu R_D/[R_D + r_d + (\mu + 1)(R_F \| R_S)]$,
$K_2 = (\mu + 1)R_D(R_S \| R_D)/R_F[R_D + r_d + (\mu + 1)(R_F \| R_S)]$. **(6.3)** THD $= 1.43 \times 10^{-3}\%$.
(6.4) $G_M = \mu K_v/[r_d + R_S(1 + \mu(1 + K_v))]$, $R_o = r_d + R_S[1 + \mu(1 + K_v)]$.
(6.6) (a) $G_M = -K_vR_2/(Z_LR_2 + Z_LR_1 + K_vR_FR_1)$, $I_1/V_1 = -R_2/R_1R_F$ as $K_v \to \infty$;
(b) $R_o = K_vR_FR_1/(R_1 + R_2)$. **(6.8)** (b) $I_L = -V_1R_4/R_1R_3 = -V_1/R_2$.
(6.10) (a) $V_o/V_1 = -\alpha K_v/\{[1 + K_v(\beta - \alpha)][s\tau/(1 + K_v(\beta - \alpha)) + 1]\}$, $V_o/V_1(0) = -\alpha/(\beta - \alpha)$,
GBWP $= \alpha K_v/\tau$. **(6.13)** (a) $R_o = h_{ie}/(1 + h_{fe}) + R_m(1 + K_d) = 1 \times 10^5 \ \Omega$;
(b) $G_M = K_d(1 + h_{fe})/\{h_{ie} + (1 + h_{fe})[R_L + R_m(1 + K_d)]\} = 0.9994 \ S$.

Chapter 7: (7.1) (a) $A_v(s) = -sRC$; (b) $V_o/V_1(s) = -1/(s10^{-9} + s1.01 \times 10^{-5} + 1)$,
$\omega_n = 3.162 \times 10^4$ r/s, $\xi = 0.1597$. **(7.2)** (a) $I_L/V_1 = -\alpha K_v/\{[R_c(1 + K_v\beta)][sL/R_c(1 + K_v\beta) + 1]\}$,
$\beta = R_1/(R_1 + R_2)$; (c) $K_v = 1.998 \times 10^3$; (d) $R_o = R_c(1 + K_v\beta)$.
(7.5) (a) $-0.01 \le \beta \le 0.08$ for stability. (b) $\omega_o = 1.732 \times 10^6$ r/s for $\beta \ge 0.08$.
(7.9) (b) $\omega_o = \omega_n = |R|$; (c) $K \ge 1$. **(7.10)** (a) $A_L(s) = -Ks^4/(s - s_1)^4$, $s_1 = -1/RC$;
(c) $\omega_o = 1/R_C$; (d) $K > 4$ for oscillations. **(7.12)** (a) $K_v < 0$ (NFB); (b) $\omega_o = a/3$;
(c) $K_v > 1/8$. **(7.14)** (b) $\beta = 10^{-3}$, $\omega_n = 7.142 \times 10^3$ r/s. **(7.16)** $V_o/V_1 = -30.831$,
$R_F = 3.0841 \times 10^4 \ \Omega$. **(7.18)** (b) $K_v\beta = 3.764 \times 10^7$ for $\xi = 0.5$.

Chapter 8: (8.2) (a) $v_{n(rms)} = \sqrt{2kT/C}$. **(8.3)** (a) $S_o = (V_s^2/2)/[(2\pi f_o\tau)^2 + 1]$ MSV;
(b) $N_o = \eta/4\tau$ MSV; (c) $f_{B(opt)} = f_o$; (d) SNR$_{(max)} = (V_s^2/2)/\eta\pi f_o$. **(8.4)** (a) $S_o = (I_s^2/2)R_F^2$;
(b) $N_o = (e_{na}^2 + i_{na}^2R_F62 + 4kTR_F)B$; (c) $B = 5$ Hz.
(8.7) (a) $v_{n(rms)} = \sqrt{N_o} = 2.729 \times 10^{-2}$ Vrms; (b) $V_S = 3.86 \times 10^{-6}$ Vpk.
(8.9) (a) $S_o = v_1^2 \ R_F^2/(R_1n)^2$;
(b) $N_o = B[i_{na}^2R_F^2 + e_{na}^2 (1 + RF/n^2 \ R_1)^2 + 4kTR_1n^2(-R_F/n^2R_1)^2 + 4kTR_F]$;
(c) $N_o = 14.8$; (d) $V_o/V_1 = -67.56$. **(8.12)** (a) $P_{o(max)} = V_2^2 \ A_{v2}^2/4R_L$;
(b) $F = 1 + e_{n1a}^2/4kTR_S + e_{na2}^2/A_{v1}^24kTR_S$. **(8.14)** (b) SNR $= v_s^2/B(4kTR_S + 2i_{na}^2R_S^2)$.
(8.18) (a) $S_{n1}(f) = (4kT_1R_1 + 4kT_2R_2)/(R_1 + R_2)^2$; (b) $S_{ni}(f) = 4kTG$.

Chapter 9: (9.2) (a) $V_o/V_1 = s^2C^2R_1R_2/(s^2C^2R_1R_2 + s2CR_1 + 1)$;
(b) $\omega n = 1/C\sqrt{R_1R_2}$, $\xi = \sqrt{R_1/R_2}$.
(9.5) $V_o/V_1 = (s^24R^2C^2 - s2RC + 1)/(s^24R^2C^2 + s2RC + 1)$,
$\omega_n = 1/2RC$, $\xi = 1/2$. **(9.6)** (a) $Z_{in}(j\omega) = j\omega(10^2)$, 100 Hy inductor;
(b) $V_1/V_s = sCG/(s^2LC + sCG + 1)$. **(9.10)** (b) $V_o/V_1(0) = -1$, $\omega_n = 5.7735 \times 10^6$ r/s, $\xi = 0.2887$.
(9.11) $V_o/V_1 = (-sC_2T_c/C_b)/[s^2(C_1C_2T_c^2/C_aC_b) + s(C_1 + C_2)T_c/C_a + 1]$,
Pk gain $= -C_2C_a/(C_1 + C_2)C_b$, $\omega_n = (1/T_c) \ C_aC_b/C_1C_2$, $Q = C_a\sqrt{C_1C_2}/\sqrt{C_aC_b} (C_1 + C_2)$;
(b) $C_a = 14.05$ nF, $C_b = 140.5$ pF, Pk. resp $= -50$.
(9.12) (b) $V_o/V_1 = -1/(s^2C_2C_5T_c^2/C_1C_4 + sC_3C_5T_c/C_1C_4 + 1)$,
dc gain $= -1$, $\xi = (C_3/2)\sqrt{C_5/C_1C_4}$, $\omega_n = (1/T_c)\sqrt{C_1C_4/C_2C_5}$.

Chapter 10: (10.1) (a) $V_o/V_1 = [sRC(1 + 16.7\,I_{ABC}R) + 1]/(sRC + 1)$.
(10.3) $V_o/V_1 = (R_2R_3/R_1R_4)(1 + sC_1R_1)/(s10C_2R_3/V_c + 1)$.
(10.5) (a) $V_o/V_1 = -(R/R_1)/(s10R_4C/|V_c| + 1)$; (b) $A_o = -R/R_1$, $\omega_b = |V_y|/10R_4C$ r/s;
(c) $R = 10$ kΩ, $R_1 = 1$ kΩ, $R_4 = 10$ kΩ; $C = 1$ μF.
(10.7) (b) $V_4/V_1 = (sC/I_{ABC}16.7 - 1)/(sC/I_{ABC}16.7 + 1)$; (c) $\theta = 180° - 2\tan^{-1}(\omega C/I_{ABC}16.7)$.
(10.9) (b) $\omega_n = V_{y2}/10RC$ r/s, $V_4/V_1\,(j\omega_n) = -1$, $\xi = 1/20V_{y1}$.
(10.12) (a) $V_4/V_1 = -1/(s10RC/|V_c| + 1)$; (b) $V_4/V_5 = -K$;
(c) $V'_1/V_5 = s(10RC/|V_c|)/s(10RC/|V_c| + 1)$; (d) $A_L(s) = +s(K|V_c|/10RC)/(s + |V_c|/10RC)^2$.

Chapter 11: (11.1) (a) $K_pK_fK_v = 2a$; (b) $\theta_{e(SS)} = V_p/K_p$.
(11.3) (a) $K_pK_fK_v = b^2/2$; (b) $\theta_o/X_m(s) = (2K_v/b)(s/b + 1)/[s^2(2/b^2) + s(2/b) + 1]$;
(e) $K_m = K_v2/b$.   **(11.5)** (a) $A_L(s) = -K_pK_fK_v/s(s + b)$; (b) $K_pK_fK_v = b^2$;
(c) $V_c/X_m(s) = (K_m/K_v)/(s^2/b^2 + s/b + 1)$; (e) $K_pK_fK_v = 9.8696 \times 10^8$.
(11.6) (a) $\tau = 8.231 \times 10^{-3}$, $K_p\,K_v = 60.75$ r/s; (b) $\omega_n = 85.91$ r/s, $K_T = 60.75$.
(11.9) (a) $K_pK_fK_v = 3a^2/4$;
(b) $\theta_o/X_m(j\omega_m) = (4K_v/3a)(j\omega_m/a + 1)/[(j\omega_m)^2/(a^23/4) + j\omega_m(4/3a) + 1]$.
(11.11) $N_o = 9a^3N/32K_v^2$, $S_o = (X_m^2/2)(K_m^2/K_v^2)/$
$\{[1 - (2\pi f_m)^2/(a^23/4)]^2 + (2\pi f_m)^2(4/3a)^2\}12)$ $\tau = 5 \times 10^{-3}$ sec., $\omega_n = 1.414 \times 10^2$ r/s.

Chapter 12: (12.1) (a) $I_{BQ} = 5$ mA; (b) $\bar{P}_{LQ} = 1$ W; (c) $\bar{P}_{max} = 1.5$ W.
(12.2) (a) $P_L = 0.875$ W; (b) $n_1/n_2 = 1/10$, $P_{L(max)} = 87.5$ W.
(12.4) (a) $V_{cc} = 31.62$ Vdc; (b) $V_o = 20.13$ Vpk, $\bar{P}_{c(max)} = 10.13$ W;
(c) $\bar{P}_{PS(max)} = 31.83$ W from each supply.   **(12.6)** (a) $K_v\beta = 2.47 \times 10^3$; (b) $K_v = 2.223 \times 10^5$,
$\beta = 1.111 \times 10^{-2}$.   **(12.8)** (a) Peak power in $R_L = 256$ W; (b) $\bar{P}_L = 128$ W;
(c) $\bar{P}_{c1} = 8.744$ W; (d) R_L sees $2h_{ie}/(h_{fe} + 1)$; (e) Efficiency = 0.785.
(12.10) (a) R_L sees $R_o = r_d/(1 + g_mr_d)$; (b) $V_o/V_1 = \mu R_L/[R_L(\mu + 1) + r_d]$; (c) $\beta = 0.1389$;
(d) $V_o/V_s = -7.773$.

References and Bibliography

Angelo, E.J., Jr. 1969. *Electronics: BJT's, FET's and Microcircuits*. McGraw-Hill, N.Y. Chs. 4–10.

Belanger, P.R., E.L. Adler, and N.C. Rumin. 1985. *Introduction to Circuits with Electronics*. Holt, Rinehart & Winston, N.Y. Chs. 4–7.

Berkovitz, R., and K. Gundry. 1973. Dolby B-type noise reduction system. *Audio* (October).

Best, R.E. 1984. *Phase-Locked Loops: Theory, Design and Applications*. McGraw-Hill, N.Y.

Blanchard, A. 1976. *Phase-Locked Loops*. John Wiley & Sons, N.Y.

Broberg, H.L. 1986. *Micro-Cap* on the IBM PC as an integral part of 1st through 3rd year analog electronics. *Proceedings, 1986 Frontiers in Education Conference*. IEEE, N.Y. Pp. 224–231.

Chirlian, P.M. 1981. *Analysis and Design of Integrated Electronic Circuits*. Harper & Row, N.Y. Ch. 16.

Data Acquisition and Conversion Handbook. 1979. Datel-Intersil, Inc., Mansfield, MA.

Dolby, R.M. 1967. An audio noise reduction system. *J. Audio Engineering Society* 15(4):383–388.

Dolby, R.M. 1972. "Compressors and Expandors for Noise Reduction Systems." U.S. Patent 3, 665, 345.

Egan, W.F. 1981. *Frequency Synthesis by Phaselock*. John Wiley & Sons, N.Y.

Franco, S. 1988. *Design with Operational Amplifiers and Analog Integrated Circuits*. McGraw-Hill, N.Y. Ch. 11.

Franklin, G.F., J.D. Powell, and A. Emami-Naeini. 1986. *Feedback Control of Dynamic Systems*. Addison-Wesley, Reading, MA. Sec. 5.2.1.

Gardner, F.M. 1979. *Phaselock Techniques*, 2nd ed. John Wiley & Sons, N.Y.

Garrett, P.H. 1981. *Analog I/O Design*. Reston, VA. Chs. 7–9.

Ghaussi, M.S., and K. Laker. 1981. *Modern Filter Design*. IEEE Press, N.Y. Ch. 6.

Ghaussi, M.S. 1985. *Electronic Devices and Circuits*. Holt, Rinehart & Winston, N.Y. Chs. 1–3, Appendix C.

Graeme, J.G., G.E. Tobey, and L.P. Huelsman. 1971. *Operational Amplifiers: Design and Applications*. McGraw-Hill, N.Y. Ch. 8.

Gray, P.R., and R.G. Meyer. 1984. *Analysis and Design of Analog Integrated Circuits*, 2nd ed. John Wiley & Sons, N.Y. Secs. 7.2. and 7.3.

Grebene, A.B. 1984. *Bipolar and MOS Analog Integrated Circuit Design*. John Wiley & Sons, N.Y. Sec. 5.1.

Holt, C.A. 1978. *Electronic Circuits: Digital and Analog*. John Wiley & Sons, N.Y. Secs. 18.1–18.3.

James, H.M., N.B. Nichols, and R.S. Phillips. 1947. *Theory of Servomechanisms*. McGraw-Hill, N.Y. Appendix.

Jung, W.G. 1977. *IC Op-Amp Cookbook*. W.W. Sams & Co., Indianapolis. Ch. 6.

Kingley, H. 1980. *The PLL Synthesizer Cookbook*. Tab Books, Blue Ridge Summit, PA.

Kitchin, C., and L. Counts. 1983. *RMS to DC Conversion Handbook*. Analog Devices, Inc., Norwood, MA.

Kuo, B.C. 1982. *Automatic Control Systems*, 4th ed. Prentice-Hall, Englewood Cliffs, N.J. Secs. 3.4–3.9

Letzler, S., and N. Webster. 1970. Noise in amplifiers. *IEEE Spectrum* (August), pp. 67–75.

Lindsey, W.C., and M.K. Simon. 1978. *Phase-Locked Loops and Their Applications*. IEEE Press, N.Y.

Middlebrook, R.D. 1963. *Differential Amplifiers*. John Wiley & Sons, N.Y.

Millman, J., and C.C. Halkias. 1972. *Integrated Circuit Electronics: Analog and Digital Circuits and Systems*. McGraw-Hill, N.Y. Ch. 11.

Millman, J. 1979. *Microelectronics*. McGraw-Hill, N.Y. Chs. 2, 3, and 8.

Millman, J., and A. Grabel. 1987. *Microelectronics*. McGraw-Hill, N.Y. Ch. 3.

Mitra, S.K. 1980. *An Introduction to Digital and Analog Integrated Circuits and Applications*. Harper & Row, N.Y. Ch. 8.

Motchenbacher, C.D., and F.C. Fitchen. 1973. *Low-Noise Electronic Design*. John Wiley & Sons, N.Y.

Ott, H.W. 1988. *Noise Reduction Techniques in Electronic Systems*, 2nd ed. John Wiley & Sons, N.Y.

PMI Telecommunications Applications Handbook. 1981. Precision Monolithics, Inc., Santa Clara, CA.

Roden, M.S. 1989. *The Student Edition of Micro-Cap II*. Addison-Wesley, Reading, MA.

Savant, C.J., Jr., M.S. Roden, and G.L. Carpenter. 1987. *Electronic Circuit Design*. Benjamin/Cummings, Menlo Park, CA.

Schauman, R., M. Soderstrand, and K. Laker, eds. 1981. *Modern Active Filter Design*. IEEE Press, N.Y.

Schilling, D.L., and C. Belove. 1979. *Electronic Circuits, Discrete and Integrated*. McGraw-Hill, N.Y. Ch. 5.

Sparkes, R.G., and A.S. Sedra. 1973. Programmable active filters. *IEEE Trans. on Solid-State Circuits* 8(2):93–95.

Spence, R., and J.P. Burgess. 1986. *Circuit Analysis by Computer: From Algorithm to Package*. Prentice-Hall International, Englewood Cliffs, N.J.

Taub, H., and D.L. Schilling. 1977. *Digital Integrated Electronics*. McGraw-Hill, N.Y. Ch. 8.

Truxal, J.G. 1955. *Automatic Feedback Control System Synthesis*. McGraw-Hill, N.Y.

van Valkenberg, M. 1982. *Analog Active Filter Design*. Holt, Rinehart & Winston, N.Y.

van der Ziel, A. 1974. *Introductory Electronics*. Prentice-Hall, Englewood Cliffs, N.J. Ch. 13.

Wait, J.V., L.P. Huelsman, and G.A. Korn. 1975. *Introduction to Operational Amplifier Theory and Applications*. McGraw-Hill, N.Y. Ch. 4.

Wojslaw, C.F., and E.A. Mostakas. 1986. *Operational Amplifiers*. John Wiley & Sons, N.Y.

Index